Essential Oil Safety

The physician must … have two special objects in view with regard to diseases, mainly, to do good or to do no harm.

Hippocrates

Authors

Robert Tisserand
Robert started practising aromatherapy in 1969 and is best known for bringing aromatherapy to the UK through the publication of *The Art of Aromatherapy* in 1977. The first book on the subject in English, it is now published in nine languages. Robert has written two further books, is on the advisory board of two publications, and lectures at international conferences. He is the Editor of *The International Journal of Aromatherapy*, is the Principal of the Tisserand Institute, and has been actively involved in establishing training standards for aromatherapy in the UK.

Tony Balacs
Tony is a lecturer in Essential Oil Science at the Tisserand Institute. He obtained his Bachelor's degree in Pharmacology in 1985 and his Master's degree in Molecular Biology in 1987. From 1987 to 1990, he was researching genetic aspects of fetal development, before concentrating on work in education. He has over 10 years' experience teaching various aspects of science to people from a wide variety of backgrounds, including at the Open University, where he is both tutor-counsellor, and the Enterprise in Higher Education Co-ordinator for London. He has lectured on the safe use of essential oils throughout the UK. His other interests include counselling, and he has a private psychotherapy practice which he runs alongside his other work.

For Churchill Livingstone:

Commissioning editor: Inta Ozols
Project development editor: Valerie Bain
Copy editor: Sue Beasley
Indexer: Tarrant Ranger Indexing Agency
Project manager: Valerie Burgess
Project controller: Pat Miller
Sales promotion executive: Hilary Brown
Design direction: Judith Wright

Essential Oil Safety
A Guide for Health Care Professionals

Robert Tisserand
Principal, The Tisserand Institute, Sussex, UK

Tony Balacs
The Tisserand Institute, Sussex, UK

Foreword by

John Caldwell PhD
Professor of Biochemical Toxicology
St. Mary's Hospital Medical School, London

CHURCHILL LIVINGSTONE
EDINBURGH HONG KONG LONDON MADRID MELBOURNE NEW YORK AND TOKYO 1995

CHURCHILL LIVINGSTONE
Medical Division of Pearson Professional Limited

Distributed in the United States of America by Churchill
Livingstone, 650 Avenue of the Americas, New York, N.Y.
10011, and by associated companies, branches and
representatives throughout the world.

First published 1995
Reprinted 1995

ISBN 0 443 05260 3

British Library Cataloguing in Publication Data
A catalogue record for this book is available from the
British Library.

Library of Congress Cataloging in Publication Data
A catalog record for this book is available from the
Library of Congress.

The
publisher's
policy is to use
paper manufactured
from sustainable forrests

Typesetting by IMH (Cartrif), Loanhead, Scotland

Printed by Bell and Bain Ltd., Glasgow

Contents

Foreword

We are all aware of the growing interest in complementary therapies from the media and, of these, aromatherapy is probably the most widely used. During the exploration of the true value of complementary therapies, it is important that the public be protected from any particular risk that these present.

My research work has concerned aspects of the safety of essential oils for many years and I am all too aware that, in this area, 'natural' certainly does not always mean 'safe', whatever the general public may think. I therefore welcome the appearance of this volume, which assembles together relevant information on the safety of a very wide range of essential oils in a fashion appropriate for the intelligent, non-specialist reader.

This collection of safety data presented at a level suitable for the general public is unique, and will help raise awareness of the risks which some of the essential oils used in aromatherapy can present, while also documenting the safety of the great majority of oils in popular use. Anyone requiring more detailed information is advised to consult the comprehensive list of references to the original scientific literature.

As a practising toxicologist, I have no position on the validity or otherwise of the claims made for aromatherapy but I do have a great concern that essential oils be used safely. I have no doubt that the work of Tony Balacs and Robert Tisserand in writing this volume represents a necessary contribution to the safety of the ever-more-widespread practice of aromatherapy.

J. C.

Preface

The many books now available on the practice of aromatherapy usually touch on the possible adverse effects of certain essential oils, while naturally concentrating on their therapeutic properties. However, there is no text written with aromatherapy in mind which is concerned specifically and in detail with essential oil toxicology. This book is designed to fill that gap.

The fragrance and flavour industries already have their own guidelines for controlling essential oil safety. These, however, are not necessarily appropriate for aromatherapy.

At the time of writing there appear to be no regulations governing the sale or use of essential oils in aromatherapy that effectively protect the consumer. The increasing availability of undiluted essential oils, some of which undoubtedly present a potential hazard, is cause for concern. In the UK and the USA at least, it is currently possible to purchase, by mail order, the majority of the essential oils which we recommend should not be available to the general public.

We suspect that in many countries, there may be too few controls on the minority of essential oils which do present a hazard. We believe that the most responsible course of action for those, such as ourselves, who have information about known, or suspected toxicity, is to make it public. Both those who sell, and those who use essential oils, will then be in a better position to take informed decisions about which oils are safe to use, and in which circumstances.

We are not simply talking about banning or restricting certain essential oils. There is also a need to improve labelling, giving warnings where appropriate, and to make packaging safer, especially with regard to young children. There is a need for a greater awareness of the potential dangers among those who package, sell and use essential oils.

In recent years there have been many 'scare stories' in the media about the dangers of essential oils. Some of these have been quite accurate, but often the information given is misleading and based more on rumour than fact. We believe that it is vital for the aromatherapy community to address safety issues, and to take responsible action, in order to safeguard its future.

We are aware that a book such as this could have the effect of presenting essential oils as generally dangerous substances – this is certainly not our intention. On the contrary, there are several instances where we have shown that supposed dangers do not in fact exist. The majority of essential oils turn out to be non-hazardous as they are used in aromatherapy. The function of this book is to reassure, when appropriate, as well as to hoist some red flags.

The same intention to inform is behind our inclusion of physiology and biochemistry. We explain how different forms of toxicity arise and why the use of certain oils is sometimes inadvisable. We believe that this is much more useful than simply presenting summaries of 'safe' or 'dangerous' oils.

There are two approaches one can take when dealing with issues of safety. The first is to assume that the materials in question are hazardous until proven to be safe. This is the approach taken when dealing with pharmaceutical drugs. The second approach is to assume that substances are safe unless proven hazardous. This line is taken in the food and fragrance industries.

In general, we have taken the approach that essential oils are safe unless proven hazardous. It seems unnecessary to treat essential oils as pharmaceuticals, especially if they are only used externally. There are cases where we have flagged an essential oil as hazardous even though absolute proof may not be available. In practice one must steer a middle course, and use all the information available, both positive and negative.

In the context of these dilemmas we have attempted to give a balanced view. We acknowledge both that animal toxicity may be relevant to the human situation, and that experimental doses almost always greatly exceed those given therapeutically. We have given detailed information concerning the relevance or otherwise of specific animal tests to humans. Toxicologists increasingly acknowledge that giving excessive doses of a substance to a genetically in-bred mouse living in a laboratory may not have great relevance to the human situation.

Our hope is that this book constitutes a practical, positive basis for guidelines, both in the essential oil retail trade and the aromatherapy profession. While it is primarily written for the aromatherapy market, it will be of interest to all those who use essential oils, whether in fragrances, flavourings, toiletries or pharmaceuticals. Pharmacists, doctors, nurses and poisons units may find it a particularly useful summary.

This book replaces *The Essential Oil Safety Data Manual* by Robert Tisserand, first published in 1985. This text was largely an extrapolation of toxicological reports from the Research Institute for Fragrance Materials (RIFM). The RIFM data still form a very important part of the current volume which, however, contains more detail about a greater number of hazardous oils than its predecessor, and a great deal more toxicological and pharmacological information.

The aim of the book remains the same: to provide information for the benefit of all who are interested in the therapeutic use of essential oils, so that aromatherapy may be practised, and products may be developed, with the minimum of risk. This can only be accomplished if all those involved, in both the aromatherapy profession and the trade, are thoroughly familiar with the hazards which do exist, and which, in a few cases, are rather serious.

R. T.
T. B.

Sussex 1995

The Tisserand Institute
65 Church Road, Hove
East Sussex, BN3 2BD, UK
Tel: +44(0)1273 206640
Fax: +44(0)1273 329811

Acknowledgements

The authors gratefully acknowledge the considerable help of the following people in giving valuable information and advice:

Dr Tim Betts (consultant neuropsychiatrist); Dr John Caldwell (toxicologist); Dr Desmond Corrigan (pharmacognocist); Tony Dann (essential oils consultant); Dr Peter Gravett (consultant haematologist); Jennie Harding (aromatherapist); Dr Sharon Hotchkiss (toxicologist); Dr Wendy Jago (educational consultant); Dr Brian Lawrence (essential oils consultant); Simon Mills (medical herbalist); Ruth H. Smith (aromatherapist).

Important notices

• All parties involved in the preparation and publication of this book have taken the greatest possible care to ensure that the information contained herein is complete and accurate. However, no liability will be accepted in respect of such information by either the publisher or the authors following any accident or injury arising from the use or misuse of any essential oils, essential oil ingredients, or of any product containing them. The information given in the text is offered only as a guide. The absence of an essential oil, or of a particular chemical substance, from any part of this book should not be taken as an indication of that oil's or substance's safety.

• Throughout the text, the cautions and contra-indications given with regard to oral administration are based on a minimum dosage per essential oil of around 0.5 ml per 24 hours. Much of the cautionary advice given in the text will not apply to dosage levels which are considerably less than this, and which may be encountered, for example, in food flavourings. The cautions and contraindications given for dermal administration are based on a minimum concentration of 1%. These may not apply to concentrations of single essential oils which are considerably lower than this, and which could be encountered, for example, in the fragrances of (aromatherapy) products such as skin creams.

Notes

- A chemical, followed by an essential oil name in parentheses, e.g. 'cinnamaldehyde (<> cinnamon bark)', denotes that the chemical is found, often as a major component, in the bracketed essential oil. The same chemical may also occur in many other essential oils. The <> sign, to denote 'found in', is our invention – it is not conventional.

- For the sake of clarity we sometimes make use of the = sign, to denote alternative terms, especially for certain botanical names, e.g.

Cinnamomum zeylanicum (= *Cinnamomum verum*).

- Throughout the book some of the plant names are followed by a superscript A (e.g. oakmoss[A]). This indicates that the plant is more commonly processed by solvent extraction than by steam distillation. Solvent extracted products are either concretes, absolutes or resinoids, and should not be referred to as essential oils.

- Chemotypes are presented, for instance: 'thyme, linalool CT'. See pages 8 and 9 for an explanation of chemotypes.

- Those who use essential oils in aromatherapy are used to measuring them in 'drops'. We have, therefore, referred to drops of essential oil in the text, even though a drop is not a precise measure. Having tested several drop-dispensers commonly used for aromatherapy oils, we have worked on the basis that 1 drop equals approximately 0.05 ml.

- Notes are indicated in the text by a superscript figure and can be found at the end of each chapter.

- At the end of each chapter, just before the notes, there is a summary which highlights the major points and conclusions of that chapter.

- We have not always given references in the text for essential oil composition. Most of this information is derived from Brian Lawrence's texts [214–217] and it seemed laborious to give the same references repeatedly.

- All of the numbered references have been read by the authors, and we have listed them in as complete a form as possible, except that we have not listed every author in cases of multiple authorship.

- Two organizations are frequently referred to in the text. These are: the Research Institute for Fragrance Materials (RIFM) and the International Fragrance Association (IFRA). RIFM funds and publishes research on the safety of fragrance materials, and IFRA formulates and publishes safety guidelines for the fragrance industry, based on RIFM's research.

- The abbreviation NCA has been used to denote oils 'not commercially available'.

- ⚠ This symbol is used in the Chemical Index for especially hazardous materials.

1

Introduction

AROMATHERAPY

While synthetic chemicals now constitute the major part of most fragrances and, to a lesser extent, flavours, true essential oils are undoubtedly being produced in greater quantities than at any time in history, such is the demand. They are also being used in a greater variety of ways than ever before. Most of the world's essential oil production is put to use as fragrance or flavouring; we estimate that aromatherapy accounts for only some 3–5% of total consumption. Nevertheless, interest in the therapeutic applications of essential oils has seen a tremendous upsurge in recent decades, accompanied by a renewed Western appreciation of traditional and unconventional forms of medicine.

Aromatherapy is probably the fastest growing of the alternative health disciplines in the UK and possibly also in several other countries. One of the most important aspects of the 'health revolution' has been the realisation that the treatment of illness and the maintenance of well-being are a personal responsibility. This has led to growth in over-the-counter sales of natural remedies, such as essential oils.

Another recent development which impinges on safety is the increasing use of essential oils, often by nursing staff, in hospitals and hospices. This has led to demands for information about, for example, drug interactions with essential oils, safety in cancer, and safety in pregnancy. One UK local authority has told its midwives not to allow aromatherapy in combination with pethidine during labour, due to lack of information.

The safety issue

Many of the oils used in aromatherapy have been employed medicinally for thousands of years, but in the form of the whole plant. When distilled from the plant, the essential oils become concentrated 100-fold or more, and so certain properties become very pronounced in the oils. Some of these properties may be desirable, but any toxicity associated with the essential oil will be correspondingly greater compared to that of the whole plant.

Most essential oils represent little or no risk when correctly used in aromatherapy. However, just as there are poisonous herbs (hemlock), berries (nightshade), woods (yew) and fungi (death-cap), there are also toxic essential oils (unrectified bitter almond oil contains cyanide). There have been a number of deaths from accidental ingestion of large quantities of oils such as wintergreen. As with any substance which comes into contact with the body, the dose is paramount; even seemingly innocuous materials like salt or sugar can be fatal if enough is ingested.

The natural origin of essential oils does not guarantee their safety; 'natural' never has been synonymous with 'safe'.

Finding toxins in essential oils is as easy as finding them in foods. In the majority of cases, however, these toxins are present in very small amounts. The difficulty comes in determining in what context (dosage level, frequency, mode of administration, etc.) the substance in question poses a threat to health. For example, rosemary oil contains 10–20% of camphor, and camphor is especially dangerous in both pregnancy and epilepsy, but only above a certain dose. In practice this means that although the amount of rosemary oil used when prescribed orally could be hazardous, the amount absorbed from aroma–therapy massage is safe.

Packaging and labelling

Bottles of undiluted essential oils are on unrestricted sale in a great many countries, and need carry no hazard warning. (This situation may change in Europe, with new European Community [EC] legislation.) We believe that, where there is a specific and serious risk associated with an essential oil, either an appropriate warning should appear on the bottle and/or its packaging or the sale of the oil should be restricted. At the same time general warnings, such as 'keep out of reach of children' may also be appropriate.

AREAS OF CONCERN

In reviewing the risks presented by essential oils, it seems that only a relatively small number are hazardous, and many of these, such as mustard and calamus, are not widely used in aromatherapy. However, some commonly used oils do present particular risks, basil, bergamot and cinnamon bark for example.

Poisoning—essential oils and children

In terms of safety, there is one issue which is of paramount concern. In the quantities in which they are most commonly sold (5–15 ml) essential oils can be lethal if drunk by a young child, and there have been many recorded fatal cases over the past 70 years. Perhaps the only reason that child fatalities have not increased with the current growth of aromatherapy is because today most essential oils are sold in bottles with integral drop-dispensers. These make it more difficult for a toddler to drink the contents of a bottle. Most urgently, we would like to see 'open-topped' bottles (i.e. without drop-dispensers) of undiluted essential oil banned, and appropriate warnings printed on labels.

Skin problems

Skin sensitisation and irritation are less emotive issues than poisoning, but there is a higher risk of misuse here because aromatherapy oils are most commonly applied to the skin. The fact that they are usually diluted in vegetable oil or some other vehicle is not an absolute safeguard because severe allergic reactions are possible after repeated contact even with minute amounts of allergen.

There are a small number of essential oils, including cinnamon bark, which present a high risk of allergic skin reaction.

Those unaware of the risks may be using hazardous oils in the firm belief that natural oils, diluted and applied externally, cannot possibly do any harm. This is not the case. In particular, caution must always be exercised when applying essential oils to damaged skin, or to areas where there is already evidence of irritation. Dermatitis may signal that the person's skin is especially liable to react adversely to essential oils.

Risks in pregnancy

A thorough search of the literature has not revealed any cases of unwanted abortion resulting from the use of essential oils. In fact most attempts by women to use essential oils to cause abortion have proven unsuccessful. Nonetheless, there is evidence that a few oils, including the notorious savin, are possible abortifacients. Dangers to the fetus are also a risk with some oils.

Oral administration, with its higher doses, is especially risky in pregnancy. This is primarily because the placenta is known to be permeable to many essential oil constituents which will therefore reach the fetus if applied to the mother. Furthermore, the fetus is likely to be far less able than the mother to detoxify these substances.

Cancer

Much research has been carried out in recent years which provides evidence for the carcinogenicity of a small group of essential oil components. This research indicates that these components are low-level carcinogens and that, while high doses of essential oils such as basil present an unacceptable risk, low doses present a negligible risk. However, in some essential oils the amounts used in massage are still too high for safety.

There is evidence that some phototoxic essential oil components are also photocarcinogenic. Many people are familiar with this risk in relation to bergamot oil. Again, only a few oils are hazardous, and in this case only if the skin to which the oil has

been applied is exposed to UV radiation within 12 hours of application.

Cancer, of course, is a very serious disease, and establishing and implementing safe levels for potentially carcinogenic or photocarcinogenic essential oils is therefore of great importance. In some cases the level of risk is such that the oil should probably not be used at all in aromatherapy.

Epilepsy

Essential oils present very little risk to people with epilepsy, in fact virtually no risk as long as the oils are not taken orally. Taken orally, however, some commonly used oils, such as rosemary, are potentially hazardous.

Oral administration

Great caution should be exercised when taking or prescribing any essential oil by mouth. Oral administration is more likely to lead to systemic toxicity problems than application to the skin, since a greater concentration (about 10 times as much) is likely to reach the bloodstream. The dangers with regard to pregnancy, cancer, epilepsy, and virtually every other area of risk, increase with oral administration.

A further consideration is that some essential oils can irritate the mucous membrane of the alimentary canal, especially if undiluted. It is our recommendation that essential oils should only be prescribed orally, for therapeutic purposes, by primary care practitioners such as medical doctors and medical herbalists.

SETTING SAFE LEVELS

This book would be of less practical use if it did not give any recommendations as to the safety or otherwise of essential oils in a given context. Inevitably, the translation of factual information into recommendations involves subjective judgement. The authors acknowledge that other interpretations are possible, particularly in the light of new information.

In recommending safe levels we have several types of data to draw on: experimental animal toxicology, experimental human toxicology, and cases of poisoning, whether accidental or deliberate. Where safe levels for dermal or oral use have been established previously we have tended to follow them, but have not done so in every instance.

The fragrance and flavour industries have their own guidelines for controlling essential oil safety. These, however, are not necessarily appropriate for aromatherapy. Some essential oils, for instance, have a maximum use level in fragrances of 0.1%, while these same oils, tea tree for example, could be used at 5% in aromatherapy. (The maximum use level referred to here is not a level set for safety reasons, but is simply the maximum amount used in any fragrance.)

Where there are no established recommend-ations, we have assumed that oils are safe when diluted for dermal use except where indirect experimental data (in vitro or in vivo) show a potential risk which we believe has not yet been appreciated. In some cases we have recommended that the oils should not be taken orally, but are safe to use topically. This is due to the higher dose levels of oral administration. In other cases we have indicated that specific essential oils should be avoided in certain vulnerable states, such as pregnancy, or that they should be used with special caution.

There are some essential oils which we have recommended should not be used in aroma-therapy. This recommendation assumes certain use levels in practice. For external use, between 1% and 5% is assumed, and dosage of between 0.5 ml and 2.5 ml per 24 hours is assumed for oral administration. An alternative approach would be to set low maximum use levels for the more hazardous essential oils. For example, it might be decided that a maximum use level of 0.2% was safe for the external use of pennyroyal oil.

There are several problems with this approach. While none of them is insurmountable, taken together, they loom large. Firstly, the maximum safe level would need to be ascertained. While this may not be impossible, it will often involve a degree of arbitrary decision-making. Secondly, there is little point in setting such levels unless they are practicable. They need to be fully understood and implemented by all those involved. In the case of aromatherapy, this presumably includes the general public. Currently there are no clear distinctions between aromatherapists and general public. The problem here is one of ensuring that guidelines are followed. Thirdly, it is questionable whether concentrations of essential oil in the region of 0.2% in carrier oil have any physiological therapeutic benefit. If they do not, there may be little point in legislating for safe levels to be accepted.

In spite of the above comments, maximum use levels of less than 1% are suggested in the text for certain potentially phototoxic essential oils. How-ever, in this instance, safe levels have been ascertained, and these levels are only important if the skin to which the oils have been applied is exposed to UV light in the following 12 hours. The simple alternative is to avoid such exposure following the use of the oils. This seems practical, especially since the dangers of sunlight are so well known, and since in any case the chances of exposure following application are low.

Maximum use levels are also recommended for a few essential oils which contain potentially carcinogenic compounds. (Contrary to popular belief, carcinogenic substances are frequently permitted in both foods and fragrances, but only at very low levels.) In doing this we have tried to be consistent with what the toxicological data show. Of the three problems outlined above, the first and third do not apply here. Safe levels are not difficult to arrive at, and they are between 1.5% and 2.5%. However, the problem of implementing them still remains.

Clearly, making decisions on restricting the availability of essential oils is not a simple matter of examining toxicological data. As we have said, different interpretations of the data are often possible, and other factors may need to be taken into consideration. In the case of pharmaceutical drugs, for instance, risks may be weighed against benefits, although very few benefits have been established for essential oils through clinical trials.

Political and sociological factors may also need to be considered. Many consider that cigarettes are potentially lethal and, in addition, addictive, and yet they are widely available to all. Paracetamol is lethal in overdose, and yet is available over the counter, with only non-specific warnings such as 'do not exceed the stated dose'. Some 200 people a year die from paracetamol overdose in the UK. Paradoxically, in Australia, both tea tree and eucalyptus oils are required by law to carry the word 'poison' on packaging, in spite of the fact that there is no reason to regard either oil as particularly toxic.

THE OILS IN THIS BOOK

A problem for us in writing this book has been the lack of a clearly defined 'list' of essential oils used in aromatherapy. In reality this is constantly growing. The common and popular ones, such as rosemary and lavender, are oils which are produced on a commercial scale. However, even here essential oils go in and out of fashion— witness the rapid growth of the tea tree oil industry in recent years.

There are also many essential oils which are on the 'fringes' of aromatherapy. They are not produced with regularity on a commercial scale, but sometimes find their way into aromatherapy literature or training seminars. This is natural, inevitable and adds interest to the subject, but it also makes 'legislation' difficult. Since it is difficult to know which essential oils to include, we have attempted to cover as many as possible.

There is a further complication. The chemical composition of most of the oils commercially available is fairly standard and well known. However, aromatherapists do sometimes obtain oils which are obscure or unusual, and which may have a composition quite different from the norm.

This gives us even more reason to attempt to cover the very obscure essential oils, and to be clear about chemotypic differences in botanically similar oils.

ANIMAL TESTING

We do not intend to argue the case against testing chemicals on live animals, but we would be unhappy if aromatherapy were to adopt vivisection as a means of safety testing for essential oils. There is, however, a large amount of animal data which has been obtained concerning skin absorption, toxicity and essential oil metabolism. Some of this data is undoubtedly relevant to humans; some of it is less so. The problem lies in distinguishing the useful from the not so useful. For instance, extrapolating animal data regarding skin absorption of essential oils to humans is probably unwise.

Within the scientific community, there are many that believe animal testing has not progressed, in terms of science and technology, over the past 50 years or so. There is currently so much public opinion against animal testing, that alternatives are now being urgently investigated.

At the time of writing there are several hundred non-animal alternatives being researched, and animal testing in EC countries may be banned altogether for cosmetics in January 1998. (The scope of this ban is not certain as yet, but skin and eye irritation tests will probably be the first to change.) It seems unlikely that animal testing in the pharmaceutical industry will be banned completely in the foreseeable future. However in all areas, including pharmaceuticals, alternatives are being increasingly used for initial screening, in order to identify those substances which are likely to present problems of toxicity, and therefore to minimise the use of animal tests.

2

Basics

ESSENTIAL OILS

Aromatic plants and infusions prepared from them have been employed in medicines and cosmetics for thousands of years, but the use of distilled oils dates back only 1000 years, which is when distillation, as we know it today, was invented [519].

Essential oils are the volatile, organic constituents of fragrant plant matter and contribute to both flavour and fragrance. They are extracted either by distillation or by cold-pressing (expression). They are not present in all plants. An essential oil should not have any of its normal components removed following extraction, nor have any other substance added. Some essential oils do not stand up to the above definition (see

---

Box 2.1   Examples of 'essential oils' which are not essential oils according to most definitions

Camphor oil (white)

This is only a fraction of true camphor oil, which is not commercially available. This is due to the removal of camphor from the raw material for commercial reasons.

Cognac oil

This is obtained from wine lees, not directly from plant material.

Cornmint oil *(Mentha arvensis)*

Commercial qualities are almost always 'dementholised', i.e. large amounts of menthol are removed from the oil, reducing the menthol content from around 80% to around 40%. Commercial cornmint oil is therefore not the oil in its natural state.

> **Box 2.2** Spurious essential oils
>
> **Marigold/calendula oil**
>
> An absolute of *Calendula officinalis* is occasionally offered, but 'marigold oil' usually turns out to be taget oil, from *Tagetes patula*, *Tagetes minuta*, or *Tagetes erecta*.
>
> **Marjoram (Spanish)**
>
> This is actually a species of thyme, *Thymus masticina*.

Box 2.1). Other oils do not come from the plants that they seem to (see Box 2.2).

Essential oils can be obtained from: flowers (rose), leaves (peppermint), fruits (lemon), seeds (fennel), grasses (lemongrass), roots and rhizomes (horseradish, calamus), woods (cedar), barks (cinnamon), gums (frankincense), tree blossoms (ylang-ylang), bulbs (garlic) and dried flower buds (clove).

Essential oils are usually liquid, but a few are solid (orris) or semi-solid (guaiacwood) at room temperature. The majority of essential oils are clear and colourless or pale yellow in colour, although a few are very deeply coloured, like German chamomile (blue) and European valerian (green).

The historical, and most widely used, method of obtaining essential oils from aromatic plants is by steam distillation. According to this method the essential oil present in the plant, which may be fresh or dried, is vaporised in the steam and then condensed by cooling. The product is a complex mixture of odoriferous, sometimes coloured and frequently biologically active compounds—an essential oil.

Fresh plant material typically yields between 1% and 2% by weight of essential oil on distillation, although rose, for example, typically yields as little as 0.015%, and is consequently highly priced.

Citrus oils are usually cold-pressed for perfumery (cold-pressed citrus oils are generally also preferred for aromatherapy) or distilled for flavour work.[1]

Extraction by organic solvents (producing concretes, absolutes or resinoids) or less frequently with liquid carbon dioxide (producing CO_2 extracts) is also used. Some absolutes and resinoids are covered in this text, but concretes have very little relevance.[2]

Composition

Essential oils are thus mixtures of organic compounds originating from a single botanical source. Most essential oils are so-called plant secondary metabolites, a term which puts them in the category of plant by-products. An essential oil may have hundreds of individual chemical components, most of which have their own characteristic smell and which may have their own biological activity once absorbed into the body.

A typical essential oil is a complex mixture of over 100 different chemical compounds, created and mixed by the parent plant. These have related but distinct types of chemical structure and give the oil its smell, its therapeutic properties and, in some cases, its toxicity. Many of them are extremely widespread throughout the plant kingdom, and many essential oils share some of the same ingredients. Some essential oil constituents are present in only trace amounts; if sufficiently potent they may still be important ingredients, either therapeutically or toxicologically.

Essential oil-producing plants come from many different botanical types and are found in all parts of the world. Even oils from the same type of plant can vary in the amounts of the different chemicals to be found. This variation may be due to factors which affect the plant's environment, including geographical location, elevation, weather conditions and soil type. It may also be due to other factors, such as age of plant, time of year harvested, time of day harvested and so on.

Variations in yield, number of harvests or flowerings can result from growing the same plant in different locations. Differences in production techniques and manufacturing equipment will be apparent in the quality and composition of the resultant oil.

Significant differences in chemical composition of oils from the same type of plant give rise to what are called 'chemotypes'. In many cases there exist more than one chemotype (CT) for a commonly-named essential oil, such as thyme.

The great majority of thyme oils on the market are those rich in thymol, a chemical with irritant properties [153, 158]. However, other chemotypes of thyme oil exist which are rich in compounds such as linalool, and contain little or no thymol. Such chemotypes would not carry the irritation risks associated with thymol-rich thyme oil. 'Safe' chemotypes exist for a few of the hazardous oils, such as thyme. In other instances safe chemotypes exist, but are not commercially available (tansy oil with little or no thujone, for instance). It might be helpful if such oils were to become available for use in aromatherapy. In yet other cases no safe chemotypes exist. Clear labelling is of great importance in distinguishing different chemotypes of essential oils with the same name.

Contamination and adulteration

For the purposes of this text we have assumed that the essential oils under discussion have not been adulterated or contaminated. Contaminants could include pesticides, and adulterants might include, for instance, other essential oils, synthetic chemicals similar to those normally present in the oil and chemicals not normally found in the oil. Contamination or adulteration could feasibly increase the potential toxicity of an otherwise safe essential oil, although we believe that the risk of this happening is, in practice, a small one.

The only absolute guarantee of the purity of an essential oil is to be present during the distillation process, and to take away the desired quantity of essential oil forthwith. There are no tests, including gas chromatography, which guarantee purity per se. Adulteration is widespread in the essential oil industry, but gross adulteration is relatively easy to detect by running the essential oil through a series of tests. Sophisticated adulteration might be detected with sophisticated testing, but the more tests that need to be performed the more expensive the process becomes. Olfactory analysis by a trained nose is often a useful adjunct to technological testing.

The best advice one can give, as far as purchasing pure essential oils is concerned, is to buy from a reputable source, which carries out its own quality control testing. No amount of testing *guarantees* purity, but the more energy a supplier puts into quality control, the purer the oils are likely to be.

With regard to pesticides and herbicides, known collectively as biocides, there are over 400 chemicals which might be used on aromatic plants. Most of these do carry over during steam distillation [228, 229, 230]. The products of solvent extraction, such as absolutes, are even more likely to retain any biocides, as are expressed oils, such as citrus and vegetable oils.

It is feasible that some of the reported allergic reactions to essential oils are caused by biocide residues, and not by the oils themselves [205]. However, skin absorption is necessary for an allergic reaction to take place, and most biocides are poorly absorbed through the skin, less than 10% being absorbed [468].

Biocides are, by definition, toxic, and even tiny amounts are not welcome in aromatherapy. However, it is worth remembering that biocides are present in many common foods. The level of biocide in essential oils is not significantly higher than that found in foods. Since essential oils are used in very much smaller quantities than foods are consumed, and since biocides are poorly absorbed through the skin, we are forced to conclude that, as far as biocide contamination is concerned, aromatherapy massage is probably much safer than eating [205].

Fabrication

Most essential oils, while they may be adulterated, are not complete fabrications. However, in a few cases fabrication is commonplace (see Box 2.3). In other cases, the natural oil is common, but there are also 'nature-identical' versions of these oils. They are made by combining the same components as are found in the natural oil. Typically, most of the trace components of the natural oil are not included. At the same time impurities, present in the synthetic chemicals used to make up the 'essential oil', will be present in trace amounts.

As with contamination and adulteration, there are probably no gross safety problems associated with fabricated oils. However, there could be some increased risks, especially in the area of skin reactions. Fabrication is generally detectable with sophisticated analysis. Technology exists which can differentiate between a chemical of natural origin, and its synthetic version.

Degradation

Chemical degradation is the process by which the quality of a chemical substance is reduced over time. In the case of essential oils for aromatherapy it is definitely undesirable, and tends to occur on prolonged storage, or under poor storage conditions.

The three major factors responsible for essential oil degradation are:

- atmospheric oxygen
- heat
- light.

Atmospheric oxygen can change the chemical composition of essential oils by combining with some of their components. This process is called oxidation and tends to occur in essential oils rich in terpenes, such as lemon and pine. Limonene and pinene, the relative major components, are both somewhat reactive terpenes. (Reactive means likely to undergo chemical reaction.)

Oxidation is speeded up by both heat and light: this is why it is important to store essential oils in dark bottles and away from sources of heat. The more air there is in a bottle of essential oil the more rapidly oxidation will take place.

One safety implication of chemical degradation is that we can not be certain of the composition of a degraded oil unless it is retested, and we may therefore be using a mixture in treatments whose composition is uncertain.

Oxidation can also affect the efficacy of an essential oil. In one recent study lemongrass oil was intentionally oxidised, and was found to have lost much of its antibacterial activity, when compared to fresh, unoxidised lemongrass oil [447]. In extensively oxidised samples antibacterial activity was completely lost.

A more serious implication of degradation is that chemical change can render essential oils more hazardous. Terpene degradation in certain oils leads to compounds being formed which make the oils potential skin sensitisers, while fresh oils are safe to use. Examples include terebinth and all pine oils [430]. Degradation in oils from the *Citrus* genus leads to the formation of chemicals which are weakly carcinogenic in rodents (see p. 95). Although it is unlikely that these can cause problems in humans, this further underlines the importance of using fresh, non-degraded oils for aromatherapy.

It is recommended that, in general, essential oils should be used within 1 year of purchase, or 1 year of first opening the bottle. The best single way to maximise the useful lifespan of essential oils is to store them in a cool place, such as a refrigerator. Oils kept in this way will keep for roughly twice as long, i.e. 2 years following purchase.

Essential oils which readily degrade should ideally be used within 6 months of purchase, or 12 months if kept cool. Note that some essential oils, when refrigerated, become very viscous, and may be difficult to pour.

Another useful way to avoid degraded oils is never to buy oils in clear glass bottles. Light rays can damage undiluted oils (especially citrus oils again) by causing the formation of free radicals. Coloured glass bottles give sufficient protection against light rays to prevent this happening.

The above guidelines apply to undiluted essential oils, but they are not generally important with regard to products made with essential oils.

ESSENTIAL OILS AND TOXICOLOGY

The subject of toxicology is a broad one. It encompasses not only acute toxicity (e.g. poisoning from drinking a whole bottle of essential oil) but also chronic toxicity. In this case a relatively small amount of a toxic essential oil, repeatedly applied, damages the body in some way.

Toxicology also includes the study of hazards in pregnancy and cancer, and threats to the skin, such as phototoxicity and allergy. Factors regarding the status of the individual may also need to be taken into account when applying toxicological information. These might include, for instance, age, blood pressure, liver dysfunction and whether or not the person suffers from particular conditions such as goitre or epilepsy. Any medication being taken might also be relevant.

Some people mount allergic reactions to specific essential oils, when the majority do not, although the reasons for this are unclear. These are known as idiosyncratic reactions. Such reactions are not easy to guard against, but patch testing can be very useful in this context (see p. 78).

In general, toxicity is dose-dependent. The greater the amount of essential oil applied, the greater the risk of harm being caused; the less used, the smaller the risk. However, there are instances where even low doses can produce severe reactions, notably in contact allergy (= skin sensitisation) and phototoxicity.

The degree of toxicity depends, to some extent, on the route of application, and oral administration of essential oils carries the highest practical risk, especially if the oils are taken undiluted. The majority of toxicity data is from oral dosing, although other methods of administration may give similar results if the same amount of essential oil enters the body.

It is important to remember that everything is potentially toxic; it is merely a question of degree. Approximately 150 very strong cups of coffee would be fatal to an adult, but nobody would ever be able to drink that much at one time, so coffee requires no warnings or restrictions. In the same way 150 ml of many essential oils would be fatal if drunk, but this quantity, in terms of aromatherapy, is as nonsensical as 150 cups of coffee; unfortunately, unlike the coffee, it is not impossible to drink.

CHEMICAL COMPOUNDS

Essential oils are composed of chemicals, and every toxic effect displayed by an essential oil has a chemical explanation. We can say, with great certainty, that the general toxicity of wormwood oil is primarily due to its high content of thujone, or that the phototoxic effect of bergamot oil is due to its (very small) content of bergapten and other furanocoumarins.

Because essential oils are highly complex chemically, it is likely that in many instances of toxicity more than one component contributes to the effect. In some cases there may be a 'toxic synergy' taking place between two or more components. In most cases the major contributor to a given toxic effect is easily identifiable.

There are a few known instances in which the toxic effect of one constituent is dampened by another component of the oil. However, these instances are the exception rather than the rule. Enough is now known about essential oil toxicity to make a reasonable prediction about the dangers of a given oil, if we know its chemical composition in detail. A few of such predictions will be quite wrong, but most will be very close to target.

Understanding the chemistry of an oil is, therefore, a very useful basis for understanding essential oil toxicity. The language of toxicity is closely allied to that of chemistry. Before we explore the various ways in which essential oils might be dangerous we will need to have a basic understanding of this chemical vocabulary.

Organic chemicals

In everyday speech, 'organic' substances are taken to be natural, untampered-with, wholesome. However, in chemistry, the word 'organic' means 'containing the element carbon'.

All essential oils are mixtures of chemically organic molecules, as are the vegetable oils with which they are often mixed. Essential oils contain atoms in addition to carbon, the most common being hydrogen and oxygen. These elements are given symbols for ease of identification: C, H and O for carbon, hydrogen and oxygen.

Natural and synthetic chemicals

Natural chemicals, as the name suggests, are those which occur around us without the intervention of humans. The term includes inanimate objects such as sand and rocks, as well as matter which has come from decaying plant and animal remains (coal and gas). In the context of aromatherapy, we are concerned with the chemical products of living things—aromatic plants and their essential oils.

Plants have a staggering capacity to synthesise chemical substances, and plant biochemistry is correspondingly complex. A single coffee bean contains 141 known constituents and the humble potato has 367 [8]. In both cases, yet more chemicals remain to be identified. Most essential oils contain between 50 and 500 different chemicals.

Synthetic chemicals are those which have been made in a laboratory or by an industrial process. They may be artificial copies of natural substances or they may be previously unknown. For instance, in the days before decaffeinated coffee became popular, the caffeine in some medicines was synthesised (none of it came from coffee). Today at least some is obtained from the process of decaffeination.

The minuscule size of molecules

To understand just how small molecules are, and how many of them there are around us, consider the fact that a drop of essential oil contains about 40 000 000 000 000 000 000 000, or forty-thousand

million million million molecules! Even if we are considering trace constituents of essential oils (perhaps present at only 0.001%) the numbers of molecules involved will still be gigantic.

On a cellular scale, therefore, amounts of chemicals which seem insignificant to us, can in fact be very significant; it is important not to see a 'drop' of essential oil as merely a 'small' amount. (A few drops of a highly toxic essential oil could be fatal to an infant if ingested.)

Molecular diagrams

It is useful, although not essential, to have an understanding of how molecules are visually represented. This will help those who refer to other literature concerning essential oil toxicology or chemistry. Molecular diagrams are usually drawn in one of two ways. Either letters denoting different types of atom are included, with lines (single, double or triple) showing the types of bond holding the atoms together (Fig. 2.1)

citronellol

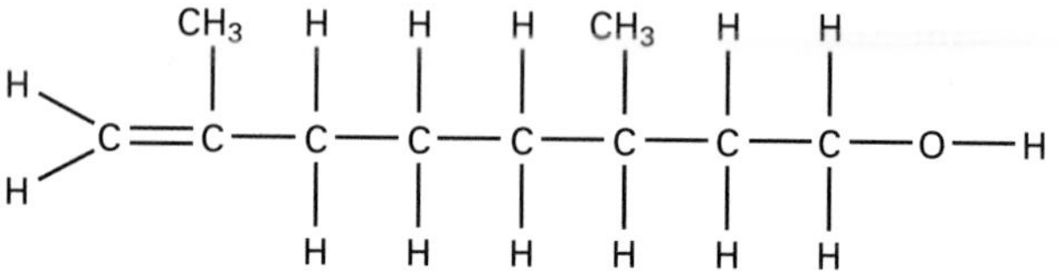

$$H_2C = CH - CH_2 - CH_2 - CH_2 - CH(CH_3) - CH_2 - CH_2 - O - H$$

Figure 2.1 Molecular structure of citronellol.

Or, a shorthand system is used, whereby the letters C (carbon) and H (hydrogen) are left out and only other types of atom (such as oxygen or sulphur) are shown explicitly (Fig. 2.2)

citronellol

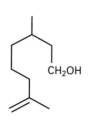

Figure 2.2 Molecular structure of citronellol.

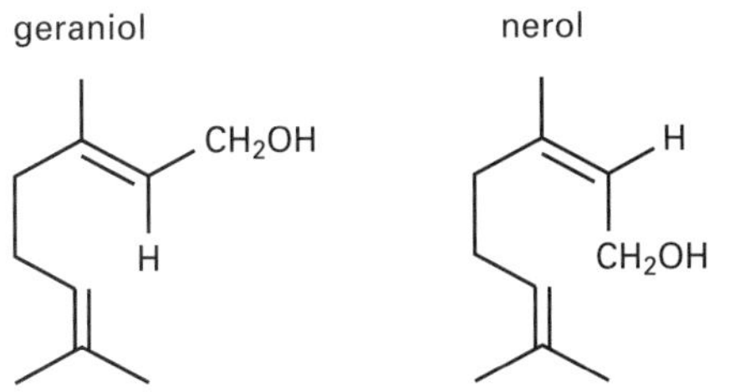

geraniol

CH₂OH

H

nerol

H

CH₂OH

Figure 2.3 Molecular structure of geraniol and nerol—isomeric alcohols.

It is possible to show the relative positions of atoms in a molecule, even though molecular diagrams are two-dimensional representations of three-dimensional objects (Fig. 2.3).

ESSENTIAL OIL CHEMISTRY

This section is concerned with important information about the chemistry and composition of essential oils. It is not intended to do the job of a textbook of essential oil chemistry or the science of biologically active plant compounds (pharmacognosy). For more information on these subjects, the reader is referred to the bibliography [4, 7].

Rather, we will explore here chemical and physical factors of direct importance to aromatherapists, as well as the confusing jungle of chemical names by which essential oil ingredients are known.

The chemistry of essential oils is complex and many of the compounds which make up the oils are hard to detect. Some of them are only present in minute quantities and so are difficult to measure; some are very similar to each other and so are difficult to distinguish with certainty.

Modern analytical methods used for essential oils include GC/MS (gas chromatography and mass spectrometry), HPLC (high performance liquid chromatography) and NMR (nuclear magnetic resonance).

A gas chromatographic trace of peppermint oil is shown in Figure 2.4. This identifies cineole (10) menthone (24) isomenthone (25) and menthol (45). Each peak generally represents a single chemical

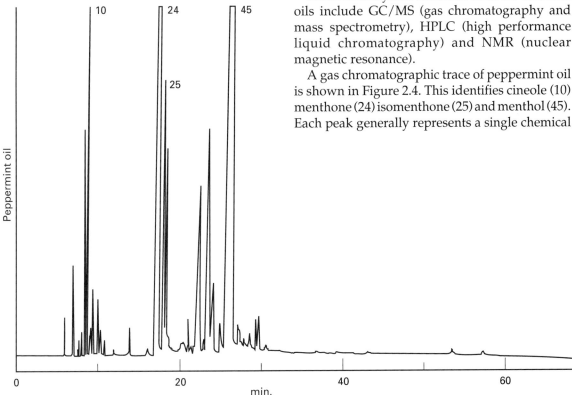

Figure 2.4 A peppermint GC trace.

compound. The size of the peak is proportional to the quantity of that compound in the essential oil.

The major components of the best-known essential oils have been known for many years [7]. The introduction of gas chromatography, a very fine separation technique ideally suited to volatile compounds, has revolutionised the detection of minor chemical constituents, especially when used in conjunction with MS and NMR spectroscopy. Today, even trace constituents, including pollutants like pesticides and adulterants such as synthetics, can be detected.[3]

Essential oil composition

Essential oils are largely composed of volatile chemicals and so tend to evaporate rapidly in a warm or breezy environment. Many are extremely sensitive to the effects of light, heat, air and moisture, and should be stored in tightly stoppered, well-filled, dark-glass bottles. Because essential oils tend to gradually change their composition during storage, it is not possible to guarantee with absolute certainty the composition of a sample without a recent analysis.

Most essential oils contain many different types of compound, but with one or two major components, which largely shape the pharmacology and toxicology of the oil. Many of the properties of peppermint oil, for instance, can be attributed to its 40% or so content of menthol.

In most cases, the level of a given compound in a given essential oil varies within a certain range. For instance, the camphor content of rosemary oil is normally somewhere between 10% and 20%. Although camphor contents both above and below these ranges are possible, they are unlikely to be found in commercially sold essential oils, as they tend to occur in plants not grown on a commercial scale. Some ranges are much narrower than the 10% seen here, while others are considerably broader.

A typical essential oil contains several hundred individual chemicals, with the great majority at levels of less than 1%. In some cases these components also figure in the action of the oil relative to the human body. For example

bergapten, which makes bergamot oil strongly phototoxic, is found at levels of around 0.3%.

Variation of essential oil composition

Widely differing oils from completely different plant species may have some ingredients in common; coriander seed and lavender oils both contain large amounts of linalool, for instance. Also, structurally similar compounds may have very different chemical properties; a common example is that of carvone, a ketone with two isomeric forms (Fig. 2.5). These are chemically structured mirror images and are identified by their ability in solution to reflect polarised light. This reflection can be measured with accuracy. The dextro-rotatory isomer, 'd' or (+) carvone, is found in caraway oil, and smells of caraway. The laevo-rotatory isomer, 'l' or (-) carvone, smells minty and is the main component of spearmint oil. An equal mixture of the (+) and the (-) isomers is called 'racemic'; 'dl' or (±) carvone has been identified in gingergrass and lavandin oils.

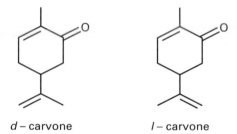

d – carvone   l – carvone

Figure 2.5 Molecular structure of d- and l-carvone.

Hydrocarbons and oxygenated compounds

The first major category of compounds for us to consider is the hydrocarbons, compounds which only contain carbon and hydrogen. From these hydrocarbons the plant makes oxygenated compounds, which constitute the second major category. In some essential oils, such as pine, the hydrocarbons predominate and only limited amounts of oxygenated constituents are present. In others, such as clove, the bulk of

Box 2.4 Classes of compounds found in essential oils

Hydrocarbons
Terpenes

Oxygenated compounds
Alcohols
Aldehydes
Ketones
Esters
Phenols
Oxides
Peroxides
Lactones
Acids
Furans
Ethers

Other compounds
Sulphur compounds

the oil consists of oxygenated compounds. A few essential oils have sulphur-containing constituents, which do not come under either of the previous categories.

The odour and taste of an essential oil is mainly determined by the oxygenated constituents; the fact that they contain oxygen gives them some solubility in water and considerable solubility in alcohol.

Box 2.4 lists the names given to the various classes of chemicals found in essential oils. Compounds such as these are found throughout nature, and are not limited to essential oils.

Terpenes

Terpenes are composed of hydrogen and carbon atoms only. All terpenes are based on the isoprene unit, an essential building block in plant biochemistry (Fig. 2.6).

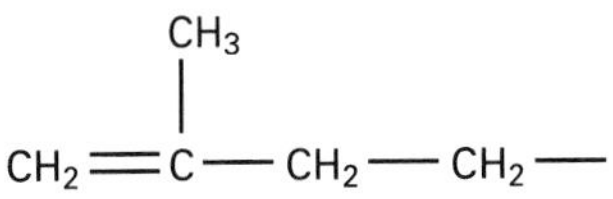

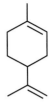

$$CH_2 = C - CH_2 - CH_2 -$$
with CH_3 above the C

isoprene unit

Figure 2.6 The isoprene unit.

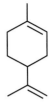

d – limonene

Figure 2.7 Molecular structure of *d*-limonene.

Monoterpenes are made up of two such units joined head-to-tail and so contain 10 carbon atoms. They are called monoterpenes, because this is the basic terpene unit as found in nature. Examples are limonene (<> lemon) (Fig. 2.7) and pinene (<> pine).

Sesquiterpenes are composed of three isoprene units and therefore have 15 carbon atoms. Monoterpenes are very common in essential oils, sesquiterpenes are less common.

The names of terpenes end in -ene (see Table 2.1).

Table 2.1 Some common terpenes

Monoterpenes	Sesquiterpenes
Camphene	Bisabolene
Carene	Cadinene
Cymene	Caryophyllene
Dipentene	Cedrene
Limonene	Chamazulene
Myrcene	Copaene
Ocimene	Farnesene
Phellandrene	Germacrene D
Pinene	Humulene
Sabinene	Selinene
Terpinene	Terpinolene

Alcohols

These are perhaps the most varied group of terpene derivatives. They are usually based on the monoterpenes and so contain 10 carbon atoms. Linalool (<> rosewood) and geraniol (<> geranium) are examples of monoterpenoid alcohols.

More rarely, plant alcohols are based on a sesquiterpene. The santalols, found in sandalwood oil, and the fusanols, obtained by distilling the

Box 2.5 Some common alcohols

Benzyl alcohol	Nerol
Bisabolol	Nerolidol
Borneol	Nuciferol
Carotol	Olibanol
Cedrol	Patchouli alcohol
Citronellol	Phenylethyl alcohol
Daucol	Pinocarveol
Farnesol	Sabinol
Geraniol	Santalol
Lavandulol	Terpineol
Linalool	Terpinen-4-ol
Menthol (Fig. 2.8)	Vetiverol
Neomenthol	Viridiflorol

Box 2.6 Some common aldehydes

Acetaldehyde	Cuminaldehyde
Anisaldehyde	Geranial
Benzaldehyde	Myrtenal
Cinnamaldehyde	Neral
Citral	Perillaldehyde (Fig. 2.9)
Citronellal	Valeranal

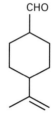

l – perillaldehyde

Figure 2.9 Molecular structure of *l*-perillaldehyde.

menthol

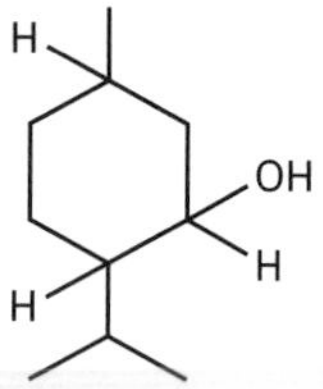

Figure 2.8 Molecular structure of menthol.

wood of *Eucarya spicata*, a small Western Australian tree, are examples of sesquiterpene alcohols [4].

The names of all alcohols end in '-ol' (see Box 2.5). Unfortunately, some other essential oil components, particularly phenols, are also known by names ending in -ol. (Ethanol, best known as an ingredient of alcoholic drinks, is not a natural component of any essential oil.)

Aldehydes

These may be considered as partially oxidised primary alcohols. Aldehydes are widely distributed as natural essential oil components. Examples are cinnamaldehyde (<> cinnamon bark) citral (<> lemongrass) and citronellal (<> citronella). Aldehydes have a slightly fruity odour when smelled on their own. They often cause skin irritation and allergic reactions. (A well-known aldehyde, not found in essential oils, is formaldehyde.)

The names of aldehydes end in -al or -aldehyde (see Box 2.6).

Ketones

These compounds are structurally similar to aldehydes. They are produced by oxidation of secondary alcohols. They are stable compounds and are not easily oxidised further. Examples are fenchone (<> fennel) carvone (<> caraway) and camphor (<> rosemary). Ketones are often relatively resistant to metabolism by the body and

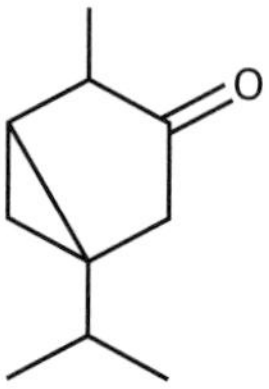

alpha – thujone

Figure 2.10 Molecular structure of *alpha*-thujone.

may be excreted in the urine unchanged. (A well-known ketone, not found in essential oils, is acetone.)

The names of ketones generally end in -one (see Box 2.7).

Box 2.7   Some common ketones

Acetophenone	Perilla ketone
Camphor	Pinocamphone
Carvone	Pinocarvone
Fenchone	Piperitone
Ionone	Pulegone
Irone	Tagetone
Jasmone	Thujone (Fig. 2.10)
Menthone	2-undecanone
Methylheptenone	Valeranone
Nootkatone	Verbenone

Esters

These compounds often have an intensely fruity odour. They are produced from the corresponding terpene alcohol and an organic acid, and highest levels are reached on maturity of the fruit/plant or on full bloom of the flower. In bergamot, as the fruit ripens, linalool is converted to linalyl acetate; in peppermint, menthol is converted to menthyl acetate.

Box 2.8   Some common esters

Benzyl acetate	Menthyl acetate
Benzyl benzoate	Methyl anthranilate
Bornyl acetate	Methyl benzoate
Bornyl isovalerate	Methyl butyrate
Butyl angelate	Methyl salicylate (Fig. 2.11)
Citronellyl acetate	Neryl acetate
Citronellyl formate	Sabinene hydrate
Eugenyl acetate	Sabinyl acetate
Geranyl acetate	Terpinyl acetate
Lavandulyl acetate	Vetiveryl acetate
Linalyl acetate	

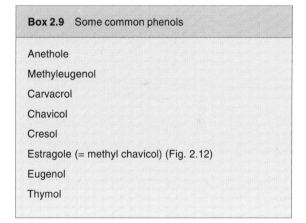

methyl salicylate

Figure 2.11   Molecular structure of methyl salicylate.

The names of essential oil esters generally follow the pattern: -yl -ate (see Box 2.8).

Phenols

Like alcohols, phenols have an –OH group. However in phenols, the –OH is attached directly to a benzene ring. This makes the –OH very reactive, and means that phenols are very likely to be irritating. Unfortunately many plant phenols

Box 2.9   Some common phenols

Anethole

Methyleugenol

Carvacrol

Chavicol

Cresol

Estragole (= methyl chavicol) (Fig. 2.12)

Eugenol

Thymol

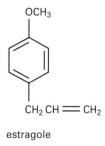

estragole

Figure 2.12   Molecular structure of estragole.

have names which sound similar to those of alcohols, but chemically, they are quite different. Thymol (<> thyme) and eugenol (<> clove) are phenols. (Phenol [= carbolic acid] is a disinfectant derived from coal tar. It is not found in essential oils and, confusingly, is an acid, not a phenol.)

The names of most phenols end in -ol or -ole (see Box 2.9).

Oxides

Organic oxides are unusual, highly reactive chemicals which decompose easily at high temperatures and on prolonged exposure to air or water. Oxides are materials where an oxygen atom in the molecule is situated between two carbon atoms: –C–O–C–. The most important oxide found in essential oils is cineole, which exists in two forms. The more abundant one is 1,8-cineole (Fig. 2.13), also known as eucalyptol when obtained from eucalyptus oil.

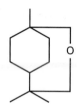

1,8 – cineole

Figure 2.13 Molecular structure of 1,8-cineole.

The names of oxides often end in -ole or 'oxide' (see Box 2.10).

Box 2.10 Some common oxides	
Bisabolol oxide	Linalool oxide
Bisabolone oxide	Rose oxide
Caryophyllene oxide	Sclareol oxide
Cineole	

Peroxides

In peroxides two atoms of oxygen are in the link between two carbon atoms: –C–O–O–C–.

The typical example is the toxic ascaridole (Fig. 2.14), found in wormseed oil. Few other peroxides exist in essential oils.

ascaridole

Figure 2.14 Molecular structure of ascaridole.

Lactones

Lactones are cyclic esters which may be simple molecules, or more complex molecules, such as bergapten, which can increase the tanning effect of UV rays on the skin. They generally have a low volatility. Costuslactone (<> costus) tends to cause skin sensitisation without the help of UV.

The names of lactones are rather variable (see Box 2.11). They are generally unrelated chemically to terpenes and sesquiterpenes.

Box 2.11 Some common lactones	
Alantolactone (Fig. 2.15)	Costuslactone
Ambrettolide	Coumarin
Bergapten	Pentadecanolide
Costunolide	Xanthotoxin

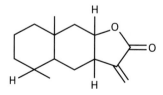

alantolactone

Figure 2.15 Molecular structure of alantolactone.

Acids

Acids are rare in essential oils, and generally have a low volatility. They usually have the COOH grouping. The names of acids take the form: -ic acid (Box 2.12). Hydrocyanic acid (= hydrogen cyanide) (Fig. 2.16) is found in bitter almond oil (it forms during distillation) and is removed before the oil is used.

Box 2.12   Some common acids
Alantic acid
Anisic acid
Benzoic acid
Cinnamic acid
Citronellic acid
Hydrocyanic acid (= prussic acid, = hydrogen cyanide)
Phenylacetic acid
Valerenic acid

hydrocyanic acid

$$H - C \equiv N$$

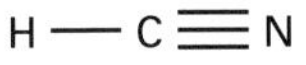

Figure 2.16   Molecular structure of hydrocyanic acid.

Furans

Furans are oxygenated compounds where the oxygen atom is part of a ring. They are only found in a few essential oils. Menthofuran (Fig. 2.17) is found in most mint oils. The names of some furans are given in Box 2.13.

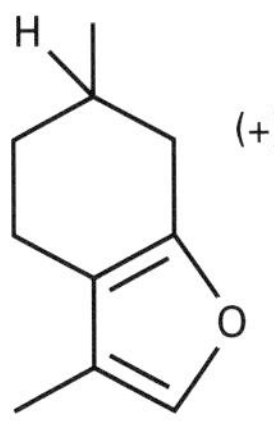

(+) – menthofuran

Figure 2.17   Molecular structure of (+)-menthofuran.

Box 2.13 Furans	
Butyl phthalide	Ligustilide
Butylidine phthalide	Menthofuran
Dihydrobenzofuran	

Ethers

Ethers are fairly unusual essential oil constituents whose molecules have simple, carbon-containing groups either side of an oxygen atom. Pandanus absolute has an ether as a major constituent, methyl-*beta*-phenylethyl ether.

Sulphur compounds

These are rather reactive molecules found in only a few essential oils. They are not derived from terpenes or sesquiterpenes and are often very pungent. Diallyl disulphide (<> garlic) is a typical sulphur compound.

Chemical names for sulphur-containing molecules usually have sulph- or -thio- in them (see Box 2.14).

Box 2.14   Sulphur compounds
Allyl isothiocyanate (Fig. 2.18)
Allylmethyl sulphide
Allylpropyl disulphide
Diallyl disulphide
Diallyl polysulphide
Diallyl trisulphide
Diallyl thiosulphinate (= allicin)
Diisopropyl disulphide
Dimethyl disulphide
Dimethyl sulphide
Dimethyl trisulphide
Methyl disulphide
Methyl allyl trisulphide
Phenylethyl isothiocyanate
Propyl disulphide

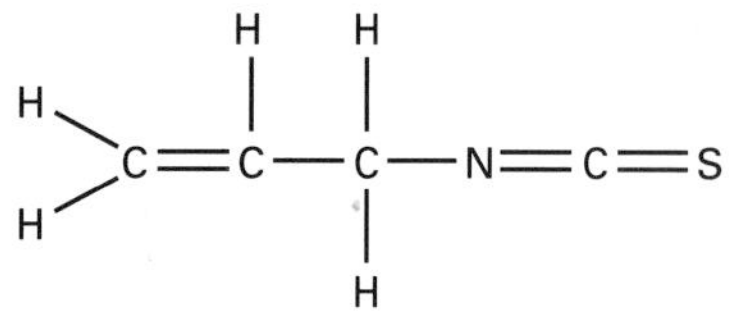

allyl isothiocyanate

Figure 2.18 Molecular structure of allyl isothiocyanate.

Trace constituents

A trace constituent is one which is present in very small amounts. For instance, 1,8-cineole is the principal component of eucalyptus oil and is usually present at a concentration of about 80%. However, in mandarin oil, 1,8-cineole has been detected at 0.002%, a 40 000-fold lower concentration than in eucalyptus oil. In mandarin oil, therefore, it is a trace constituent. Mandarin oil comprises one major component, d-limonene which, along with other terpenes, accounts for some 95% of the oil. There are then the remaining constituents, the remaining 5%, which comprise at least 74 individual chemicals.

SUMMARY

• A typical essential oil is a complex mixture of over 100 organic chemical compounds. These have common types of chemical structure and give the oil its smell, its therapeutic properties and, in some cases, its toxicity.

• Some essential oil constituents are present in only very small amounts; if sufficiently potent they may still be important ingredients, either therapeutically or toxicologically.

• Oils from the same species of plant can vary in the amounts of the different chemicals to be found. This variation may be due to many factors, most of them relating to either the plant's environment and growing conditions, or to harvesting and distillation techniques.

• Oils from the same type of plant, but with quite different major components, are called chemotypes.

• Essential oils are either distilled or, in the case of citrus oils, cold-pressed. Other forms of aromatic extract are: concretes, absolutes, resinoids and CO_2 extracts.

• Essential oils can be adulterated, contaminated or fabricated. All are, in theory, detectable by laboratory analysis which, however, is expensive. As far as we know, none of these is likely to present serious safety problems. Biocide contamination in essential oils is no more of a problem than biocide contamination in foods.

• Essential oils are very sensitive to the effects of light, heat, air and moisture, and should be stored in tightly stoppered, dark-glass bottles.

• Degradation can lead to increased hazards. The oxidation of some terpenes, for instance, makes them more likely to cause skin sensitisation.

• To avoid degradation, essential oils should be stored in a cool place, such as a refrigerator, in dark bottles, and should be used within 1 year of purchase or first opening.

• In general, toxicity is dose-dependent. The more of a substance that is used, the greater the potential for harm.

• Patch testing is a good way of predicting allergic reactions to essential oils.

• Toxic reactions depend not only on the amount of oil used but also on the method of administration, and the physiological status of the individual.

• Most toxic effects of essential oils are attributable to known chemical components.

• All essential oils are mixtures of chemically organic molecules. That is, all the chemicals in essential oils contain carbon.

• A typical essential oil contains several hundred individual chemicals, with the great majority at levels of less than 1%.

• The types of chemical compound found in essential oils include: terpenes, alcohols, aldehydes, ketones, esters, phenols, oxides, peroxides, lactones, acids, furans, ethers and sulphur compounds.

Notes

1. It is important, for reasons of clarity, to distinguish between the various types of oils and extracts, and not

all of them should be referred to as 'essential oils'. Unfortunately, however, there is no single word to describe the whole family of aromatic extracts, especially since for many people the word 'extract' connotes a material which is specifically not an essential oil. We have, therefore, taken liberties with our book title (somehow, 'Aromatic Material Safety' does not have the same ring to it).

2. CO_2 extracts are relatively new and little used, and consequently there is little or no toxicological data on them. However, they are likely to be used in aromatherapy, as are the even newer 'phytols', extracted by a process known as 'advanced phytonics'.

Both CO_2 extracts and phytols give more complete and 'truer' products than essential oils, but are more costly. Toxicologically they are likely to be similar to essential oils from the same plant, although in a few cases there may be significant differences.

3. Sophisticated analysis is very costly, and the full range of equipment required runs into hundreds of thousands of pounds. The problem of detecting sophisticated essential oil adulteration in the laboratory is similar to drug-testing in athletes. Technicians may well discover adulterants (or drugs) if they know exactly what they are looking for. Clever adulterants (or drugs) are, of course, those which are difficult to detect.

3

Administration

In this chapter we explore the methods used to administer essential oils to the body; their frequency, dosage, and safety implications. Emphasis is given to dermal application, since this is the most commonly used route, in both therapy and retail products.

In an aromatherapy session, essential oils are usually applied to the skin diluted in a vegetable oil vehicle, in the classical 'aromatherapy massage'. Clinical experience in aromatherapy suggests that clients gain benefit in three ways: from absorbing the essential oil through the skin; from inhaling the vapour; and from the massage itself. Commercial aromatherapy products also operate via skin application and/or inhalation.

Some practitioners, notably a group of French doctors, favour the oral administration of essential oils for the treatment of certain conditions. Oral administration is considered later in the chapter.

It is important to know whether, and under what conditions, essential oils enter the body's circulation, as well as where they go once absorbed and how the body eliminates them. In this chapter, we consider the passage of oils through the skin, absorption by inhalation, and oral, rectal and vaginal administration. The following chapter covers distribution and elimination.

DOSAGE

In toxicology, dosage is everything. Extremely small amounts of 'toxic' substances can be ingested without causing any harm, while very large amounts of common foodstuffs could, in theory, be fatal. A number of homoeopathic remedies are prepared from toxic substances, arsenic for instance, but the amount of arsenic actually ingested is so very small (or even non-existent) that it is harmless. Traces of cyanide are present in apple pips, and eating a few of them causes no harm. However, in the late 1970s an Englishman died after eating a bowlful of apple pips. He had saved them up for a feast because he liked them so much.

A great many food substances naturally contain very small amounts of toxic chemicals. Raw cabbage contains allyl isothiocyanate, the ingredient in mustard essential oil that makes it so toxic and violently irritant. However, the amount of raw cabbage you would have to consume at one sitting to commit 'cabbage suicide' would be many thousands of times more than it would be possible to eat. Cabbage, therefore, is not considered toxic.

Although mustard essential oil is used in food flavours, the quantities are very small indeed (equivalent to about one hundredth of a drop of mustard oil in 50 g of pickle). However, in aromatherapy, mustard oil should be regarded as toxic. Even if mustard oil is diluted down to 2% in vegetable oil, this still represents a concentration of essential oil some 2000 times higher than in the pickle.

In general, the amounts of essential oil which aromatherapists use during massage bring greater quantities into contact with the body than does the consumption of foodstuffs or the use of perfumes. Essential oils which are not regarded as hazardous in those contexts may be dangerous when used in aromatherapy if insufficient regard is given to dose and to the condition of the client.[1]

Amounts of essential oil used in aromatherapy

Throughout the book we refer to 'the amounts used in aromatherapy massage' as being either safe or not in a given instance, as we do with oral dosing. It is important, therefore, to clarify what these amounts are.

The total quantity of essential oil absorbed into the body from an aromatherapy massage varies according to:

Important:
- The percentage dilution of the essential oil
- The total quantity of oil applied
- The total area of skin to which the oil is applied

Less important:
- The particular essential oil(s) used
- The particular vegetable oil (or other vehicle) in which the essential oil is dispersed
- The part(s) of the body to which the oil is applied
- The temperature and moisture content of the skin
- The health and integrity of the skin
- The absorptive capacity of the skin
- The extent to which the skin is covered after the massage
- How soon the skin is washed following the massage.

In spite of all these variants, the range of quantities that might be absorbed is easy to calculate. The percentage dilution used is normally between 2% and 3%, but we can take a minimum and maximum range of 1% and 5% for massage over a large area of skin. The usual vehicle is vegetable oil, and experience shows that the total quantity of oil applied will vary between a minimum of 5 ml and a maximum of 25 ml for a full-body massage. We will assume that the oil is spread evenly and thinly over the area to be massaged, and that the maximum number of full-body applications in 24 hours is one.

These same parameters could also apply to aromatherapy products. Although more than one application per 24 hours is possible, full-body application, with the skin-absorption assistance of massage is very unlikely.

Referring to Table 3.1 we can see that the smallest quantity of essential oil likely to be used in practice is 0.05 ml, and the largest quantity is 1.25 ml. We know that, after application to uncovered skin, between 4% and 25% of the essential oil is absorbed [53, 91]. If we take 4% of 0.05 ml, and 25% of

Table 3.1 Amount of essential oil applied to the skin, in millilitres

Concentration in oil or product applied	Total quantity of oil or product applied				
	5 ml	10 ml	15 ml	20 ml	25 ml
1%	0.05	0.1	0.15	0.2	0.25
2%	0.1	0.2	0.3	0.4	0.5
3%	0.15	0.3	0.45	0.6	0.75
4%	0.2	0.4	0.6	0.8	1.0
5%	0.25	0.5	0.75	1.0	1.25

1.25 ml we end up with a minimum of 0.002 ml and a maximum of 0.3 ml. This last figure is equivalent to approximately 6 drops of essential oil in a 24-hour period for our 'worst scenario'.

4% of 0.05 ml = 0.002 ml = 1/20th drop
25% of 1.25 ml = 0.3 ml = 6 drops

In most cases the amount actually absorbed is likely to be in the $\frac{1}{2}$ drop to 2 drops (0.025–0.1 ml) range.[2]

For oral dosing the quantity taken within a 24-hour period will vary depending on the approach of the prescribing practitioner, but the range is 0.5 ml (10 drops) to 2.5 ml (50 drops). This gives us a maximum eight times greater than the maximum for percutaneous absorption. The more usual oral dosage range is 0.5 ml to 1 ml (10–20 drops), about 10 times greater than the usual massage range. (We have assumed that 100% of any oil administered orally is absorbed. Although this is unlikely in every instance, it is appropriate for a worst-case scenario.)

If oral dosage is 8–10 times greater than for massage, this gives us a reasonable basis for making a clear distinction between the two in terms of safety. Absorption into the bloodstream after dermal application is much slower than after oral dosing. Concentrations will not build to high levels because the oil is continually being removed from the bloodstream. At any one time the amount present in the blood after dermal application will be relatively low.

The amounts absorbed from baths, inhalation and vaporisation are relatively small, and will certainly not exceed the maximum for massage.

The amount of essential oil used in pessaries, douches and suppositories varies, and could go up into the oral dosage range. The safety of individual essential oils in these cases will depend on the oil and the quantity administered.

In the practice of aromatherapy, dosages and dilutions are not standardised, and are not always measured precisely. However, as with herbal medicine, the amounts of pharmacologically active substance applied to the body are, in general, significantly smaller than the dosages of orthodox drugs.

That the dose of a substance makes a difference to its therapeutic and toxic effects is beyond question. Determining how these effects translate into clinical experience with essential oils still requires a lot of study.

Frequency of application

If dosage is the most important factor in toxicology, frequency of application comes a close second, and in terms of chronic (long-term) toxicity frequency is all-important. A relatively small quantity of a toxic essential oil, if applied daily for several months, could cause minor tissue damage in the liver or kidneys. The effects of chronic toxicity may not be recognised as such, because the symptoms produced are relatively minor and very common. See page 47 for a full explanation of chronic toxicity.

Babies and children

The dosage of a drug is normally reduced for children in proportion to body weight, so that approximately the same amount of drug per kilogram of body weight is administered. So, for a child of 20 kg the oral dose would be 0.15–0.3 ml (3–6 drops) in 24 hours. We do not recommend oral dosing in children smaller than 20 kg. Our recom-mendation that only primary care practitioners such as medical doctors and medical herbalists should prescribe essential oils for oral administration applies equally, if not more so, to children.

For dermal application, great caution is certainly in order for very small infants. This is because a baby's skin is especially thin, and is therefore both potentially more sensitive, and more permeable to essential oils. A baby is also less equipped to deal with any adverse effects than is an adult. These cautions apply even more to premature babies, and here it might be prudent to avoid the external use of essential oils altogether, unless there are very important benefits to the baby. Even vaporising oils into the air might be best avoided.

The total amount of essential oil absorbed will in any case be less than in adults, because of the smaller body size. The essential oils used will also make a difference to any potential skin reactions. It would be prudent to completely avoid the oils, at least for children under 2 years, which are potentially problematic in this regard (see p. 229). It is easier to list those oils which are unsafe to use than it is to list safe ones for infants. However, as examples of those oils which would be considered safe, see the list of oils safe in pregnancy (p. 111). Recommendations for the maximum concentrations of essential oil that can be used for dermal application in children are given in Table 3.2.

It is recommended that children under about 2 years of age should not be given baths with essential oils unless the oils have been previously dispersed in a water-soluble medium. This is to guard against skin irritation from improperly dispersed oils, and could be applied to any age group.

Instillation into the nose, or introducing by drops, results in a combination of oral ingestion and inhalation. There are reports from Belgium of non-fatal, but serious toxicity in children who have had solutions containing either menthol (4 cases) or cineole (9 cases) instilled into their noses [250]. The ages ranged from 1 month to 3 years and 9 months. The effects of poisoning included irritated mucous membranes, tachycardia, dyspnoea, nausea, vomiting, vertigo, muscular weakness, drowsiness and coma [5, 250]. The most serious symptoms (including coma) were seen in a child of under 2 months, who had 1 ml of a menthol solution instilled into his nose. Some of the other cases suffered no more than mucous membrane irritation, but some of these were given gastric lavage.

In most of the cases the drops were given accidentally, in mistake for another, safer preparation. No details were given regarding the amounts of menthol or cineole administered, so it is difficult to extrapolate to essential oils. Clearly, peppermint and eucalyptus oil would be implicated, and these are among the primary oils used as decongestants. However, many other essential oils, administered in this way, could cause similar problems. The seriousness of these cases leads us to recommend that essential oils should not be instilled into the noses of children under the age of 5.

BIOAVAILABILITY

In this text we have used the term 'bioavailability' to indicate the proportion of a substance which reaches the systemic circulation after administration. It is lower after dermal or oral dosing than after intravenous injection which gives, by definition, 100% bioavailability.

Essential oils are complex mixtures of chemical substances; each of these will have its own bioavailability. Strictly speaking, an essential oil component is not bioavailable if it is bound to plasma albumin (see p. 37). Since the true bioavailabilities of most essential oil ingredients by different routes are not known, we have used the term as defined above. Also, bioavailability is subject to biological variation; individuals will have their own bioavailability profile related to their own physiology and metabolism. The choice

Table 3.2 Recommended maximum concentrations of essential oil for dermal administration to children

Age	Maximum concentration
Up To 6 months	1%
6–24 months	2%
2–10 years	3%
10+ years	5%

The guidlines here are not research based, and should be taken as helpful suggestions rather than absolute rules. The particular oils used, and the individual they are used on are invariably important factors.

and dose of therapeutic agents are, ideally, tailored to the individual patient.

Looking at Table 3.3 we can see that the amount of essential oil reaching the bloodstream at any given time depends not only on how much is administered, but on the route of administration. In dermal application, for example, the amount of essential oil applied may be relatively high, but because absorption is slow the levels in the blood are much lower than those from oral or mucous membrane administration.

In a recent human study, peak plasma concentrations of lavender oil components were detected 20 minutes after the oil had been applied by massage; after 90 minutes concentrations had fallen close to zero (see Fig. 3.1) [444]. In this study 1.5 g of massage oil (2% lavender oil in peanut oil)

Table 3.3 Bioavailability and likely dose assumed to apply to essential oils

Route	Bioavailability	Likely dose
Skin	Variable	Variable
Nose	High	Low
Lungs	High	Low
Mouth	Variable	High
Rectal/vaginal	High	Variable
Intravenous	100%	High

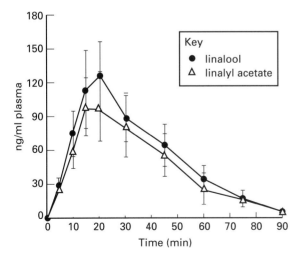

Figure 3.1 Blood levels of linalool and linalyl acetate after the application of lavender oil by massage (with permission from [444]).

was gently massaged into the abdomen for 10 minutes, and blood samples were drawn from the arm 0, 5, 10, 20, 30, 45, 60, 75, and 90 minutes after finishing the massage. Plasma concentrations of up to 120 ng/ml are extremely low. The two components measured were linalool (24.8% of the oil) and linalyl acetate (29.6% of the oil).

DERMAL ADMINISTRATION

To what extent do essential oils and vegetable oils penetrate human skin and find their way into the circulation? Which factors might aid penetration and which hinder it? If essential oils do pass through the skin, how do they do so? Surprisingly, these questions have not been much addressed in the aromatherapy literature, although dermal absorption is studied in great detail by toxicologists.[3]

For many years biologists believed that the skin formed an impervious barrier to the outside world, but we now know that this is not the case.[4] Most chemicals which are applied to the skin are absorbed to some degree. However, the skin is still an important protective barrier, limiting the rate at which potentially toxic substances gain entry into the body. As well as its outer, horny layer the skin has other means of protection, including sweating, enzymes and certain immune mechanisms [468].

The structure of the skin (Fig. 3.2)

The skin is the largest organ in the body. It is about 3 mm thick, and one of its functions is to keep undesirable chemicals from entering the body. Its outer layer, which consists of dead, epidermal cells is known as the stratum corneum. Below this lies the remainder of the epidermis (the living, or 'viable' epidermis) consisting of living cells arising in the deep epidermis, which become flatter as they rise to the surface. Below the epidermis is the much thicker dermis, containing nerves, sweat glands, sebaceous glands, hair follicles, blood vessels and lymph vessels. Beneath the dermis lies subcutaneous fat.[5]

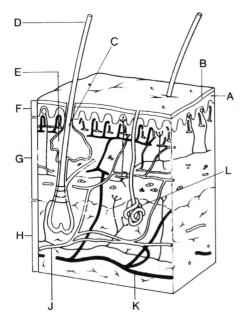

Figure 3.2 Cross-section of skin. (A) Viable epidermis
(B) Stratum corneum (C) Hair follicle (D) Hair
(E) Sebaceous gland (F) Epidermis (G) Dermis
(H) Subcutaneous tissue (J) Vein (K) Artery
(L) Sweat gland.

Percutaneous absorption

Since the cells of the stratum corneum are not alive they are incapable of reacting to chemicals. Therefore, before an essential oil that has been applied to the skin can cause a toxic response in the skin, or indeed anywhere else in the body, it must first cross the stratum corneum. This probably takes place by a combination of mechanisms including entry through the ducts of the sweat glands and the hair follicles [93] and around the cells of the stratum corneum [53].

The essential oil will then reach the viable epidermis, and it is here that toxicity may first take place. If the oil reaches the upper dermis it enters the capillary circulation, and will be carried to distant sites [468].

In contrast to sweating, which is an active, energy-requiring process, the passage of molecules inwards through the skin occurs quite passively; the dermal cells do no inward pumping, and passage is by simple diffusion only [10]. This means that substances which can enter the body through the skin will do so if they come into contact with it for more than a few moments. In an aromatherapy session, almost the whole body, representing about 1.8 square metres of surface area, is generally massaged with diluted essential oil.

In 1940, the kinetics of essential oils were investigated by a researcher called Straehli. He found that all the oils tested appeared in the breath, following absorption through the skin, after time intervals which differed with each oil.

Turpentine, 1,8-cineole and *alpha*-pinene took 20 minutes; eugenol, linalool, anethole, linalyl acetate, geranyl acetate and methyl nonyl ketone took 20–40 minutes; anise, bergamot, lemon and methyl salicylate took 40–60 minutes; citronella, pine needle, lavender, geranium and cinnamaldehyde took 60–80 minutes; coriander, rue, peppermint, citral, citronella and geraniol took 100–200 minutes [12]. Other workers came to similar conclusions, finding in addition that skin penetration of terpenes is increased by putting them in hot bath water [156]. There is no discernible pattern in the results Straehli obtained for the rates of absorption of essential oils and their constituents.

Chemical structure and percutaneous absorption

Because essential oils *diffuse* through the skin, larger molecules pass through more slowly than smaller ones. Essential oil constituents are smaller than those in vegetable oils and so diffuse more rapidly through the skin. Substances of molecular weight greater than 500 have great difficulty passing through the skin at all. All essential oil constituents have molecular weights well below 500, and so pass through with reasonable ease.[5]

It is generally thought that lipophilic (fat-soluble) compounds cross the skin better than hydrophilic (water-soluble) ones. However, this principle is too general to be really useful taken on its own. The problem is that the stratum corneum is partly hydrophilic and partly lipophilic—it has both characteristics. Something which is highly fat-soluble will have trouble getting through the watery parts of the stratum corneum, whereas a molecule which is highly water-soluble will be unable to pass through lipid-rich regions.

It is really a simplification to talk of absorption of an essential oil, since the various components of a single oil may be absorbed at different rates. This probably results in the composition of the oil changing while it is on the surface of the skin [53]. Essential oil constituents have both water- and fat-solubilities, with fat-solubility predominating, which make them ideally suited to passing through the skin.[6]

Factors affecting absorption

The rate of essential oil absorption through the skin, and the total amount absorbed, depend on several factors. Because they are volatile, essential oils evaporate rapidly when applied to warm, uncovered skin, and so the amount applied to the body surface is substantially more than the amount absorbed, especially if large regions of the body remain uncovered for a long period.

When the skin is damaged or diseased, the rate of percutaneous absorption can greatly increase [15]. The risk of skin reactions also increases and essential oils should always be applied with caution to damaged skin. Anecdotal reports indicate that the percutaneous absorption rates of oils varies between individuals.

Viscosity tends to impede absorption through the skin. For instance olive oil and almond oil, which are relatively viscous, are slowly absorbed through rat abdominal skin whereas linseed oil, which is much less viscous, is rapidly absorbed [98]. The rate of absorption of the vehicle will probably affect that of the essential oil to some degree. The viscosity of most essential oils is very low, although that of absolutes and other solvent-extracted materials is often much higher.

In general, the more volatile components are absorbed more easily than less volatile ones. However, they will also evaporate more readily away from the skin. These two effects tend to cancel each other out when oils are applied to skin which remains uncovered. However, covering the skin may particularly enhance the absorption of volatile compounds.

It is clear from toxicity studies that essential oils are absorbed into the circulation when applied undiluted on to animal skin [98]. Subsequent studies have corroborated this finding both in animals and humans. The vehicle in which fragrance is delivered may also influence skin absorption by decreasing volatility [91]. Viscosity has been shown to slow absorption from the gastrointestinal tract [14].

Temperature

The absorption of methyl salicylate is enhanced by a fairly modest rise in temperature, such as would be encountered in a not particularly hot bath [99]. Most probably, the absorption of other essential oil ingredients would alter in the same way.

A rise in temperature of 10°C in the oil or oil suspension in which volunteers' hands are immersed increases the rate of percutaneous absorption several-fold, no doubt because of enhanced capillary circulation in the area [88].

In aromatherapy massage, anything which warms the client's skin would be expected to enhance absorption. Warmth will also increase evaporation, particularly of the more volatile components. This will decrease the total amount of essential oil available for skin absorption.

Water

The presence of water has been shown to influence the rate of absorption of some essential oil constituents. Methyl salicylate absorption is greatly enhanced by waterlogging the skin of the hand by prolonged prior immersion in hot water [99]. Ethanol, which comprises the greater part of perfumes, also enhances percutaneous absorption.

Hydration of the stratum corneum, which happens during a bath or shower, facilitates the penetration of essential oils [16]. Taking a hot bath increases the amount of blood circulating in the dermis, which will enhance the absorption of essential oil. The oiliness of the skin's surface encourages oils to adhere to it, meaning that a high proportion of the bath oil which has not evaporated will come into contact with the skin.

The temperature of a hot bath will cause essential oils to evaporate quickly so this method of application greatly increases the proportion of oil available for inhalation. Clean skin probably absorbs better than dirty skin.

Covering the skin

If the skin is occluded (covered) with a non-permeable material after application of an oil, absorption into the bloodstream is greatly increased [90, 91]. Experiments using isolated rat skin sections have shown that occlusion greatly enhances the absorption of benzyl acetate (<> jasmine[A]) when applied in an ethanol vehicle [88] and that this model is a good system for predicting absorption through rat skin in vivo [89].

Occlusion changes the temperature and hydration of the skin, and these physical factors affect absorption. It is believed that occlusion is the single most effective way to enhance absorption [90]. In an American study, it was found that about 75% of an applied dose of fragrance is absorbed through human skin when the site of application is covered, regardless of the particular fragrance, whereas when the skin is uncovered, the proportion of the total dose absorbed is only 4% [91].[7]

There is some suggestion, however, that skin occlusion may be less important than was previously thought, at least for some non-volatile substances [88] despite considerable data showing that occluding some common fragrance materials on the skin can increase absorption [86].

Other factors which influence absorption

Many substances can increase the permeability of the skin on contact. They need not cause any great damage to the skin, only altering the transport properties of the stratum corneum temporarily. In in vitro tests on human epidermis, lipid solvents, water, soaps and detergents have all been found to increase permeability [92, 94]. Detergents increase epidermal permeability reversibly after mild treatment and irreversibly after prolonged contact [96]. A thorough wash with soap and water prior to an aromatherapy massage would be expected to increase skin permeability.

Greasy ointments provide an occlusive layer which promotes skin hydration by slowing the evaporation of water away from the skin. However, since these ointments are not well absorbed, they tend to hold back essential oils, and are not appropriate if absorption is desired [16].

Oil-based creams and lotions facilitate the penetration of molecules into the skin [16]. Their fat-solubility enables essential oils to overcome the barrier effect of the skin's sebum more easily [86] and their water-solubility enables essential oils to penetrate the skin's horny surface layer [99].

Dermal metabolism

The skin contains many important enzymes that can break down or inactivate toxic chemicals. Esterase enzymes, for instance, may make certain pesticides less toxic. For some chemicals the enzyme activity is so efficient that the chemical is completely metabolised during skin absorption [468].

There are other enzymes in the skin that may activate chemicals, making them *more* toxic. These include cytochrome P_{450} enzymes, which are known to convert a small number of essential oil constituents (safrole, estragole and perhaps others) into potentially carcinogenic compounds (see pp.93–94). However, it is not known how efficient dermal cytochrome P_{450} is in this regard. The more efficient it is the greater the risk will be from estragole-rich essential oils such as basil when they are applied to the skin.[8]

Passage into the blood

Once an essential oil has been absorbed, the epidermis acts as a reservoir, where some of the oil will remain for a time before it enters the bloodstream. This period will vary with different essential oil constituents, but is likely to be measured in hours or even days [53]. Benzyl acetate (<> jasmine[A]) is rapidly absorbed into human skin, with a large reservoir in the epidermis [56].

Essential oil components, then, are gradually 'time-released' from the epidermis into the dermis. This means that the oils will enter the bloodstream more slowly than orally administered oils. Any acute toxicological effects are therefore likely to be less pronounced following dermal application compared to ingestion.

Once in the dermis, the molecules enter its blood capillaries, and are carried away. This process tends to happen easily because the dermis is more-or-less freely permeable and the capillaries let small molecules pass through their walls [17]. When the area on to which the oil is applied is massaged, the rate of systemic absorption increases due to the effect on blood flow, which is increased by massage. There is evidence that massage enhances the absorption of at least some essential oil constituents [99].

INHALATION

Aromatherapy invariably results in odoriferous molecules being inhaled. Because the lungs have a large surface area, the amount of essential oil which reaches the bloodstream will be increased by deep or rapid breathing.

The main stream of inhaled air does not impinge directly on to the olfactory epithelium, but air currents carry odoriferous molecules to it. The olfactory epithelium only covers an area of about 5 square centimetres and so the surface available for absorption of molecules into the circulation is some 4000 times smaller than that of the skin. Nevertheless, the lining of the nose is extremely thin, and being well supplied with capillary blood, a high percentage of the molecules which come into contact with the nasal mucosa are absorbed into the general circulation [13].

Inhaled substances pass down the trachea into the bronchi, and from there into finer and finer bronchioles, ending at the microscopic, sac-like alveoli of the lungs, where gaseous exchange with the blood takes place. These alveoli are extremely efficient at transporting small molecules, such as essential oil constituents, into the blood.

The efficiency with which molecules pass from the alveoli into the blood increases with the rate of blood flow through the lungs, with their fat-solubility [66, 259] and with the rate and depth of breathing. The exertion required to give a massage is one of the reasons why the aromatherapist will often absorb more essential oil by this route than the client. Carbon dioxide stimulates respiration and levels can build up in a stuffy room; such an environment might cause more essential oil to be absorbed via the lungs than in a well-aired room.

In orthodox medicine, the inhalational route is usually favoured only if the compound's intended site of action is in the respiratory tract. Nasal decongestants and bronchodilators for the treatment of asthma would be drugs in this category. Because of the relative lack of interest in the nose as a means of introducing molecules into the body, many essential oil components have not been studied for their behaviour after inhalation.

However, when human volunteers were exposed to high air concentrations of d-limonene they absorbed up to 70% of the dose into their bloodstream over a 2 hour period, during which time they were at rest or slightly active. Only 1% of the d-limonene was exhaled unchanged, and 0.003% was eliminated unchanged in the urine. Almost all of the excreted dose had been metabolized by the liver and the long blood half-life of d-limonene indicated that much of the dose had partitioned to adipose tissue. The authors suggested that it might take 3 days for the highest dose to be eliminated entirely. The doses of d—limonene used ($10\text{-}450$ mg/m^3/2h) were equivalent to evaporating $1\text{–}40$ g in a 100m^3 room ($2\text{m} \times 5\text{m} \times 10\text{m}$) [1].

Central nervous system effects from inhalation

The nose and surrounding facial bones are close to the brain, and substances absorbed via the nasal mucosa have easy access to the central nervous system (CNS). It is very likely that essential oil molecules absorbed through the lining of the nose will reach the brain. It is also unlikely that they will be absorbed in quantities likely to be dangerous, unless concentrated vapours are inhaled for very long periods of time.

One theory holds that essential oil molecules exert their mood-altering effects by triggering the olfactory nerves, but without being absorbed into the CNS. In this model, stimulation of olfactory nerve endings would cause messages to be sent ultimately to the brain's limbic area, the seat of memory and emotion. Another theory holds that essential oils influence the CNS by being absorbed into the bloodstream via the olfactory membranes.

The frequent anecdotal reports of mood altering effects after inhalation do suggest that many, if not all essential oil components gain rapid entry to the CNS after inhalation. Cineole, for example, has been shown to reach the arterial circulation in mice after the inhalation of rosemary oil and produces increased physical activity whether given to the mice orally or by inhalation [55].

Other studies suggest that essential oil constituents may reach the circulation in subpharmacological amounts after inhalation of even quite high concentrations, for instance in mice exposed to coumarin, *alpha*-terpineol or sandalwood oil vapours [2].

If essential oils are readily absorbed into the CNS via the olfactory membranes, this may have safety implications, especially if potentially neurotoxic essential oils are being inhaled in concentration. There might be particular risks for people with CNS problems, such as epilepsy (see p. 69).

There are very few reports of problems following inhalation of essential oils. A 13-year-old boy who inhaled 5 ml of Olbas Oil, instead of the recommended few drops, experienced double vision and ataxia (loss of control over voluntary movements). He was described as euphoric, barely rousable and talking gibberish. 12 hours later he had fully recovered, and there were no neurological signs [141]. Olbas Oil is a commercial preparation containing 35.5% peppermint oil, 35.5% eucalyptus oil, 18.5% cajuput oil, 4.1% menthol, 3.7% wintergreen oil and 2.7% juniperberry oil. The report surmises that the symptoms experienced were due to the menthol content of the product.

Inhalation is an important route of exposure because of the role of odour in aromatherapy, but from a safety standpoint it presents a very low level of risk to most people. Even in a relatively small closed room, and assuming 100% evaporation, the concentration of any essential oil (or component thereof) is unlikely to reach a dangerous level, either from aromatherapy massage, or from essential oil vaporisation.

The only likely risk would be from prolonged exposure (perhaps 1 hour or more) to relatively high levels of essential oil vapour, such as could occur when directly sniffing from a bottle of undiluted oil. This could lead to headaches, vertigo, nausea and lethargy. In certain instances more serious symptoms might be experienced, such as incoherence and double vision.

ORAL ADMINISTRATION

The oral route is the preferred one for dosing with medicines; it is both convenient and economical. Most oral preparations can be formulated so that they have little or no taste and, with many, gastrointestinal irritation is minimal or nonexistent.

Some medical practitioners do prescribe essential oils to be taken orally, but great care must be exercised if prescribing in this way. The bioavailable dose of an essential oil given orally is significantly higher than after inhalation or topical application. Oral administration is certainly a useful way of getting large amounts of essential oil into the body. Those medical practitioners who favour it are frequently treating severe infectious diseases which require heavy dosing. However, in general, any hazards are also magnified proportionately.

All recorded cases of serious poisoning with essential oils have occurred after the ingestion of relatively large amounts of essential oil. Only in very few cases were the oils being taken for therapeutic purposes, and in these few instances the person was generally self-administering rather than following the advice of a practitioner. It is recommended that essential oils to be taken orally for medicinal purposes are only prescribed by primary care practitioners such as medical doctors or medical herbalists who have an intimate knowledge of essential oil toxicology.

One disadvantage of oral dosing with essential oils is that some of their ingredients might irritate the gastrointestinal mucosa. Except for a few highly irritant oils, this effect is likely to be unpredictable, with some people tolerating an oil well and others reacting badly to it. Such irritation will also vary according to dosage and dilution. If essential oils are taken by mouth, it is recommended that they are first dispersed in copious amounts of water, or dissolved in some other suitable medium before being swallowed.

Oral administration always carries the potential for inducing nausea and vomiting, and for causing irritation of the gastrointestinal tract. Digestive enzymes can destroy some types of essential oil constituents, and the presence of food has extremely unpredictable effects on absorption into the bloodstream.

After absorption from almost all regions of the gastrointestinal tract, most substances pass directly to the liver, where the great majority will be deactivated, and some, paradoxically, will be made more toxic.

RECTAL ADMINISTRATION

Suppositories are used as an alternative to oral dosing where there is a likelihood that the essential oil would be substantially broken down in the gastrointestinal tract or by the liver, and where high systemic concentrations are desired.

The principal advantage of rectal dosing is that it largely bypasses the portal circulation, resulting in the passage of compounds around the body before metabolism by the liver can commence. Another advantage is that it is the most efficient way to administer a remedy directly to the lower colon.

The main considerations from the point of view of safety are similar to those for oral administration. Being lined with mucous membrane, the rectum is highly sensitive to irritation, especially if the essential oil is unevenly dispersed, and the absorption of large amounts potentiates any toxicity.

VAGINAL ADMINISTRATION

Pessaries are placed into the vagina for absorption. In orthodox medicine they are usually in tablet form. Because essential oils are not currently available in this form, tampons are sometimes used to carry the oils. Douches have a higher water content than pessaries and take the form of a few drops of essential oil in warm water.

Pessaries and douches may be recommended for vulval and vaginal infection or irritation. The main considerations from the point of view of safety are similar to those for oral and rectal administration. Being lined with mucous membrane, the vagina is highly sensitive to irritation, especially if the essential oil is unevenly dispersed, and the absorption of large amounts potentiates any toxicity.

SUMMARY

- Toxicology is as much concerned with dosage, frequency and route of administration as it is with the toxicity of the substance administered.
- The amount of essential oil absorbed from an aromatherapy massage varies according to many factors, but will normally be 0.025 ml–0.1 ml (roughly $\frac{1}{2}$ to 2 drops).
- The amount of essential oil absorbed from oral dosage in a 24-hour period is 8–10 times greater than in massage.
- Depending on the method of administration, a different percentage of the oil applied will reach the bloodstream.
- The safety guidelines are based on the assumption that the oils will be used at between 1% and 5% dilution for external use, and between 0.5 ml and 2.5 ml (around 10 to 50 drops) per day for oral use.
- The amount of essential oil used in pessaries, douches and suppositories varies, and could go up into the oral dosage range. The safety of individual essential oils in these cases will depend on the oil and the quantity administered.
- Frequent use of small amounts of toxic essential oils can give rise to chronic toxicity, the effects of which are often not recognised.

- The stratum corneum acts as a reservoir where some essential oil components can remain for many hours.
- In general, the skin is highly permeable to primarily fat-soluble but partially water-soluble molecules such as are found in essential oils.
- The more volatile components are absorbed more easily than less volatile ones. However, they will also evaporate more readily away from the skin. These two effects tend to cancel each other out.
- Warmth and massage both enhance absorption, as does hydrating the skin before oil application.
- Clean skin probably absorbs better than dirty skin.
- Covering the skin following massage aids absorption.
- Damaged skin is more permeable than undamaged skin, and so essential oils should be applied to it with caution.
- Any acute toxicological effects are likely to be less pronounced following dermal application than after oral administration.
- The amount of essential oil which reaches the bloodstream from inhalation will be greatly increased by deep or rapid breathing.
- Substances absorbed via the nasal mucosa have easy access to the CNS.
- From a safety standpoint, inhalation presents a very low level of risk.
- Essential oils taken by mouth are likely to have been significantly altered by the time they reach the general circulation.
- Large amounts of essential oil can be absorbed when they are taken orally in comparison to inhalation or dermal application. Caution is therefore indicated when prescribing essential oils to be taken by mouth.
- Oral, rectal and vaginal dosing carry some risk of mucous membrane irritation.

Notes

1. For instance, Steffen Arctander, writing about the use of thuja oil in perfumery comments: 'The chief constituent, thujone, is a ketone which is considered skin-irritant and toxic to a certain degree. At the normal use-level of cedarleaf (thuja) oil in perfumes, however, it seems inconceivable that the oil could be responsible for any

harmful effects.' [155]. This may be so, but for aromatherapists, thuja oil should be considered toxic.

2. Discounting skin reactions, dermal application is probably the safest way of administering essential oils, since they are absorbed so slowly into the bloodstream. We do not know what levels of essential oil become bioavailable following dermal application, but they are likely to vary a great deal. For example, only about 3% of d-limonene is absorbed through uncovered human skin [468]. This may be linked to the fact that d-limonene has a high volatility.

3. St Mary's Hospital Medical School in London is pioneering a way of studying skin absorption using fresh, healthy human skin taken from a person undergoing surgery, usually mastectomy or cosmetic breast surgery. A section of skin, including the dermis, is removed and kept at 32°C (normal skin temperature) in a special chamber. The underside of the skin is continually bathed in a solution which mimics the flow of blood. This keeps the skin alive and metabolically active for at least 24 hours by providing nutrients and oxygen and carrying away waste products. After applying a chemical to the surface of this skin, the rate and degree of absorption can be measured.

4. Some chemicals can have disastrous effects when absorbed into the body. For example, in the 1970s hexachlorophene was used as an antiseptic in baby soaps and talcs, causing brain damage and even death in some babies after it penetrated their skins.

5. There is good evidence that many poisons can be absorbed through the skin to lethal effect, including arsenic, cyanide, phenol and toxic plant alkaloids. It seems as though the vehicle in which the poison is administered can have a profound effect on the rate at which it produces toxicity. Not surprisingly, heavy, viscous fats such as lard and wool-fat retard percutaneous absorption [98]. It seems likely that the use of varying oil bases in aromatherapy would have an effect on the rate of absorption of essential oils.

6. It has been shown that monoterpenes and sesquiterpenes, which are hydrocarbons, are more easily absorbed through rat skin than are the terpenoid alcohols and esters [97]. There is further evidence showing that water-solubility is the most important factor in determining how easily a substance will pass from the dermis into the bloodstream [95]. Those substances which pass most readily from the surface of the skin into the bloodstream are therefore those which have both water- and lipid-solubility [95].

7. We are *not* suggesting that aromatherapists should wrap their patients in a non-permeable material after a massage to maximise absorption. However, the wearing of normal clothing post-treatment surely aids absorption in a safe and acceptable manner.

8. In 1775, an English doctor named Percivall Pott noticed an increase in scrotal cancer in chimney sweeps, due to skin contact with soot. Studies this century have established that the polycyclic aromatic hydrocarbons found in soot are themselves harmless. However, the cytochrome P_{450} enzymes in the skin convert them into reactive compounds that can damage cellular DNA in a way which can lead to cancer [468].

4

Metabolism

In the previous chapter we discussed different methods of administration, and their safety implications. We also described the dermal metabolism of essential oils. This chapter is concerned with the fate of a compound once it has been absorbed into the circulation, and with the consequences of this metabolism for toxicity. In any such discussion, there are three major processes which need to be considered:

- Distribution
- Biotransformation
- Excretion.

It is important to recognise that metabolism of essential oils occurs however they are administered. In most aromatherapy situations the oil will be applied to the skin and/or inhaled. Essential oil components may be metabolised differently depending on the route of administration.

Because most essential oils are highly fat-soluble, they will initially tend to behave like fat-soluble drugs in terms of their distribution—where they go in the body.

Once in the body, many different kinds of transformation are possible. As essential oil components pass through the liver, they are changed by enzymes into molecules which are more water-soluble. Any foreign molecule which enters the body will undergo changes such as this. While they are still predominantly fat-soluble, essential oil components will have affinities for different types of tissue than they will after being made predominantly water-soluble.[1] Figure 4.1

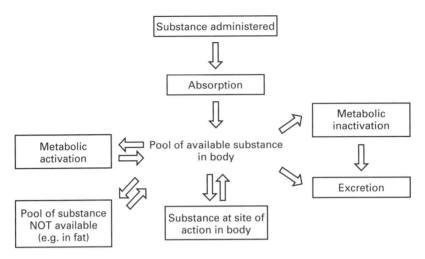

Figure 4.1 Absorption, distribution, metabolism and excretion.

illustrates the possible routes and outcomes for a substance that has entered the body.

DISTRIBUTION

The distribution of a substance, once inside the body, depends to a large extent on the solubility that it has in various types of tissue. For instance, the brain is very rich in fats, and so fat-soluble molecules will be taken up very efficiently by the brain. Conversely, highly water-soluble compounds will tend to linger in the watery environment of the blood. Organs such as the adrenal glands, the kidneys and exercising skeletal muscle, which have a very high blood flow, will receive a high dose of water-soluble compounds. Resting skeletal muscle has a relatively low blood flow and so will fill up with the substance more slowly [151]. However, skeletal muscle is often active.

Fat-soluble substances usually pass readily into the CNS and the liver, more slowly into muscle and very slowly into fat tissue, which has a very low blood flow. However, fat acts very much as a reservoir, building up a concentration of lipid-soluble compounds like essential oils slowly, but hanging on to them very tightly. Compounds which end up here could remain for hours or days before being completely eliminated. In fact, fatty

tissue effectively drags highly fat-soluble compounds out of other tissues [152]. This happens with short-acting barbiturates, for instance.

While they are fat-soluble, essential oils probably have a fairly short lifespan in the bloodstream, being redistributed first to muscle, because of its high blood flow, and then over a longer period of time to fat. They would probably stay there largely unchanged, because fat has a low metabolic rate. Some highly fat-soluble drugs can be detected in fat weeks or even months after last use. The same may be true for the most fat-soluble essential oil components such as terpenes. Once lodged in fat, however, most substances are inactive.

In experimental animals, thymol, carvacrol, eugenol and guaiacol have all been found to redistribute rapidly to the blood and kidneys following oral administration [18]. Citral is seen to be rapidly and completely absorbed from the gastrointestinal tract and then redistributed equally to all the tissues [17].

BIOTRANSFORMATION
Metabolism

The liver is by far the most important organ for metabolising chemicals which have found their way into the body. Nonetheless, the skin, nervous tissue, kidneys, lungs, intestinal mucosa and blood plasma also have this ability. The aim of such 'detoxification' reactions is to make the compound less fat-soluble and more water-soluble. This makes excretion in the urine much easier.

Binding to plasma albumin

Many substances which enter the bloodstream bind well to plasma albumin, a soluble protein which is present in very high concentration in the blood. Binding to albumin is usually reversible, but the large amount of it in circulation means that its effect can be large: if a small amount of a substance with a high affinity for albumin enters the bloodstream, it will effectively be mopped up, leaving little or none free in the circulation.

Ketones, esters, aldehydes and carboxylic acids, all of which are found in essential oils, tend to bind to plasma albumin. It is not known for certain whether essential oil components react this way, but from the knowledge we have of how drugs behave, we would expect some of them to do so. The importance of such plasma protein binding is that pharmacologically active molecules are inactive while bound to proteins in the blood.

Essential oils are often electrically charged at body pH. This means that they can easily stick to electrically charged molecules like proteins, in much the same way as a rubbed balloon will stick to a wall. This is why they would be expected to bind, at least to some extent, with plasma albumin. This will have the twin effects of prolonging the time they remain in the bloodstream and of reducing the amount of free oil available.

There may be a case for reducing the dosage of essential oil given orally to patients with kidney disease or cirrhosis of the liver since they have low plasma albumin levels: such people will have less protein available to bind the oil and so the concentration of oil in the bloodstream may be higher than usual [152].

Drugs which bind to the same plasma proteins can interact with each other by interfering with each other's binding. Because essential oils may bind to plasma proteins, there is a theoretical possibility that they could interact with some drugs (Fig. 4.2).

After external application, most essential oil compounds almost certainly find their way into the bloodstream in very much lower concentrations than do drugs. Significant interactions between essential oils and drugs, at the level of plasma binding, are therefore unlikely unless the oils have been given by mouth, when the quantity reaching the bloodstream is likely to be much larger.

Phase I and Phase II reactions

Phase I reactions are those in which a compound is chemically altered as a prelude to making it water-soluble. They usually involve oxidation, reduction, or splitting by hydrolysis [61]. Esters are typically metabolised by hydrolysis, and esters found in essential oils, such as linalyl acetate (<> lavender) are probably metabolised in this way. The oxidation, reduction and

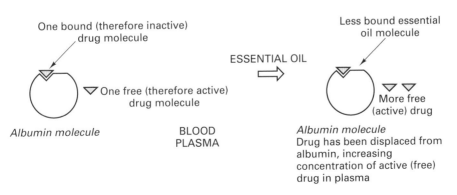

Figure 4.2 Possible interaction between oils and drugs at level of plasma albumin binding.

hydrolysis reactions go on mainly, but not exclusively, in the liver.

Some ingredients of toxic essential oils can increase the activity of liver enzymes. Safrole (<> sassafras) increases the activity of many liver enzymes and this action underlies its carcinogenicity [26]. Citral and linalool also increase the activity of rat liver enzymes [24] as does 1,8-cineole [31] but the significance of this for humans is unclear.

Phase II reactions are those in which substances which have undergone Phase I are joined to other molecules in order to make them finally ready for excretion. This process is known as 'conjugation'. Most drugs undergo reactions of this type and it seems likely that many essential oils do also [152]. Eugenol and *trans*-anethole have both been found to increase the levels of liver enzymes responsible for conjugation [314]. This is unlikely to cause any toxicity because Phase II metabolism enables toxic chemicals to be excreted more easily.

Oxidations (Phase I)

These are oxidation reactions carried out by the liver which use molecular oxygen for the direct oxidation of the drug or essential oil component. Oxidations are also carried out in the lung, kidney, skin, placenta and small intestine, but the liver is the most important site of oxidation. The crucial group of enzymes in oxidation are the cytochromes P_{450}. These can oxidise an extremely wide range of foreign molecules and certainly do metabolise essential oil constituents.[2]

Oxidations are reactions used very widely by the liver to prepare molecules for conjugation. They often detoxify molecules, rendering them inactive, but they can also activate some molecules to highly reactive metabolites capable of seriously damaging the liver or other organs.

Glucuronide conjugation (Phase II)

Glucuronides are formed by foreign molecules containing an -OH group, such as alcohols and phenols. Many essential oil components fall into this category and so are probably eliminated from the body as glucuronides. These conjugates are excreted in the urine if the molecular weight of the parent drug (or essential oil constituent) is 300 or less. (The molecular weight of terpenoids is about 150 and of sesquiterpenoids, about 225.) Other substances are conjugated by reaction with amino acids or with sulphate.

Glutathione detoxification

The liver contains a chemical called reduced glutathione, which mops up reactive molecules (e.g. free radicals) before they can damage DNA or protein. However, the liver only contains so much glutathione, and once it has been completely depleted of the chemical, reactive molecules are then able to attack liver cells. It is very unlikely that any essential oils could be absorbed dermally in sufficient quantities to deplete glutathione, but they could be from oral administration.

As we have seen, many substances which are introduced into the body, or which find their way there accidentally, undergo biotransformation. If they are pharmacologically active, this process usually results in loss or reduction of their activity—but not always; some substances are actually activated by biotransformation, often into toxic metabolites [118].

For instance, *d*-pulegone (<> pennyroyal) is metabolised in the liver to menthofuran via a highly reactive metabolite. This metabolite irreversibly binds to the components of liver cells in which metabolism takes place, quickly destroying the liver [119]. Menthofuran is only

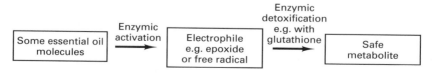

Figure 4.3   Electrophile detoxification.

one of several metabolites of *d*-pulegone [52] but is able to produce its toxic effects at very low concentrations. A similar process of bioactivation is responsible for the hepatotoxicity of paracetamol (= acetaminophen) and for the liver toxicity of estragole (<> basil) [62].[3]

The metabolism of substances to highly reactive intermediates (electrophiles) which must then be detoxified is illustrated in Figure 4.3.

It appears that several essential oil constituents, such as eugenol and cinnamaldehyde, may be able to decrease the level of glutathione in the liver (see p. 61).

Induction of liver enzymes

Some drugs can increase the activity of the hepatic enzymes which metabolise them. There is the possibility that some essential oil constituents may also be able to do this, in which case they would increase the metabolism of certain drugs. A chemically-induced increase in enzyme activity is called induction.

Essential oils probably do not reach the tissues in high enough concentrations to affect drug metabolism after aromatherapy massage. However, there are studies which have shown that certain essential oil compounds can affect drug metabolism under certain circumstances, and these are detailed below.

Safrole, an hepatocarcinogen, has been shown to induce reductase and hydroxylase enzymes in the liver as well as UDP-glucuronyltransferase and cytochrome P_{450} [26]. Geraniol and linalool have both been shown to induce liver cytochrome P_{450} activity after oral administration [41] as has citral [24, 43] in contradiction to earlier findings that linalool and nerol have no such properties [43]. Cedarwood [47] as well as limonene, borneol and terpineol are also known to induce cytochrome P_{450} [43]. Cedarwood oil (type unspecified) can decrease the sleeping time of rats given pentobarbital by enhancing the latter's metabolism [47].

Linalool has been shown to increase the activity of cytochrome b_5 [40] which is part of the enzyme chain by which cells respire. It can also induce *beta*-oxidation in the liver, which is a part of the process by which fat molecules are metabolised [24].

Cineole has been found to enhance the drug metabolising activity of enzymes even when given by aerosol inhalation [31]. This raises the possibility that some other essential oil components could be equally potent inducers [44]. It is unclear how cineole produces its effects, as the concentrations of enzymes measured in liver fractions do not increase. Possibly it enhances the activity of other enzymes or co-enzymes associated with cytochrome P_{450} activity [39, 46]. The cytochrome P_{450}-inducing activity of cineole is not seen in lung tissue after aerosol administration, possibly because the compound is cleared too rapidly from lung tissue to affect enzymes found there [42].

Diallyl disulphide, the major component of garlic oil, induces changes in the enzymes of suckling rats [374]. However, marked protective activity of garlic oil against hepatotoxic enzyme inducers has been demonstrated, apparently via an ability to inhibit free radical production [375].

Inhibition of liver enzymes

The *d*-isomer of pulegone (<> pennyroyal) has been shown to rapidly destroy cytochrome P_{450} in laboratory animals [22]. This effect probably underlies its toxicity to humans. Cytochrome P_{450} activates pulegone to toxic metabolites, which are then in place to destroy the P_{450} which has just made them. Because the activation of *d*-pulegone occurs via cytochrome P_{450}, pretreatment with drugs which induce cytochrome P_{450} greatly enhances pulegone's hepatotoxicity [25]. Since such drugs are common, this makes pulegone-rich essential oils particularly hazardous, especially in oral doses. The same applies to *beta*-asarone, the genotoxicity of which is due to a novel, but similarly P_{450}-dependent activation mechanism [445].

Bitter almond oil has been found to inhibit liver aldehyde dehydrogenase; both bitter almond and anise oils also inhibit liver alcohol dehydrogenase [23]. Both enzymes are essential for the metabolism of alcohol. However, this effect was obtained with dosage levels of 1.6 g/kg/day and 3.2 g/kg/day, and so the data have no relevance to aromatherapy.

There is no evidence that the changes in enzyme activity detailed above occur following essential oil administration in aromatherapy. The examples do illustrate, however, that essential oil constituents can be biochemically active, at least under experimental conditions.

EXCRETION

For many substances, the only important organ of excretion is the kidney, the body's filtration system. Essential oils are extremely volatile and, like alcohol, small quantities are excreted simply by being breathed out. Nonetheless, the kidney is the body's main organ of excretion and any essential oil in the bloodstream will have to be filtered by it. (Binding of a molecule to plasma protein will slow its filtration.)

Terpenes

When *d*-limonene is given to rabbits by oral administration, 72% of it is excreted in the urine within 72 hours, with 7% appearing in the faeces, because of biliary excretion and non-absorption from the gastrointestinal tract. No unchanged *d*-limonene is found in the urine; the substance is completely metabolised to oxidation products and water-soluble conjugates [49]. Similar products are seen if *d*-limonene is given orally to rats [48]. It is likely that humans metabolise *d*-limonene in a similar fashion to these two species.

Myrcene is a major constituent of juniper and hops and is used in the manufacture of alcoholic beverages, especially beer. It is metabolised via similar pathways to elemicin and *iso*-elemicin, (see 'Alkenylbenzenes', p. 41) but there is no evidence that any of its metabolites is hepatotoxic [54]. Terpenes have to be converted to alcohols before they can be oxidised further, for instance, to carboxylic acids.

Alcohols

Geraniol is known to be metabolised to Hildebrandt acid by the liver, and excreted as such in the urine [67]. In contrast, linalool is much more difficult to oxidise. It has been found that linalool is rapidly absorbed from the stomach after oral administration, and that a significant proportion of the administered dose is excreted in the bile and appears in the faeces [50]. Almost all of the linalool is removed from the body within 72 hours, that which remains being located mainly in the skeletal muscle [50].

Linalool is oxidised by cytochrome P_{450} before being conjugated to a glucuronide. The activities of many enzymes involved in its metabolism are increased after an oral dose in rats, but it is not clear what, if any, the significance of this is for human use [328].

Aldehydes

Cinnamaldehyde can be oxidised to cinnamic acid before being conjugated with glucuronic acid or glycine and excreted in the urine, but much of an oral dose is excreted as hippuric acid, a normal component of urine [28]. Citronellal, hydroxycitronellal, citral, perillaldehyde and cuminaldehyde are all oxidised and converted to carboxylic acids by the liver of mammals [35]. The rat and mouse can eliminate all of a single oral dose of citral (up to 1 g/kg) within 72 hours and 120 hours, respectively [17].

Esters

After dermal application of benzyl acetate, virtually all the absorbed dose is excreted within 24 hours, with water-soluble metabolites (glucuronide conjugate, benzoic acid and hippuric acid) being detectable in the urine [51].

Oxides

Camphor has been shown to be capable of being both oxidised and reduced (to borneol) in liver preparations from the dog and rabbit [72]. However, there are distinct inter-species differences, and it is not clear whether similar reactions occur in humans.

Alkenylbenzenes

There has been considerable interest in the metabolism of *trans*-anethole (<> fennel) which is much used as a food additive in a variety of baked goods, sweets and beverages [68]. A study using doses of *trans*-anethole close to those found in the diet has discovered that the major route of excretion for the compound is via the urine; it is also excreted via the expired air, as carbon dioxide. *Trans*-anethole produces nine urinary metabolites, all oxidation products. Estragole is chemically related to *trans*-anethole, and is metabolised in a similar fashion [30, 71]. Most of the metabolites found in urine have also been detected in bile, suggesting that biliary excretion is an important route for the removal of *trans*-anethole [71].

Trans-anethole has been shown to be distributed and metabolised similarly in the rat, rabbit and human [29] with most of an orally administered dose being removed from the body within 48 hours [33]. However, anethole does not partition uniformly in the body; after intravenous dosing in the mouse, most of the administered anethole is accumulated by the liver, lungs and brain. In some studies, anethole appears to be very poorly absorbed from the gastrointestinal tract, with most of an administered dose remaining in the mouse stomach [34, 68]. In another, however, very good absorption after oral dosing is seen [65]. These differences may be due to the methodologies chosen.

Elemicin and *iso*-elemicin are constituents of nutmeg oil and, with myristicin, may be partially responsible for nutmeg's hallucinogenic properties [61]. Elemicin is also found in elemi oil. Both compounds are metabolised by oxidation, hydroxylation and epoxidation; the latter process may give rise to a risk of liver toxicity and carcinogenicity [61].

Safrole (<> sassafras) is metabolised to a carcinogen [59] but is itself thought to be partly responsible for the liver carcinogenicity of sassafras [32]. Safrole is definitely metabolised via a reactive intermediate which is sufficiently stable in the blood to reach all regions of the body [72]. The carcinogenic metabolites of safrole have been shown to bind to DNA, RNA and protein in the liver; such binding probably underlies the carcinogenicity of safrole and its metabolites [59]. Safrole can be metabolised to a further compound which binds to the iron-containing part of cytochrome P_{450} [45].

INTERACTION WITH DRUGS

It is difficult to say how likely is drug interaction with essential oils. In theory it could occur in several ways: at the level of plasma albumin binding; at cell surface receptors or at intracellular receptors; or physiologically, where a drug and oil may act through very different mechanisms but produce similar or opposing physiological effects.

The plasma binding mechanism is, in our view, likely to be quite unimportant, as essential oils scarcely reach the circulation in sufficient quantities after aromatherapy massage or inhalation. It is quite likely that essential oil molecules and drugs could interact at cellular receptors, although the details can at present only be guessed at—we know next to nothing about the receptor activity of essential oil constituents. The physiological mechanism is also probably important, in those instances where essential oils have powerful effects on function, such as alertness, blood pressure and metabolism.

Enhancing transdermal penetration

Both essential oils and essential oil components have been shown to enhance the transdermal penetration of drugs. In various animal studies, both eucalyptus oil and camphor enhanced nicotine absorption [Nuwayser et al, cited in 522], terpineol enhanced prednisolone absorption [Yamahara et al, cited in 522] and limonene increased indomethacin absorption [Okabe et al, cited in 522]. In a study using excised human skin, four essential oils enhanced the dermal absorption of 5-fluorouracil to varying degrees; anise oil 2.8-fold, ylang-ylang oil 7.8-fold, wormseed oil 33-fold, and eucalyptus by 34-fold [522].

Essential oils are rarely applied to the skin undiluted, and it is not known to what extent dispersion of an essential oil in a vegetable oil, for instance, would affect the above data. Few drugs are applied to the skin, but even a two-fold increase in transdermal penetration could have undesirable results in some instances, such as nicotine patches.

Blood clotting

Two cases demonstrate the potentiating effect of topically applied salicylates on the action of warfarin, and the kind of symptoms this can cause. An 85-year-old woman was admitted to hospital with acute cardiovascular problems, and was prescribed heparin, followed by warfarin. Some weeks after leaving hospital she returned, suffering from gross vaginal bleeding, ecchymoses on the left cheek, the right flank, and perineum, and orthostatic hypotension. Her blood haemoglobin had fallen from 117 g/l to 80 g/l. It turned out that the previous week she had liberally applied a preparation containing menthol and methyl salicylate to her arthritic joints [521].

In a second case, a 68-year-old man was put on warfarin, and his prothrombin time stabilised at 1.3 to 1.5 times the control value. On routine follow-up, the prothrombin time had risen to 2.5 times the control value, considerably more prolonged than it had been for several months. The patient had liberally applied topical trolamine salicylate to his neck and shoulders on several recent occasions because of pain. Within 1 week his prothrombin time had fallen to 1.3 times the control value, after doing nothing but ceasing to apply the trolamine salicylate [521].

Another salicylate, acetylsalicylic acid (aspirin) has been known to increase the risk of haemorrhage in patients on warfarin. Animal models have demonstrated that high salicylate levels decrease hepatic synthesis of vitamin-K-dependant coagulation factors [521]. It is possible that methyl salicylate would cause problems with anticoagulant drugs other than warfarin, aspirin and heparin for example, but this has not been established.

MAO inhibition

Myristicin has been shown to be an inhibitor of monoamine oxidase (MAO) in rodents [367]. Myristicin is found in significant quantities in essential oils of parsnip, parsley leaf and parsleyseed, and in smaller amounts in mace and nutmeg oils. Nutmeg oil was less potent an inhibitor of MAO than myristicin [367]. MAO inhibitors should not be given in conjunction with pethidine. 'Very severe reactions, including coma, severe respiratory depression, cyanosis and hypotension have occurred in patients receiving monoamine oxidase inhibitors and given pethidine' [5]. Oral dosages of the above oils, in conjunction with pethidine, would be inadvisable. The safety of non-oral dosages is uncertain. Phenelzine sulphate is an MAO inhibiting antidepressant, and its action might be potentiated by oral doses of myristicin-rich essential oils.

Cytochrome P_{450} induction

Cytochrome P_{450}, an important detoxifying enzyme, is induced by certain drugs (see Box 4.1). Any essential oil which also induces this enzyme may potentiate the toxic effect of a drug, if taken at the same time in oral doses. The destructive effect of pulegone on cytochrome P_{450} was described on page 39. In rats, phenobarbitone (either 40 or 80 mg/kg/day for 4 days) potentiated the hepatotoxicity caused by pulegone given orally at 0.4 g/kg once daily for 5 subsequent days [25].

Box 4.1   Substances which induce cytochrome P_{450}

These should not be combined with oral doses of essential oils rich in either *beta*-asarone or *d*-pulegone, as this will probably increase liver toxicity. (Such oils are, in any case, contraindicated due to their toxicity.)

Nicotine

Ethanol (present in alcoholic drinks)

Progestogens (found in combined and mini contraceptive pills)

Diphenhydramine (antihistamine)

Nitrazepam (= Mogadon, tranquilliser)

Pethidine (narcotic analgesic)

Carbamazepine (anticonvulsant)

Phenobarbitone (anticonvulsant)

Phenytoin (anticonvulsant)

Glutathione depletion

The important protective role of glutathione was explained on page 38, and the depletion of glutathione by essential oil components is detailed on pages 61–62. It is very likely that essential oils rich in glutathione-depleting constituents would increase the hepatotoxicity of glutathione-depleting drugs, such as paracetamol, if the oils were taken in oral doses. However, such an interaction has not been recorded clinically.

Table 4.1 Drugs and essential oils: established incompatibilities, probable incompatibilities, and possible incompatibilities

Drug	Essential oil
Established incompatibilities, even when the essential oils are applied topically	
Warfarin	Birch (sweet), wintergreen
Probable incompatibilities if the essential oils are taken in oral doses (listed as cautions in Chs 10 and 13)	
Acetaminophen	See paracetamol
Aspirin	Bay (W. Indian), betel leaf, cinnamon leaf, clove bud, clove stem, clove leaf, garlic, *Ocimum gratissimum,* onion, pimento berry, pimento leaf
Heparin	Bay (W. Indian), betel leaf, cinnamon leaf, clove bud, clove stem, clove leaf, garlic, *Ocimum gratissimum,* onion, pimento berry, pimento leaf
Paracetamol (=acetaminophen)	Anise, basil, bay (W. Indian), betel leaf, buchu *(B. crenulata),* calamus, camphor (brown), camphor (yellow), cassia, cinnamon bark, cinnamon leaf, clove bud, clove stem, clove leaf, fennel, *Ocimum gratissimum,* pennyroyal, pimento berry, pimento leaf, sassafras, star anise, tarragon, tejpat leaf
Pethidine	Parsley leaf, parsleyseed, parsnip
Phenobarbitone (=phenobarbital)	Buchu *(B. crenulata),* calamus, pennyroyal
Warfarin	Bay (W. Indian), betel leaf, cinnamon leaf, clove bud, clove stem, clove leaf, garlic, *Ocimum gratissimum,* onion, pimento berry, pimento leaf

Possible incompatibilities if the essential oils are taken in oral doses (not listed in Chs 10 and 13)

The following list is very much cautionary rather than definite, and none of the interactions listed here has been confirmed by clinical experience

Aminopyrine (= amidopyrine)	Eucalyptus
Carbamazepine	See cytochrome P_{450} inducers
Diphenhydramine	See cytochrome P_{450} inducers
Hexobarbitone (= hexobarbital)	Cedarwood
Nitrazepam (= Mogadon)	See cytochrome P_{450} inducers
Cytochrome P_{450} inducers	Buchu *(B. crenulata),* calamus, pennyroyal
Pentobarbitone (= pentobarbital)	Cedarwood, eucalyptus
Pethidine	Mace, nutmeg, and see cytochrome P_{450} inducers
Phenelzine sulphate (MAO inhibiting antidepressant)	Mace, nutmeg, parsley leaf, parsleyseed, parsnip
Phenytoin	See cytochrome P_{450} inducers
Progestogens (in many contraceptive pills)	See cytochrome P_{450} inducers
Zoxazolamine	Eucalyptus

Table 4.1 summarises information on incompatibilities between drugs and essential oils.

THOSE MOST AT RISK

As used in aromatherapy or as eaten, essential oils usually enter the body in tiny amounts, at least when compared to many drugs. The processes which have been described almost certainly apply to all essential oils, but it is important to realise that only oils which powerfully alter the activity of enzymes, or which produce highly reactive chemicals once biotransformed, are likely to present any hazard.

Nevertheless, it is always wise to bear in mind that certain groups of patients—the seriously ill, the old, the very young, and those with kidney and liver disease—are less able to handle any foreign molecules entering the body. They should be closely monitored for strong or unusual reactions.

SUMMARY

• Most essential oils will behave like fat-soluble drugs in terms of their initial distribution.
• The metabolism of essential oils in humans is sometimes similar to that which occurs in animals, sometimes different.
• Most essential oil ingredients are metabolised by enzymes before being excreted.
• Essential oils probably have a fairly short lifespan in the bloodstream, being redistributed first to muscle, and then over a longer period of time to adipose tissue.
• Essential oils probably bind to plasma albumin. There may be a case for reducing the dosage of essential oil given orally to patients with kidney or liver disease who have low plasma albumin levels.

• After an aromatherapy massage, most essential oil compounds find their way into the bloodstream.
• Significant interactions between essential oils and drugs are unlikely unless the oils have been given in oral doses.
• The three most important types of reaction which essential oils undergo in the liver are:
—Oxidation
—Glutathione detoxification
—Glucuronide conjugation.
• A number of essential oil components have effects on liver enzymes. Small amounts are unlikely to have worrying effects, but large oral doses could do.
• Most of the essential oil in the body is excreted via the urinary tract, although some is also excreted in the breath. Small amounts are also excreted in the faeces and through the skin.

Notes

1. The metabolism of substances in animals often reflects that in humans. However, there are instances in which it does not. An example is the metabolism of phenylethyl alcohol, a component of rose oil which is toxic when given orally to rats, but which is metabolised in humans to phenylacetic acid, a naturally occurring, non-toxic material [160].
2. If two substances oxidised by cytochrome P_{450} reach the enzyme at the same time, both their rates of metabolism may be reduced. Since most good substrates are highly fat-soluble, essential oil constituents, especially hydrophobic ones like terpenes, may interfere with the metabolism of some drugs if the concentrations of both reaching the liver are sufficiently high. It is, however, unlikely that the amounts used externally in aromatherapy will be sufficient to induce changes in cytochrome P_{450} activity.
3. Dangerously reactive molecules are made in the liver as an unfortunate but necessary step in their metabolism. However, the liver is vulnerable to attack unless they can be removed extremely rapidly. An enzyme, glutathione-S-transferase, catalyses the reaction with glutathione by which the molecules are made safe.

5

Toxicity

TOXICITY TESTING

Virtually every substance used in foods, medicines, cosmetics and toiletries, including most essential oils, has been tested for toxicity on animals. Tests may not have been carried out by the company manufacturing the product; in fact this is the current trend. However, toxicology testing will almost certainly have been carried out by someone else at some point in time. (Due to the pressure of public opinion, the extent of animal testing is now under scrutiny. Already the number of animal tests required for cosmetic ingredients is beginning to fall in many countries.)

There are two broad categories of toxicity—acute and chronic. Acute toxicity describes the result of the short-term administration of a substance, and in testing usually involves a single, high dose. Chronic toxicity describes the result of the long-term use of a substance over weeks, months or years.

Both aspects of toxicity can be tested using different methods of administration. So there are 'acute oral toxicity' tests, 'acute dermal toxicity' tests, administration by subcutaneous (s.c.) injection, or injection into the abdominal cavity (intraperitoneal, or i.p.).

Acute oral toxicity

The traditionally accepted method for acute toxicity testing is known as the LD_{50} test. This involves giving different doses of the test substance to matched groups of animals, usually rats or mice, in order to identify the dose that kills 50% of a group. LD_{50} stands for lethal dose 50%, and the acute oral toxicity of valerian oil, for instance, is expressed as: 'acute oral LD_{50} 15 g/kg'.

The lethal dose of a substance varies according to the size of the person (or animal) taking it. So the lethal dose for a small baby may be about 10 times less than for an adult. Toxicity, therefore, is not usually measured in terms of an actual lethal dose, but by calculating the number of grams of substance *per kilogram of body weight* required to cause death.

For example, if you weigh 50 kg, and the lethal dose of essential oil X is 50 g, this means that 1 g of essential oil X per kg of body weight is a lethal dose (lethal dose 1.0 g/kg). The more toxic an essential oil is the lower the figure will be, because it indicates that only a small amount is toxic. The most toxic essential oil, boldo leaf, has a lethal dose of 0.13 g/kg. Probably the most toxic component of any essential oil, hydrocyanic acid (<> unrectified bitter almond oil) has a lethal dose of 0.005 g/kg. Essential oils with a lethal dose of roughly over 1.0 g/kg could be regarded as non-toxic in aromatherapy. To convert g/kg values into typical adult human dose, see Table 5.1.

Table 5.2 Extrapolating rodent toxicity to humans

Toxicity	Substance	LD_{50}	Equivalent lethal dose for a 65 kg adult
Highly toxic	Hydrocyanic acid	0.0001 g/kg	0.0065 g
Definitely toxic	Boldo oil	0.13 g/kg	8.45 g
Marginally toxic	Cornmint oil	1.25 g/kg	81.25 g
Very non-toxic	Valerian oil	15.0 g/kg	975.0 g

The average human adult weighs around 65 kg, so a lethal dose of cyanide would be 0.005 g/kg x 65 kg, or 0.325 g. One of the least toxic essential oils, valerian, has a toxicity level of 15 g/kg. For a 65-kg person, this is equivalent to a lethal dose of 975 g, almost 1 kilo of essential oil. This is around 2000 times the amount normally used in aromatherapy massage (15 ml of oil with 3% of essential oil).

It is important to understand that the LD_{50} value is not an absolute. LD_{50} values for the same substances frequently vary between the types of animal used for testing, different laboratories, and methods of administration (e.g. oral, dermal or by injection). Nor is the LD_{50} a complete key to toxicity as it does not, for example, tell us anything about carcinogenesis, or risks in pregnancy. It is, however, a rough guide to acute toxicity, and is an

Table 5.1 Converting g/kg values to typical human dose

g/kg value	Equivalent dose for 65 kg human adult
0.1 mg/kg	0.0065 g
0.5 mg/kg	0.0325 g
1.0 mg/kg	0.065 g
5.0 mg/kg	0.325 g
10 mg/kg	0.65 g
15 mg/kg	1.0 g
50 mg/kg	3.25 g
0.1 g/kg	6.5 g
0.5 g/kg	32.5 g
1.0 g/kg	65.0 g
5.0 g/kg	325.0 g

1 g of essential oil is very roughly equivalent to 1 ml, which is approximately 20 drops. g/kg or mg/kg values are mentioned many times in the text. Less often, ml/kg values are given, which are approximately the same. This table is a guide to converting these values into an actual dose for an average, adult human. The table is valid when human tests have been done, but should be treated with some circumspection when animal tests are involved.

Table 5.3 Differential between supposed lethal (adult) human dose and maximum amount used in aromatherapy

Oil	Lethal dose	Multiplication factor to convert likely maximum dose to lethal dose	
		Oral use	Dermal absorption
Hydrocyanic acid	0.0065 g	0.0026	0.021
Boldo oil	8.45 g	3.4	27.0
Cornmint oil	81.45 g	32.0	256.0
Valerian oil	975.0 g	390.0	3120.0

These figures mean, for example, that around three times the maximum likely oral dose of boldo oil might be fatal—an unacceptably close differential. The figures assume that essential oils applied orally in a dose of 2.5 ml are 100% bioavailable, and that 1.25 ml applied dermally is 25% bioavailable. Such number-crunching can only give an approximation of the real situation, and this table presents the worst-case scenario.

important factor in making a decision about the safety or otherwise of essential oils. Note that the figures in Tables 5.2 and 5.3 assume that human toxicity is similar to rodent toxicity for essential oils.

Acute dermal toxicity

Acute dermal toxicity tests are conducted in a similar way to oral LD_{50} tests, except that the oil is applied to the skin, instead of being given orally. (The kind of toxicity we are discussing here is systemic toxicity, not toxicity to the skin itself.) The Research Institute for Fragrance Materials (RIFM) has routinely conducted acute dermal toxicity tests in rabbits. We have not included these data in this volume for the following reason.

Common sense tells us that an essential oil cannot cause systemic toxicity when applied dermally, unless it is readily absorbed through the skin. The acute dermal LD_{50} values, therefore, depend as much on the ability of the essential oil to penetrate rabbit skin as they do on the toxicity of the oil if and when absorbed. So long as human skin and rabbit skin absorb the same oils at the same rates, the tests should provide useful information. However, there seems to be no evidence to this effect, and it is believed that, in general, the absorption of chemicals is significantly higher through animal skin [468].

In tests with six substances (see Fig. 5.1) including caffeine and cortisone, rabbit skin absorption rates were notably higher than human ones [506]. Although none of the substances tested were essential oils they represent a wide range of chemical types, and it is very likely that essential oils will behave in a similar fashion.

We cautioned earlier against attaching too much importance to acute oral LD_{50} figures, although adding that they provide a useful rough guide. Acute dermal LD_{50} figures should be regarded with perhaps even more circumspection, as we have very little idea how well the animal data would extrapolate to the human situation. They probably give no useful indication of the toxicity of dermally applied essential oils in relation to aromatherapy.

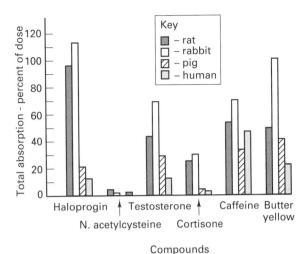

Figure 5.1 Percutaneous absorption of several compounds in rats, rabbits, pigs and humans (with permission from Williams and Wilkins [506]).

Chronic toxicity

In this context, chronic toxicity refers to adverse effects produced either in the skin or elsewhere in the body, by the repeated use of essential oils, either internally or externally. The doses involved may be significantly smaller than those responsible for acute toxicity. However, toxicity is a risk because of the possibility of cumulative effects, where damage will increase after each small dose.

Chronic toxicity is, like acute toxicity, dose-dependent, but it will also depend on frequency and total length of time of application. There is a link with acute toxicity in the sense that small, frequently applied amounts of essential oils which are acutely toxic are very likely to be chronically toxic. There is another link in that there are frequently similar causes of death in both cases—most commonly damage to liver and kidneys.

The 'maximum tolerated dose' (MTD) is the nearest equivalent to an LD_{50} value for chronic toxicity. Table 5.4 shows the MTD values established for seven essential oils, compared with LD_{50} values established in other studies. The essential oils were administered orally for 8 weeks [414].

Comparing the figures for acute and chronic toxicity, we find that the MTD is between 3 and 10.7 times less than the LD_{50} value, and that the

Table 5.4 Comparison of MTD and LD_{50} values for some essential oils

Essential oil	Oral LD_{50} (g/kg)	MTD (g/kg/day)	LD_{50}/MTD ratio
Mustard	0.15	0.05	3.0
Sage	2.52	0.4	6.3
Clove	3.73	0.7	5.3
Cassia	5.2	0.75	6.9
Chamomile	8.56	0.8	10.7
Angelica	11.56	1.5	7.4
Valerian	15.0	2.3	6.5

The MTD values show the highest dose administered without causing any noticeable injury.

average ratio is 6.6. What this seems to indicate is that a dosage of 10 times less than the oral LD_{50} is safe to use, even chronically. However, 10 times is not a very comfortable margin, and a more realistic safe chronic dosage, at least for humans, is more likely to be 100 times less than the oral LD_{50}. The point of the table is to show that there is an orderly relationship in essential oils between acute and chronic toxicity.

It is very likely that the effects of chronic toxicity from essential oils in humans will not be noticed, or at least will not be identified as such. Unlike acute toxicity, where the effects are obvious, chronic toxicity may only produce minor symptoms, which could equally be caused by many other factors. Such symptoms might include headaches, loss of appetite, minor skin eruptions, nausea, lethargy and so on. These problems may not be recognised as relating to the regular use of a particular substance in relatively small amounts.

ESSENTIAL OIL TOXICITY

Table 5.5 shows essential oils divided into four categories, according to their level of acute oral toxicity in animals. The most toxic oils are found in category A, and these are all oils that we recommend are not used in aromatherapy. Many of them are also restricted in use as flavour or fragrance ingredients due to their toxicity, and some carry other hazards, such as carcinogenic or irritant properties. Full details on each will be found in the essential oil profiles in Chapter 10.

Actual figures are only given for the A and B categories.

B category oils are on the borderline between safe and unsafe in terms of aromatherapy usage, and these are some of the oils which may present difficulties in terms of safety classification. We have very good reason to believe, for example, that hyssop (neurotoxic) and sassafras (carcinogenic) are not very safe oils to use, while we have no reason to suspect any problems with myrrh or tea tree.

Generally speaking, the oils in both the C and D categories can be regarded as safe in terms of acute toxicity. However, oils in these categories may present other hazards. For example, costus can cause severe allergic skin reactions and taget is powerfully phototoxic.

Some of the C and D category oils may even be more toxic than they would appear from the LD_{50} tests. Sage oil, for instance, found in the C group, contains about the same amount of thujone (a toxic component) as oils of tansy, thuja, wormwood and armoise, all of which are dangerously toxic. Why sage does not come out as being more toxic in LD_{50} tests is not clear, but it might be unwise to regard it as a safe oil to use in aromatherapy purely on the basis of these tests.

It should be obvious from these observations that we cannot simply play about with numbers, and say that essential oils above a certain level are dangerous, and those below it are safe. When other factors are also taken into account, however, it becomes easier to make an informed judgement on toxicity.

Human and animal toxicity data

Most essential oil toxicity testing is carried out using rats or mice. The oils are either administered by mouth, or they are applied to areas of shaven skin, either undiluted or diluted in a vehicle.

Oral toxicity data are of potential use, firstly to aromatherapists who prescribe essential oils to be taken by mouth, and secondly in assessing the risk involved when an essential oil is accidentally swallowed. These data also give a rough indication of general toxicity, regardless of route of application.

Table 5.5 Toxicity ratings based on RIFM LD_{50} tests

Oral toxicity A: LD_{50} up to 1 g/kg			Oral toxicity B: LD_{50} between 1 g/kg and 2 g/kg	
Oil	LD_{50} (g/kg)	Toxic component	Oil	LD_{50} (g/kg)
Boldo leaf	0.13	Ascaridole 16%	Methyl salicylate (wintergreen)	1.20
Wormseed	0.25	Ascaridole 60–80%	Cornmint	1.25
Mustard	0.34	Allyl isothiocyanate 99%	Savory (summer)	1.37
Armoise	0.37	Thujone 35%	Clove leaf	1.37
Pennyroyal (Eur.)	0.40	Pulegone 55–95%	Basil	1.40
Tansy	0.73*	Thujone 66–81%	Hyssop	1.40
Thuja	0.83	Thujone 39–80%	Almond (bitter) FFPA†	1.49
Calamus	0.84	Asarone 45–80%	Sassafras (Brazilian)	1.58
Wormwood	0.96	Thujone 34–71%	Myrrh	1.65
Almond (bitter)	0.96	Prussic acid 3%	Birch (sweet)	1.70
Artemisia arborescens	?‡	*Iso*-thujone 30–45%	Bay leaf (W. Indian)	1.80
Buchu (*B. crenulata*)	?	Pulegone 50%	Oregano	1.85
Horseradish	?	Allyl isothiocyanate 50%	Sassafras	1.90
Lanyana	?	Thujone 4–66%	Tarragon	1.90
Pennyroyal (N. Am.)	?	Pulegone 60–80%	Tea tree	1.90
Southernwood	?	Thujone (major component)	Buchu (*B. betulina*)	?
Western red cedar	?	Thujone 85%	Savin	?

Oral toxicity C: LD_{50} between 2 g/kg and 5 g/kg

Ajowan ?	Ho leaf
Anise	Lovage
Bay leaf (laurel)	Mace
Cajeput	Marjoram (sweet)
Camphor (brown)	Neroli
Camphor (yellow)	Nutmeg
Caraway	Parsley leaf
Cassia	Parsleyseed
Cinnamon bark	Peppermint ?
Cinnamon leaf	Perilla
Clove bud	Peru balsam
Clove stem	Pimento berry ?
Coriander	Pimento leaf
Costus	Rosewood
Cumin	Rue
Dill seed	Sage (Dalmatian)
Dill weed	Snakeroot
Elecampane ?	Spike lavender
Elemi	Star anise
Eucalyptus	Taget
Fennel (bitter)	Thyme
Fennel (sweet)	Yarrow

Oral toxicity D: LD_{50} over 5 g/kg

Abies alba (cones)	Davana	Marjoram
Abies alba (needles)	*Eucalyptus citriodora*	May chang
Angelica root	Fir needle (Canadian)	Orange (bitter)
Angelica seed	Fir needle (Siberian)	Orange (sweet)
Benzoin (resinoid)	Frankincense ?	Palmarosa
Bergamot	Galbanum	Patchouli
Bergamot mint	Geranium (Algeria)	Pepper (black)
Birch tar	Geranium (Réunion)	Petitgrain
Cabreuva	Geranium (Morocco)	Pine (Scotch)
Cade	Ginger	Pine (dwarf)
Camphor (white)	Grapefruit	Rose[A]
Cananga	Guaiacwood	Rose (otto)
Cardamon	Gurjun	Rosemary
Carrot seed	Hibawood	Sage (Spanish)
Cascarilla	Immortelle	Sandalwood
Cedarwood (Atlas)	Jasmine[A]	Schinus molle
Cedarwood (Texas)	Juniper	Spearmint
Cedarwood (Virginian)	*Juniperus phoenicea*	Spruce
Celery seed	Labdanum	Tangerine
Chamomile (German)	Lavandin	Terebinth ?
Chamomile (Roman)	Lavender	Valerian
Citronella	Lemon	Verbena
Clary sage	Lemongrass	Vetiver
Copaiba	Lime	Violet leaf[A]
Cubeb seed	Linaloe	Ylang-ylang
Cypress	Mandarin (expressed)	

*Average of 0.30 g/kg (dog) and 1.15 g/kg (rat)
†FFPA – free from prussic acid
‡The queried oils have been categorised according to their chemical composition, due to the absence of RIFM data on these particular oils.

The data from dermal testing may indicate the risk of an adverse skin reaction in humans following dermal application and, in the case of very toxic oils, they may indicate the dermal dose needed to produce systemic toxicity. (The trend now is towards using only humans for dermal irritation and sensitisation testing.)

There is one major problem with using toxicity data obtained in rodents. Both the skin and metabolism of rodents are different to those of humans. Therefore, the results obtained by animal testing, while often having considerable relevance, cannot be assumed to be the same as results which human testing would give. It is interesting to compare rodent toxicity data with reports of adverse reactions or poisonings in humans. Comparison indicates that, in some instances, rodent toxicity data give a poor indication of human toxicity.

Taking six cases of wintergreen oil poisoning in adult humans, three people died from ingestion of 15 ml, 30 ml and 80 ml; three survived after ingesting 6 ml, 16 ml and 24 ml [290]. None of these cases received any medical intervention. If we take the lowest dose which was fatal (15 ml) and the highest dose which was non-fatal (24 ml) this gives us a human lethal dose of between 0.23 g/kg and 0.37 g/kg, assuming the individuals were of average weight. This compares with an LD_{50} of 1.2 g/kg for methyl salicylate in rodents [424]. 4–8 ml of methyl salicylate is considered a lethal dose for a child [424]. If these figures do represent reality, this means that wintergreen oil is three to five times more toxic in humans than it is in rodents.

Wintergreen is a useful oil to make comparisons with, because there are many recorded cases of poisoning to draw on compared to other essential oils, and because chemical variation between different wintergreen oils is minimal. The variation in toxicity seen between the cases cited above may be due to differences in body weight and/or differences in individual health status. Four of the cases cited were between 21 and 28 years old, one was 55, and one was adult, age unknown.

Eucalyptus oil seems to be fatal to humans in amounts between 30 ml and 60 ml [261]. If true,

this would make eucalyptus oil some four times more toxic in humans than in laboratory animals.

Camphor (the chemical) is thought to be significantly (perhaps 70 times) more toxic in humans than in rodents. The probable oral lethal dose in humans is reckoned to be 0.005–0.5 g/kg, compared to an animal oral lethal dose of 0.5-15 g/kg [153]. The fatal dose of camphor for children is reported as being about 1 g [272].

RIFM gives the rat oral LD_{50} for pennyroyal oil as 0.4 g/kg [420] equivalent to about 30 g in an adult, or 35 ml. One ounce (about 30 ml) of penny-royal oil produced a fatal outcome when ingested by an adult [168]. In this instance the animal data seem to correlate well with human toxicity.

It may be that discrepancies between rodent toxicity and apparent toxicity to humans are greatest in the case of oral dosing. Also, it is possible that wintergreen, eucalyptus and camphor are exceptions, and that variations between rodent and human toxicity are not significant for the majority of essential oils.

Nevertheless, the foregoing examples illustrate that circumspection is needed when applying animal data to the use of essential oils in humans. It is certainly worth noting that, in some recorded cases of poisoning in humans, the oils seem to be more toxic than animal testing would indicate.

CASES OF POISONING

Much of the information available on the biological effects of essential oils and their ingredients has come from experimental work in animals, because vivisection can produce data which could never be obtained by doing experiments in humans. However, there have been hundreds of recorded cases of poisoning from essential oils, many of them fatal. These cases, although extremely regrettable, provide us with useful information about the toxicity of essential oils in humans.

All cases of serious poisoning from essential oils arise after oral ingestion. Most cases involve ingestion of the undiluted essential oil, in a few cases the oil was diluted in a household product.[1]

In every case, the amounts taken were very much higher than normal therapeutic doses. Most importantly, in adults, all the deaths were associated with oral self-administration, and never with aromatherapy. A few appear to be suicides. (Since death from excessive essential oil ingestion commonly takes from 1 hour to 3 days to occur, it is perhaps not an ideal method.) Of the many non-fatal accidents, only two cases (one wintergreen and one wormseed) report any long-term ill effects.

Certain oils appear quite frequently in cases of poisoning, and these are covered below. The fact that an essential oil appears in this chapter does not necessarily mean that it should be regarded as particularly toxic. It is very likely that any essential oil, drunk in sufficient quantity, will cause serious problems.[2]

Camphor

There are a great number of cases where camphorated oil (20% pure camphor in cottonseed oil) has caused serious poisoning, usually in young children. Although camphorated oil is not an essential oil these cases are included because they provide useful information about the likely toxicity of camphor-rich essential oils.

In 1973, the USA saw 500 cases of camphor intoxication reported to the medical authorities, with probably many more having gone unreported [256]. Doctors have in the past recommended that the sale of camphorated oil should be restricted [243]. In 1924 a fatal case was reported in which one teaspoon of camphorated oil had been consumed by a 16-month-old boy [244]. The effects of the intoxication included frequent fits, constriction of the pupils, rapid pulse and an extremely high respiratory rate. The child died 7 hours after ingesting the oil [247] and gross damage to the CNS, especially to the brain, was noted on autopsy [244].

Accidental ingestion of 60 g of camphorated oil by a pregnant woman probably caused the death of her full-term fetus [241]. The woman was intoxicated by the camphor but recovered fully, in contrast to her baby. In another, very similar case, the baby survived [240]. After crossing the placenta, camphor evidently passes into fetal lung, liver, brain and kidney tissue. It also rapidly destroys the placenta, causing haemorrhage [257].

Camphorated oil has often been given to children, in error, by their parents when castor oil should have been administered. Reports of such accidents have come both from the USA [272] and from Britain [269]. Toxic effects are initially those of CNS stimulation (delirium, convulsions) followed by depression (incoordination, coma). Some authors recommend gastric lavage with water as treatment but suggest avoiding oils and alcohol, as these may increase absorption into the bloodstream.

Cinnamon

Cinnamon oil (type not known) caused poisoning after the ingestion of approximately 60 ml by a 7-year-old boy. (He drank the oil when dared to by a friend.) Symptoms included a burning sensation in the mouth, chest and stomach, dizziness, double vision and nausea. There was also vomiting and later collapse. The doctors involved were of the opinion that had vomiting not occurred the dose could have been fatal. In this case, there were no serious consequences [251]. (There are two types of cinnamon oil; cinnamon bark oil contains 55–75% cinnamaldehyde while cinnamon leaf contains 70–90% eugenol.)

Citronella

As little as 15 ml was fatal to a child of 21 months when taken orally. This is probably the only recorded case of fatality from citronella oil. Signs of poisoning included vomiting, shock, frothing at the mouth, pyrexia, deep and rapid respiration, cyanosis and convulsions. Autopsy revealed numerous haemorrhages in the white matter of the brain, collapse of the lungs, intense congestion of the gastric mucosa with superficial necrosis and detachment of the lower part of the mucous membrane of the oesophagus [264].

Clove

There are two recent reports of non-fatal poisoning with clove oil. In 1991 a 7-month-old child was given one teaspoon of clove oil (equivalent to 0.5 g/kg of eugenol). Supportive care and gastric lavage were sufficient for total recovery following the resultant severe acidosis, CNS depression and urinary abnormalities (the presence of ketones in the urine) [399].

The second case, reported in 1993, involves near fatal poisoning of the paracetamol type after ingestion of 5–10 ml of clove oil by a 2-year-old boy (equivalent to 0.3–0.7 ml/kg of eugenol). Acidosis, deteriorating liver function, deep coma, generalised seizure and unrecordably low blood glucose were all noted. Heparin (anticoagulant) was given due to the possible development of disseminated intravascular coagulation. The child was fully conscious by day 6 and eventually made a full recovery [402].

Eucalyptus

There have been several instances of eucalyptus oil poisoning in children in the UK this century; in some, death has been recorded as an outcome [269, 278], in others, the poisoning has been survived [277, 279, 280].

There are three common manifestations of eucalyptus poisoning: a depressant effect; ab–normal respiration; and pinpoint pupils. A toxic dose of eucalyptus usually leads to drowsiness in a few minutes, and the patient may be deeply unconscious within 15 minutes. There is fre–quently vertigo and incoordination, but these are indications of relatively mild poisoning, as they would otherwise be rapidly superseded by stupor.

There are also likely to be the following: epigastric pain, weakness in the legs, cold sweats and headache. Respiratory effects are rather variable, possibly because eucalyptus stimulates respiratory tract mucus production at low doses and suppresses it at high doses.

Some cases show shallow, others laboured, breathing [269]. Usually, the pupils are highly constricted [278] a sign which could lead to confusion with opiate poisoning, although the smell of eucalyptus on the patient's breath should make the diagnosis obvious to someone familiar with its odour.

Under 5 ml of eucalyptus oil has apparently been fatal in an adult, although this is atypical.

Death is usual after 30 ml, following severe cardiovascular, respiratory and central nervous effects [261]. A 3-year-old boy has survived the ingestion of about 10 ml of eucalyptus oil [255] and an adult has survived ingestion of 60 ml [261]. In 1893, a 10-year-old boy died within 15 hours of ingesting about 30 ml of eucalyptus oil. Signs included shortness of breath and vomiting coming on in minutes [165].

Although there are many different types of eucalyptus oil the great majority of those purchased for household or medical use are chemically similar, being cineole-rich. Cineole has occasionally been reported as causing serious poisoning when accidentally instilled into the nose (see p. 26) [250].

Hyssop

Hyssop oil has been reported as a cause of neurotoxicity. It brought on convulsions in a 6-year-old child given 'half a coffee spoon' (perhaps 15–20 drops?), in an 18-year-old girl, who drank 30 drops, and in a 26-year-old woman who took 10 drops on each of 2 consecutive days, and suffered a seizure on the second day [140, 234]. (See p. 67 for more details on these cases.) The toxic ingredient of hyssop is probably pinocamphone [234].

Nutmeg

There do not appear to be any recorded cases of poisoning from nutmeg oil ingestion. However, it is worth noting that ingestion of one whole ground nutmeg can cause serious toxicity, manifesting as narcosis, collapse and a feeble pulse [448, 449]. It may be that this toxicity is not directly related to the essential oil which, however, does appear to be psychotropic in elevated doses (see p. 69).

Parsley/apiol

Both apiol and various preparations of parsley have been used for many years to procure illegal abortion in Italy [500, 502, 503]. There are several recorded cases of fatal poisoning from apiol ingestion [501, 502, 503]. In one case, a woman in her seventh month of pregnancy ingested 14 capsules of apiol (0.3 g of apiol per capsule) over 2 days. On the third day, she experienced violent abdominal pains with diarrhoea, vomiting and vaginal bleeding. The fetus was expelled with severe menorrhagia. Her cardio-renal signs deteriorated rapidly, with signs of nephritis. The patient's cardiac function decreased, she became comatose and died [530].

The lowest total dose of apiol causing death is 6.3 g (2.1 g/day for 3 days) [530]; the lowest fatal dose is 0.77 g, which was taken for 14 days [533]; the lowest single fatal dose is 8 g [500]. At least 19 g has been survived [530]. Common symptoms of apiol poisoning are fever, severe abdominal pain, vaginal bleeding, vomiting and diarrhoea [501, 530]. Post-mortem examination invariably reveals considerable damage to both liver and kidney tissue, often with gastrointestinal inflammation and sometimes damage to heart tissue [501, 503, 533]. In one case, contamination of apiol with triorthocresyl phosphate is believed to have contributed to the death of the patient [532].

Pennyroyal

One of the earliest recorded cases of pennyroyal poisoning was published in *The Lancet* in 1897. A 23-year-old woman ingested one tablespoon of pennyroyal oil, which was followed sometime later by vomiting. She was admitted to hospital 4 days later with symptoms of acute gastritis, and died later that day. Post-mortem examination showed congestion of the abdominal organs [390].

She had taken the oil to 'bring on menstruation which had been in abeyance for six months'. At the inquest, the pharmacist who sold her the oil stated that, in 30 years, he had never heard of a case of pennyroyal oil poisoning. It would seem that pennyroyal oil was readily available over the counter at that time, and was regarded as a safe menstrual stimulant (and abortifacient). This belief in the safety of pennyroyal almost certainly contributed to its further use, and to further accidents.

In 1906, the case of a woman who took half an ounce (about 15 ml) of 'pennyroyal essence' was reported in the *British Medical Journal*. ('Pennyroyal essence' could have been concentrated essential oil, or a 10% solution by volume.) The woman was 1 week past her expected menstrual period. She became extremely ill, the effects of the oil starting only 10 minutes after ingestion. Within 2 hours she was unconscious but had recovered the next day. Her symptoms had included tingling and numbness of the extremities [166].

The same journal reported a similar case 7 years later; the woman concerned was febrile, delirious, vomiting and suffering from involuntary twitching. The effects had commenced within 30 minutes of ingesting the oil. She had recovered by the next day [167]. In another similar case a woman ingested 15 ml of American pennyroyal oil in an attempt to stimulate menstruation [253]. She experienced minor symptoms of acute poisoning, vomited many times, and survived. She was not pregnant.

A much more serious case was reported in *The Lancet* in 1955. It was not possible to determine how much oil was taken, but the effects were abortion, vaginal bleeding, haemolytic anaemia (destruction of the red blood cells) and rapid destruction of the kidney tubules, with death following massive urea leakage into the blood [169].

Four cases of pennyroyal poisoning were reported in Colorado in 1978. In the first case a 21-year-old woman, who was less than 1 month pregnant, had taken a quarter-ounce of pennyroyal oil. She presented with nausea, numbness and tingling of the extremities and dizziness. Her pregnancy did not abort and she recovered the same day. She later had a legal abortion [168].

The second case was of a 24-year-old woman who had also ingested a quarter-ounce of pennyroyal oil, taken in an attempt to induce abortion. She was dizzy and nauseated, but her pregnancy survived. She was discharged the following day [168].

In the third case a 22-year-old pregnant woman ingested 10 ml of pennyroyal oil in an attempt to induce abortion. She experienced no symptoms other than dizziness, and did not miscarry [170].

The fourth case was of an 18-year-old, frightened of being pregnant but actually not, depressed and, on her own estimation at least, having a tendency to irregular menses. She had taken 1 ounce of pennyroyal oil and presented with rash, abdominal pain and frequent vomiting of blood. These effects persisted over the following 12 hours and she began to bleed from the vagina as well as from injection sites.

24 hours after ingestion, liver function tests had become abnormal, she lapsed into coma, developed fluid on the lungs and died on the sixth day in hospital. The cause was brain stem dysfunction due to liver damage [168, 170]. The dose of pennyroyal oil in this case was four times that in the first two.

Sage (Dalmatian) and thuja

Both of these oils have produced neurotoxicity (epileptiform convulsions) in a few cases (see pp. 67–68) [234]. The clinical picture is similar to that caused by hyssop intoxication, but in the case of sage and thuja oils, the major toxic ingredient is thujone. Sage also contains the convulsant camphor.

Sassafras

Sassafras oil has apparently been fatal in an adult after the ingestion of one teaspoon (about 5 ml) [cited in 269]. In another case, a 47-year-old ingested 5 ml and survived, suffering only vomiting and 'feeling shaky' [252]. Five children all survived following sassafras oil ingestion, although serious symptoms of poisoning occurred in each case. Their ages ranged from 15 months to 2½ years, and the amounts of oil ingested from 4 ml to 60 ml [269].

Sassafras oil appears to depress the CNS so that respiration and the blood circulation become inadequate [269]. Symptoms first appear within 10–90 minutes of taking the oil and the clinical picture resembles that of eucalyptus poisoning, except that vomiting and shock are more common with sassafras, and pinpoint pupils are not seen [252].

Wintergreen/methyl salicylate

Commercial methyl salicylate is almost always synthetically produced and is frequently sold as 'wintergreen oil'. Although the natural essential oil does exist, the great majority of the wintergreen oil sold is synthetic methyl salicylate. It is difficult to distinguish the two by smell alone.

Methyl salicylate, present in sweet birch and wintergreen oils at around 98%, has been a frequent cause of serious poisoning in children [265]. In the years 1926, 1928 and 1939–1943, 427 deaths were reported in the USA from methyl salicylate poisoning after swallowing wintergreen oil, or products containing it [237]. The toxicity of methyl salicylate is probably comparable to that of bleach.

Signs of intoxication include vomiting, fever, rapid and laboured breathing, tachycardia and respiratory alkalosis. There is often enlargement of the heart, congestion of the lungs, liver and kidneys and generalised lymphoid hyperplasia [232]. Cases of methyl salicylate poisoning have been notorious for their poor record of outcome [272]. In one report from 1937, 25 of 43 cases in the USA (59%) proved fatal [272]. In British studies of accidental poisoning (1953) methyl salicylate (36 deaths) featured far more frequently than camphor (12 deaths) citronella or eucalyptus (1 death each) [268, 270].

Three cases of wintergreen oil poisoning are detailed in a 1937 paper by Charles Stevenson, a pathologist, two of which are summarised here [290]. A 2-year-old child recovered after consuming 15 ml of the oil, but only after a severe illness involving a rapid pulse, low blood pressure, muscle twitching, cyanosis (bluish coloration) and rapid breathing.

A 1-month-old child was accidentally given 5 ml of wintergreen oil. The infant vomited a milky fluid 30 minutes later which smelled strongly of wintergreen and she appeared sleepy. 2 hours later she vomited again, and was given peppermint water and castor oil. Her breathing was rapid and deep; her breath acidotic and she was pale but not cyanotic. She died about 24 hours after the poisoning.

Stevenson also gives summarised information of 41 cases of wintergreen oil poisoning, between 1832 and 1935. Including the two cases mentioned above, these comprise 14 cases who survived and 27 who died. Of those who died, 17 were under 3 years of age. Two of the adult cases appear to have been suicides, since 200 ml and 250 ml of wintergreen oil were ingested. One 2-year-old child died after ingesting 4 ml, while another survived after ingesting 15 ml. One adult died after ingesting 15 ml, while another survived after ingesting 45 ml [290].

One teaspoon of wintergreen oil proved fatal to a child aged 22 months [260, 273]. Vomiting soon after ingestion was followed by drowsiness, laboured breathing, a craving for water, and evidence in the tongue and skin of marked dehydration. Glucose and saline infusions had no effect, the child became cyanotic and hyperpyrexic, and eventually died. On autopsy, the greatest abnormalities were found in the heart muscle, the liver and the kidneys [273]. It has been suggested that such poisoning can be effectively treated by exchange transfusion [268]. There may be great individual variation in the ability to handle the poison: adults are possibly more resilient since the ingestion of 10 ml of methyl salicylate has been survived [263].

A 1985 paper reports rapid recovery in a 21-month-old infant following ingestion of (presumed 4 ml) wintergreen oil. The child developed lethargy, vomiting and rapid breathing, but recovered fully and rapidly [293].

Methyl salicylate toxicity has been implicated as a cause of topical necrosis and kidney damage after application of a preparation containing it to the skin. In this case, a 62-year-old man presented with large blisters on his arms and thighs. He had been using an ointment containing wintergreen oil and menthol, and applied a heat pad on top of the ointment.

The clinical picture was initially that of acute eczema, such as occurs in contact dermatitis. But over the following 3 days, it became clear that the full thickness of skin was necrotic and that the necrosis extended even into the muscle layers. The patient had a high temperature, appeared toxic and complained of muscle weakness. His skin lesions were treated by grafting and he was only able to leave hospital after nearly 1 year. Residual kidney damage was still evident 2 years later [100].

Wormseed

Wormseed (= *Chenopodium*) oil, although one of the most toxic essential oils, has been used as a treatment for intestinal worms. It is not widely available in the UK and its availability in Europe has declined. Earlier in the century there were several reports of fatal poisoning in children due to wormseed oil being accidentally ingested or deliberately prescribed [271, 276].

A 2-year-old child died after being given 8 drops of wormseed oil, followed by a similar dose 1 week later. Autopsy showed broncho-pneumonia, liver and kidney damage, and oedema of the brain. Although this child had sickle cell anaemia, comparable doses of wormseed oil have been fatal in other children [342].

30 drops were fatal in a 12-year-old child, causing degeneration of the liver and kidneys, and oedema of the brain; 48 drops were fatal in a 4-year-old child, causing convulsions, bowel inflammation and generalised oedema; a 3-year-old child died after being given 9 drops of wormseed oil, 1 week after having received a similar dose [342]. It has been speculated that the damage to mucous membrane caused by the first dose allows the second dose to be more easily absorbed.

In one instance of poisoning in a 14-month-old child, less than 5 ml of wormseed oil was fatal [271]. A girl of 19 months died 24 hours after being given three 1-drop doses of wormseed oil 12 hours apart [438]. There is one instance of between 10 ml and 12 ml of wormseed oil having been survived in a male adult [274]. In this case the symptoms included slurring of speech, dullness of wits, impeded motor function, and periodic lapsing into coma. A nurse was fortunately present at the time, and administered gastric lavage which undoubtedly saved his life. He did not fully recover from the episode for 6 months, and his mental faculties are said to have remained dulled, possibly due to permanent CNS damage.

In most cases, a major finding on autopsy has been considerable damage to the CNS, particularly the brain. According to a 1939 report, wormseed oil is powerfully neurotoxic, resulting in visible oedema of the brain and meninges; the oil also affects the liver and kidneys. The fatal dose is very close to the therapeutic dose, although there is a great variation in individual sensitivity [438].

POISONING IN CHILDREN

There are many recorded cases of poisoning by essential oils in young children, often between 1 and 3 years of age. Parents, and consumers in general, need to be much more aware of the risks than they currently are. Young children are most at risk for three reasons:

- Their oral inquisitiveness leads them to try out materials by putting them in their mouths
- A liquid which is being examined by a child will probably be swallowed rather than sipped
- Children are smaller than adults and so smaller amounts of poison will do them harm [248].

Some unfortunate infants have died because a parent administered the essential oil by mistake, thinking it was castor oil, olive oil, or some other relatively innocuous oil. Some died because the essential oil was intentionally administered to them, whether by an ignorant parent, or an ignorant doctor. Many died because a bottle of essential oil was within their reach, and they were able to open it.

Essential oil poisoning in children is not a new problem. In 1953 Craig & Fraser [cited in 269] reported that, of 502 cases of accidental poisoning in children (1931–1951, Aberdeen and Edinburgh) 74 were due to essential oils. Of 454 deaths from accidental poisoning in childhood which occurred in Britain during a similar period, 54 were caused by essential oils [269].

Also in 1953, Craig (Royal Hospital for Sick Children, Edinburgh) commented that:

The problem is not to make the oils unpalatable or difficult of access, but to keep them away from children altogether. It is necessary to educate the public... Some curb should be put on these oils, which can all be sold across the counter of a chemist's shop. It seems desirable to remove such highly dangerous substances as oil of wintergreen. [269]

In over 40 years little has changed to improve the situation.

Statistics compiled for the USA show that in 1973 there were reports of 530 ingestions of camphor-containing products, 415 in children under 5 [257]. In all of these cases the products involved were over-the-counter preparations, the majority being camphorated oil. One report tells of a 10-month-old infant who stood up in her crib, reached for a bottle of camphorated oil, removed the cap, and drank approximately 1 ounce [437]. Although camphorated oil is not an essential oil, it contains similar quantities of camphor to several essential oils, and the risk to young children from these oils is a very similar one.

In 1985, Howrie and colleagues (Children's Hospital of Pittsburgh) reported on a case of wintergreen oil poisoning in a 21-month-old child. They comment:

Several aspects of this case are of great concern. First, there appears to be a lack of understanding and appreciation of the toxic potential of oil of wintergreen in the community... It is also important to recognise that oil of wintergreen is sold in volumes that may be fatal to young children if part or all of the product is consumed. The use of child-resistant safety packaging is required by the Poison Prevention Packaging Act for liquid products containing more than 5% methyl salicylate by weight. However, as noted in this case, a young child may still defeat those safeguards. [293]

In 1993, in a case report of clove oil poisoning, the authors commented:

Packaging and labelling of clove oil and other potentially toxic essential oils needs to be urgently reviewed. If all these oils were sold in bottles that had child resistant tops and were in restricted flow bottles, then even if a child managed to remove the top he/she would only be able to drink a very small amount. Also, the bottles need to be clearly labelled to indicate that the contents, if ingested, can be very toxic and urgent medical help should be sought if ingestion occurs. [402]

Serious poisoning has occurred with oils of citronella, clove, eucalyptus, pennyroyal, sassafras, wintergreen and wormseed. The point here is not just that these particular oils are hazardous to small children, but that a whole bottle of many essential oils almost certainly will be, if drunk. It is clearly very important for essential oil bottles to be kept tightly closed and well out of children's reach.

REDUCING THE RISKS

We do not know the consequences of ingesting large amounts of most essential oils. There are undoubtedly some oils which present negligible risk, but it is worrying that so many oils are sold in open-mouthed bottles in shops where there is no suitably trained supervision. We would like to see appropriate training for all retail staff who sell essential oils, even where packaging and labelling of the products sold is of a good standard.

Information on essential oil toxicity is available from specialist poisons units. In the UK, the largest of these is at Guy's Hospital; it advises doctors about substances which patients have swallowed or allowed to come into contact with the skin. Only a tiny fraction of enquiries to the Units are essential-oil-related, but the number of cases may be expected to increase as aromatherapy gains popularity unless proactive safety measures are taken.

There are complications to assessing the potential toxicity of essential oil products brought into the home:

• Not all essential oils are undiluted or unadulterated.
• Many essential oil bottles, as sold at retail outlets, are inadequately labelled.
• Some aromatherapy clients bring home mixtures of essential oils in bottles which carry insufficient information about their contents. Percentage dilution (or ingredients, if the mixture is undiluted) should be listed.

The two most important changes that both retail suppliers and aromatherapists could make which would significantly decrease the risk of serious accidental poisoning by essential oils are:

• To label clearly the contents and concentrations in every bottle of essential oil dispensed,

mentioning whether the oils are pure or diluted in vegetable oil, and if so, by how much. (Clear labelling is especially important to poisons units at hospitals. Their job is made much more difficult if preliminary tests have to be done to find out what has been ingested.)
• Only to buy, sell or dispense undiluted essential oils in bottles which are fitted with integral drop-dispensers. (Any design of bottle which leaves a wide open bottle-neck, such as a regular dropper with bulb, is relatively unsafe, because it allows large volumes to be poured out at once.)

SUMMARY

• The LD$_{50}$ value is not in itself a complete key to toxicity.
• The lethal dose of a substance varies according to the size of the person (or animal) taking it.
• Virtually every substance used in foods, medicines, cosmetics and toiletries is tested for toxicity on animals, although this situation is now beginning to change.
• Circumspection is needed when applying animal data to the use of essential oils in humans. Some essential oils may be more toxic to humans than they are to animals.
• So far, all cases of serious poisoning from essential oils have arisen after oral ingestion. Aromatherapy massage is very unlikely to give rise to such poisoning.
• There are many cases of poisoning by essential oils in very young children. Parents and consumers in general need to be much more aware of the risk to young children than they currently are.
• Both professional aromatherapists and retailers should ensure that all undiluted (pure or mixed) essential oils are adequately labelled, and are in bottles with integral drop-dispensers.
• We would like to see appropriate training for all retail staff who sell essential oils.

Notes

1. Aromatherapy massage is very unlikely to cause serious poisoning and there are no reports of this. Relative to

oral dosing, much smaller amounts of essential oil are used, and the oils enter the bloodstream at a slower rate, so concentrations in the blood and tissues will be considerably lower.

2. The absence of an essential oil from this chapter in no way suggests that it carries no poisoning risk. If an oil of unknown toxicity is ingested accidentally, use the Safety Index to help a doctor identify its constituents.

6

Tissues, organs, systems

MUCOUS MEMBRANE

Mucous membrane is the name given to the lining of the alimentary, respiratory and genito-urinary tracts. The cells forming the membrane secrete mucus, which helps protect the membrane from mechanical and chemical injury. However, because it is thinner, and more fragile than skin, mucous membrane is more sensitive to insult, and is also more permeable to essential oils.

Mucous membrane lacks the protective keratinised cell layer of skin. Undiluted essential oils should not be applied to mucous membrane (mouth, vagina, rectum) and should not be applied to or very near the eyes.

Essential oils come into contact with mucous membrane when ingested, used in a mouthwash or gargle, inhaled, or when used in pessaries, douches or suppositories. It is inadvisable to apply undiluted essential oil to mucous membrane, as this carries a high risk of irritation. Essential oils such as clove, myrrh and tea tree are sometimes used undiluted in the mouth for problems such as toothache, mouth ulcers and thrush. Clove oil is in fact very likely to irritate mucous membrane.

There are very few references to essential oil irritation of mucous membrane in the literature, but one material which has been tested is eugenol. Applied undiluted for 5 minutes to the tongues of dogs, eugenol caused erythema and occasionally ulcers [Lilly et al 1972, cited in 421]. In a 1978 study, the application of undiluted eugenol to rat labial mucosa, observed for up to 6 hours, resulted in swelling, cell necrosis and vesicle formation [73]. The authors commented that eugenol progressively destroys the cells of the mucosal epithelium, and causes an acute inflammatory response. Even when eugenol was left for 1 minute only, and then removed as far as possible, a severe reaction resulted. The effect of eugenol on mucous membrane was said to be 'more penetrating and destructive [to mucous membrane] than previously thought.'

Apart from the above, all the information in this section is derived from tests performed on one of us (RT) and two other volunteers. Arctander refers to rue oil as an irritant: 'For most people rue oil is harmful to the mucous membranes' [155]. Our tests found rue oil to be non-irritant. Since these tests were conducted on a very small number of people, the results (see Box 6.1) may be less reliable than those from the RIFM monographs.

Testing method

0.02 ml of undiluted essential oil was applied to a small area on the inside of the mouth, which was kept open to prevent dilution of the oil with saliva. Subjective observations were made for up to 3 minutes following application. Irritation generally become apparent 15–30 seconds after application.

Irritation is very much a dilution-dependent hazard. The more an essential oil is diluted, the less irritating it is.

If we look at the principal mucous membrane irritants from this study a very clear picture emerges regarding the components probably responsible. In most cases the irritant oils contain one of three phenols—thymol, carvacrol or eugenol. These phenols occur in large amounts, in every instance constituting between 38% and 95% of the essential oil. There seems little doubt that

---

Box 6.1   Mucous membrane irritants

A: severely irritating
Recommendation—do not use on mucous membrane.

Horseradish	Mustard

B: strongly irritating
Recommendation—do not use at more than a 1% concentration on mucous membrane.

Clove leaf	Massoia	Pimento leaf	Thyme
Clove stem	Oregano	Savory	Thyme (wild)

C: moderately irritating
Recommendation—do not use at more than a 3% concentration on mucous membrane.

Bay leaf (W. Indian)	Cinnamon bark*	Cornmint	Pimento berry
Caraway	Cinnamon leaf	Laurel	Spearmint
Cassia*	Clove bud	Peppermint	

* These oils, while rated here for their irritant effects, can also cause powerful allergic reactions in some people.

D: non-irritant
Caution—some oils may occasionally cause irritation if used undiluted.

Angelica root	Clary sage	Geranium	Nutmeg	Sandalwood
Anise	Citronella	Grapefruit	Orange	Sassafras
Basil	Coriander	Hyssop	Parsley herb	Tarragon
Bergamot	Cumin	Immortelle	Parsleyseed	Tea tree
Buchu	Cypress	Juniper	Petitgrain	Vetiver
Cardamon	Dill seed	Lavender	Pine (dwarf)	Wintergreen
Carrot seed	Eucalyptus	Lemon	Rose[A]	Yarrow
Cedarwood (Atlas)	Fennel (bitter)	Lemongrass	Rose otto	Ylang-ylang
Cedarwood (Virginian)	Fennel (sweet)	Marjoram (sweet)	Rosewood	
Chamomile (German)	Frankincense	Myrrh	Rue	
Chamomile (Roman)	Galbanum	Neroli	Sage (Dalmatian)	

these phenols are responsible for causing the irritation. Certainly we have not found any essential oils containing significant amounts of them which do *not* cause irritation.

The remaining mucous membrane irritant oils all contain cinnamaldehyde, carvone, massoia lactone, menthol or allyl isothiocyanate. These are presumably responsible for their irritating effects. There is, not unexpectedly, a strong correlation between mucous membrane irritant oils and skin irritant oils.

As with skin irritation, there is little risk so long as essential oils are appropriately diluted. Exactly what constitutes appropriate dilution may need more research, and will depend on the essential oils used. In the absence of such data the recommendations given above will hopefully serve as a useful interim guideline.

THE LIVER

The liver is the largest internal organ of the body. It is a remarkable structure in many respects, not least in its ability to carry out so many life-giving functions. It makes enzymes and plasma proteins, it recycles red blood cells and stores their iron, it is the main heat producer of the body, it carries out an initial metabolism on almost every nutrient material absorbed from the alimentary canal, it is involved in vitamin manufacture and it detoxifies drugs and other foreign substances.[1]

Glutathione

The liver contains a substance called reduced glutathione, which mops up reactive molecules (electrophiles, free radicals) before they can damage DNA or protein. However, the liver only contains so much glutathione, and if this is completely (albeit temporarily) depleted, reactive molecules are able to attack and seriously damage liver and blood cells, before the glutathione is replaced. If the damage is serious enough, as in paracetamol (= acetaminophen) overdose, the result can be fatal, and is due to liver failure and/ or haemolytic anaemia.[2]

It is unlikely that any essential oils could be absorbed in sufficient quantities to deplete

glutathione through dermal application. However, oils administered orally probably could do so. This effect has never been demonstrated in humans, but it would be prudent to avoid oral administration of certain essential oils (see Table 6.1) in people taking drugs such as paracetamol, which rapidly use up liver glutathione [20].

Eugenol (<> clove) has been shown to cause liver damage in mice whose livers have been experimentally depleted of glutathione [400]. Although the oral doses of eugenol used were high (5 ml/kg) the data suggest that clove oil and others rich in eugenol should be used with caution in those with impaired liver function and that they should not be administered to anyone taking paracetamol (= acetaminophen).

Work with rat liver cells in culture gives evidence that eugenol is metabolised to toxic compounds in a paracetamol-like manner, and supports the previous paper. As in paracetamol poisoning, pretreatment with *N*-acetylcysteine prevents glutathione loss and hence cell death [401].

Cinnamaldehyde (<> cinnamon bark) has been found to depress rat liver glutathione levels [475]. *Trans*-anethole (<> anise) shows a dose-dependent cytotoxicity to rat liver cells in culture, and causes

Table 6.1 Components, and corresponding essential oils, which have a potential for liver toxicity

Component	Oil
Trans-anethole	Anise, fennel, star anise
Cinnamaldehyde	Cassia, cinnamon bark
Estragole	Basil, *Ravensara anisata*, tarragon
Eugenol	Bay (W. Indian), betel leaf, cinnamon leaf, clove bud, clove stem, clove leaf, *Ocimum gratissimum*, pimento berry, pimento leaf, tejpat leaf
Methyleugenol	*Melaleuca bracteata*, snakeroot, tarragon (Russian)
d-pulegone	Buchu (*B. Crenulata*), pennyroyal
Safrole	Camphor (brown), camphor (yellow), sassafras
Unknown	Savin

It is recommended that these essential oils be avoided orally in people with liver disease, such as cirrhosis and hepatitis, people with alcoholism, and in anyone concurrently taking paracetamol (= acetaminophen).

glutathione depletion [452]. The mechanism of its action is unknown, but it is not due to a DNA-damaging effect. A primary metabolite of *trans*-anethole, anethole 1,2-epoxide, also depletes glutathione [504].

The symptoms described for nutmeg poisoning resemble those produced by paracetamol-type liver toxicity so closely that it is likely that excessive doses of ground nutmeg can be metabolised to highly reactive intermediates which damage the liver [321]. However, this is speculation and, even if true, may not apply to the essential oil.

Pulegone (<> pennyroyal) has been shown to produce hepatotoxicity of the paracetamol (= acetaminophen) type. A toxic metabolite of pulegone, menthofuran, is produced by the liver's enzyme systems [112, 396]. In rats, pennyroyal oil produces acute liver and lung damage, the active constituents being identified as pulegone, *iso*-pulegone and menthofuran [110]. Pulegone produces generalised toxic effects in the liver [22]. Both pulegone and its metabolite, menthofuran, destroy the enzyme cytochrome P_{450} [111, 117, 118, 119]. The ingestion of 30 ml of pennyroyal oil by an 18-year-old girl resulted in massive hepatic necrosis and death [170].

Similar studies to those conducted for pulegone have demonstrated that neither menthone nor carvone is hepatotoxic [111, 117]. It is possible that the toxicity of apiol (see p. 52) is due to glutathione depletion.[3]

The formation of abnormal DNA derivatives

A group of essential oil constituents (alkenylbenzene derivatives) seems able to produce DNA abnormalities in mouse liver. Some of these compounds are associated with the development of liver cancers in experimental animals [122]. Methyleugenol, estragole and safrole are most highly implicated, myristicin and elemicin less so.[4] Dill and parsley apiols, and anethole, all bind weakly to DNA and are not associated with tumour development.

Safrole has been shown to produce liver enlargement and to induce liver microsomal enzymes in the rat after chronic administration

[131]. The pattern of liver toxicity caused by safrole and sassafras oil is consistent with safrole being a weak hepatocarcinogen [343].[5]

Menthol hepatotoxicity

Large doses of menthol (above 0.2 g/kg) can produce long-term change in the appearance of hepatocytes (liver cells) when given orally to rats [329]. This is a very high dose, and the finding has no implications for the use of mint oils in aromatherapy. However, menthol-rich essential oils are likely to cause liver problems in people deficient in a particular enzyme.

Menthol has been shown to provoke severe neonatal jaundice in babies with a deficiency of the enzyme G6PD (glucose-6-phosphate dehydrogenase).[6] Usually, menthol is detoxified by a metabolic pathway involving G6PD. When babies deficient in this enzyme were given a menthol-containing dressing for their umbilical stumps, menthol was found to build up in their bodies [109]. This is a fairly common inheritable enzyme deficiency, particularly in Chinese, West Africans, and in people of Mediterranean or Middle Eastern origin [109]. G6PD deficiency is sex linked and more males are affected than females. For instance, 12% of male African Americans show G6PD deficiency, but only 3% of females [523].

Menthol should clearly be avoided, at least in oral doses, by anyone with a deficiency of this enzyme. Such people can be recognised because they will characteristically have had abnormal blood reactions to at least one of the following drugs, or will have been advised to avoid them:

- antimalarials
- sulphonamides (antimicrobial)
- chloramphenicol (antibiotic)
- streptomycin (antibiotic)
- aspirin.

Coumarin hepatotoxicity

Coumarin is a lactone found in quantity in several absolutes, and in very small amounts in a few essential oils. Chronic oral administration of coumarin produced severe liver damage at a level of 2500 p.p.m. in the diet of rats and 100 mg/kg per day for dogs [347]. This hepatotoxicity was confirmed by subsequent research [336, 469] and one of these studies reported finding bile duct carcinomas in coumarin-fed rats [469]. Coumarin shows a high toxicity in several animals; its acute oral LD_{50} is 0.2 g/kg in both guinea pigs and mice [420], and 0.68 g/kg in rats [338]. Based on all the animal data, the FDA prohibited the use of coumarin in food in 1954; both the UK and EC 'standard permitted proportion' of coumarin in food flavourings is 0.002 g/kg [455, 456].

It was noticed by several workers that there was a significant interspecies difference in the way coumarin was metabolised [108, 525, 526, 527]. For example, the percentage of orally administered coumarin excreted in the urine as 7-hydroxycoumarin is 1% in the guinea pig, 3% in the dog, 19% in the cat, and 60% in the baboon [527]. In the 1970s the relevance of the animal data to humans was being seriously questioned. One study, giving coumarin to baboons at 50 p.p.m. and 100 p.p.m. (3 weeks at each dosage level) did find histological damage in the liver [527]. In another study, coumarin was administered to baboons for between 16 and 24 months at 2.5, 7.5, 22.5 or 67.5 mg/kg/day [528]. At the highest dose level only, an increase in liver weight was noted, and ultrastructural examination of the liver revealed dilatation of the endoplasmic reticulum.

In clinical trials using coumarin as an anti-cancer agent only 0.37% of patients developed (reversible) abnormal liver function. The majority of 2173 patients received 100 mg/day of coumarin for 1 month, followed by 50 mg/day for 2 years. (The other patients received between 25 mg/day and 2000 mg/day.) Of the total, only eight patients developed elevated liver enzyme levels, which returned to normal on stopping the coumarin [520]. On this evidence, coumarin cannot be regarded as hepatotoxic in humans.

The second baboon study indicates a no-effect level of 22.5 mg/kg/day, which equates to around 1.5 g/day for an average (65 kg) adult human. This dosage level is 30 times higher than the 50 mg/day received by many of the cancer patients for 2 years.

THE KIDNEYS

The urinary system comprises two kidneys, the ureters, which carry urine from the kidneys to the bladder, the bladder itself, and the urethra. Its functions are to form, store and release urine, an aqueous solution of the waste products of metabolism, plus the metabolic products of drugs and other foreign molecules.[7]

Most drugs and their metabolites are excreted via the kidneys. Many water-soluble molecules are secreted into the kidney tubules and reach the urine; the more water-soluble essential oil constituents, such as terpenoid alcohols, aldehydes and esters are probably removed in this way, at least in part. Some such drug molecules are reabsorbed actively into the bloodstream from the kidney tubules; it is not known whether essential oil molecules are actively reabsorbable.

Many lipid-soluble molecules are known to diffuse passively back through the tubule walls into the bloodstream, which is why the liver makes such an effort to render fat-soluble substances more water-soluble in readiness for excretion. The relatively fat-soluble terpenes and terpenoid ketones undoubtedly come into this category.

Because the kidneys receive the peak plasma concentrations of all substances present in the blood, and because they concentrate a filtrate of the blood, kidney tubule cells are exposed to higher levels of chemicals than are other body cells. Consequently, the kidney is extremely vulnerable to damage caused by foreign molecules, whether drugs or essential oil constituents.

Acute toxicity resulting from apiol ingestion (at close to the oral dosage equivalent of an apiol-rich oil) frequently includes severe damage to kidney tissue (see p. 52). Apart from this, there is no evidence that essential oils can cause damage to *healthy* human kidneys, at the doses in which they are given in aromatherapy. Because of their probable general toxicity, it is recommended that apiol-rich essential oils (see Table 6.2) are not given orally at all.

There are two issues to bear in mind when considering kidney disease. Firstly, a diseased kidney will almost certainly be less able to handle

Table 6.2 Apiol-rich essential oils

Oil	% apiol
Indian dill seed	20–30
Parsley leaf	<18
Parsley seed	21–80

large quantities of any essential oil, and there may be a case for reducing the dosage of essential oil given orally to people with kidney disease (see p. 37). Secondly, in kidney disease the blood's ability to coagulate is reduced, and anticoagulant essential oils might exacerbate the situation, perhaps resulting in internal haemorrhage. It is recommended, therefore, that such oils should not be given orally in kidney disease (see p. 66).

It has been claimed that 'juniper' can irritate the kidneys, and that it should be contraindicated in kidney disease; 'juniper oil' is specified in a few instances [9, 389, 394, 395, 397, 398, 446, 447]. While other preparations made from juniper may (or may not) be problematical in kidney disease, no evidence could be found to support this idea as far as the essential oil is concerned (see juniper profile). Certainly there is nothing in the composition of juniper oil which would seem to give cause for concern.

In adult male rats, d-limonene has been shown to produce renal accumulation of a soluble protein, alpha$_{2u}$-globulin with associated microscopic abnormalities [125, 127]. However, this toxic effect is so specific to male rats, and so conspicuously absent in studies on humans and other mammals, that there is clearly no caution required in using limonene-rich oils in aromatherapy [124, 126, 128].

Methyl salicylate has been shown to produce nephrotoxicity in rats during gestation, retarding kidney development but not usually causing permanent abnormality [129]. High doses (4 ml/kg) of methyl salicylate, applied to the skin of rabbits, 5 days a week for up to 96 days, caused early deaths and kidney damage; lower dosage levels (0.5, 1.0, or 2.0 ml/kg) caused a higher than normal incidence of 'spontaneous' nephritis [415]. Safrole and sassafras oil produce kidney damage in the rat after toxic doses [343].

THE CARDIOVASCULAR SYSTEM

The cardiovascular system comprises the heart and the blood vessels through which blood flows. The heart is a biological pump whose function is to ensure that the body receives adequate blood to supply its need for oxygen, glucose, fats, amino acids and other nutrients. In addition, the rate of blood flow out of the tissues should be sufficient to remove waste products and toxins.[8]

Blood pressure

The fact that arterial blood is under pressure, even between heartbeats, means that the heart has to do considerable work against this pressure simply to get the blood moving. Any abnormally prolonged increase in the resting pressure is therefore undesirable because it increases the work of the heart. A condition of permanently raised blood pressure is referred to as hypertension. There is general agreement that a resting pressure which is consistently above 95 mm of mercury may indicate a tendency to gradually worsening hypertension.

Any toxic material, for instance an oral essential oil in overdose, is likely to produce hypertension due to reflex stimulation of the sympathetic branch of the autonomic nervous system, unless the oil in question has a markedly depressant action on the heart or a marked dilatant effect on small arteries.

One way of influencing the smooth muscle of blood vessels directly is by altering the flow of calcium ions into the muscle cells. Also, heart activity is dependent upon the movement of calcium into and out of cardiac muscle cells.

Essential oils which can influence calcium movement across cell membranes may have a direct action on cardiac tissue if they reach it in large enough quantities. This may happen after, for instance, oral ingestion of peppermint oil, but is unlikely after dermal application. Bisabolol, eugenol, d-carvone, d-menthol and possibly *trans*-anethole, inhibit cardiovascular calcium channels, giving the possibility of a depressant effect on the heart if absorbed from essential oils in sufficient quantities [355].

There is an indication that rosemary and peppermint oils may interfere both with calcium influx into myocardial cells and with the release of intracellular calcium stores. At present there is no direct evidence that they present any health risk, nor of what dosage might present a risk [404, 405]. Menthol appears to be peppermint's active ingredient [406, 407].

Calamus oil is believed to depress the myocardium because of its asarone content [350]. Taget oil is hypotensive in rats, reducing blood pressure at a dose of 0.05 g/kg [352]; hyssop oil is hypotensive at very high doses [353].

Garlic oil (18 mg) has a hypotensive effect in humans after oral administration [366]. Geranium and lavender oils demonstrate a weak hypotensive action in animals [225]. Essential oil constituents with a hypotensive action, which is probably caused by vasodilation, include:

- Linalool
- Citronellol
- Nerol
- Geraniol
- Terpineol
- Cineole.

These are in decreasing order of effectiveness, from 9 to 31 mg/kg producing a 25% fall in blood pressure in the dog [387]. No such hypotensive effects have been demonstrated in humans.

Synthetic camphor and Japanese camphor (true camphor) can cause a transient fall, followed by a rise, in blood pressure when given to the cat [377]. They also produce a rise in heart rate in the isolated rabbit heart, via a direct effect on heart muscle [377]. In humans, the predominant effect on the heart appears to be depressant. However, at least one report suggests that camphor can sensitise the heart to adrenaline [378].

An antihypertensive and hypotensive effect of carrot seed oil in the dog has been demonstrated; this is probably due to a vasodilatant action on arterioles [383].

Zanthoxylum budrunga oil has been found to cause a transient fall in blood pressure at intravenous doses of above 2.5 mg/kg [223].

At a dose of approximately 1 g/kg intravenously, clary sage oil produces a small increase

Box 6.2	Hypotensive essential oils and constituents	
Oils		**Constituents**
Calamus ?	Lavender	Cineole
Carrot	Peppermint ?	Citronellol
Garlic	Rosemary ?	Geraniol
Geranium	Rosewood ?	Linalool
Hyssop	Taget	Nerol
		Terpineol

in arteriolar diastolic and systolic pressure [224]. It is unlikely that this has any relevance to the use of clary sage in aromatherapy, since nowhere near this dose would reach the tissues after normal therapeutic doses, whether oral or dermal.

Lists of essential oils and essential oil constituents with hypotensive properties are given in Box 6.2.

There is no evidence that essential oils have an adverse effect on the control of blood pressure in humans, and the data from animal tests are inconclusive. We therefore consider that there is no need for contraindication of essential oils in either hypertension or hypotension, by any route of administration. Oils of garlic and taget are possibly the most likely to demonstrate a hypotensive effect, but there is no reason to believe that either oil would exacerbate an already established hypotension.

The heart

Inhalation of *l*-menthol has been shown to dilate capillaries in the nasal mucosa; this effect correlates with the reported ability of *l*-menthol to dilate systemic blood vessels after intravenous administration [222]. Additionally, mentholated cigarettes and peppermint confectionery have been responsible for cardiac fibrillation in patients prone to the condition who are being maintained on quinidine, a stabiliser of heart rhythm [236]. Bradycardia has been reported in a person addicted to menthol cigarettes [388]. Menthol-rich oils (cornmint, peppermint) should probably be avoided altogether in cases of cardiac fibrillation.

Citral, administered to rabbits subcutaneously or by mouth, causes damage to the vascular endothelium even at a dose of 0.005 mg/kg, as does cinnamaldehyde [324]. The effect is probably due to an anti-vitamin-A property of citral [340] and is very likely to be specific to rabbits, since the metabolism of vitamins by other animals is often different to that of humans. We therefore feel that there is no cause for concern here.

Cinnamon bark oil (0.5 g/kg) and cinnamaldehyde (0.1 g/kg) cause serious cardiovascular toxicity in mice leading to insufficiency, pulmonary congestion and death [343]. The dose for cinnamon bark oil is extremely high compared to those used in aromatherapy. No such effects have been seen after ingestion in humans, and there is no evidence that cinnamon bark oil has any adverse effect on the human heart.

A depressant effect on the dog heart has been shown by carrot seed oil [383]. When given in large doses, a depressant effect on the heart has been shown for linalool [328], sage [224], geraniol, thymol and (-)-carvone [387]. These findings are unlikely to be relevant to aromatherapy because they imply human doses of 0.15–0.5 g/kg but they could be relevant to oral overdosage.

Blood clotting

Both garlic and onion oils demonstrate anti-platelet activity [363, 365, 366, 372]. (Platelet activity is essential for blood clotting.) Onion oil is up to 10 times more potent than garlic oil [363]. The active component in garlic is believed to be methyl allyl trisulphide [373]. Tests with the synthetic compound showed it to be active at a concentration of 10 µmol/l in human platelet-rich plasma [363].

Anti-platelet activity is also demonstrated by eugenol and *iso*-eugenol, and is due to an anti-prostaglandin action [370, 434]. It would be prudent to avoid oral administration of garlic oil, onion oil, and all eugenol-rich oils (see Table 6.3) in those with blood-clotting problems. This group includes people with haemophilia, liver disease, kidney disease, prostate cancer and systemic lupus erythematosus (SLE). It would be prudent for anyone taking anticoagulant drugs, such as

Table 6.3 Eugenol-rich essential oils

Oil	% eugenol
Bay (W. Indian)	38–75
Betel leaf	28–90
Cinnamon leaf	70–90
Clove bud	70–95
Clove stem	70–95
Clove leaf	70–95
Ocimum gratissimum	12–20
Pimento berry	67–83
Pimento leaf	60–95
Tejpat leaf	75–80

aspirin, heparin and warfarin, to avoid oral use of the same group of oils.

THE RESPIRATORY SYSTEM

Perilla oil contains perilla ketone, and sometimes related substituted furans, egomaketone and *iso*-egomaketone. These are all considered to be potent agents which can cause fluid build-up in the lungs of animals [333]. The toxicity is of most concern to farmers whose cattle or sheep may graze on the plant. Pulmonary toxicity has also been induced in laboratory rodents [333].

Perilla oil is not acutely toxic to either rat or mouse [428] and there are no reports of it causing any problems in humans [332]. Perilla oil is extensively used in food flavourings in Japan. We are of the opinion that perilla oil should be used with caution in aromatherapy. There is no indication that any other essential oil might be toxic to the respiratory system.

THE CENTRAL NERVOUS SYSTEM

The CNS is uniquely sensitive to chemicals and toxicological agents of all types. Because nervous tissue is largely unable to renew itself, and because the functioning of the CNS is so exquisitely sensitive to neural damage and to alterations in neurotransmitter function, it is quite possible that some essential oil constituents have CNS toxicity, at least when given in high doses. Many types of essential oil compound, being fat-soluble, can probably gain rapid entry into the CNS through the blood–brain barrier.[9]

Convulsant effects

Convulsant effects are one example of 'neurotoxicity'. A neurotoxic substance is one which has a toxic or destructive effect on nervous tissue. Essential oils are unlikely to cause physical damage to nervous tissue, other than in extreme circumstances, such as those described in fatal cases of poisoning from ingestion of wormseed oil (see p. 55). Fenchone, pinocamphone, camphor and thujone have all been found to have convulsant activity [134, 135]. However, the doses used in these particular studies were extremely high, and have no relevance to the amounts used in aromatherapy.

Hyssop

Hyssop is convulsant, probably because of its pinocamphone and *iso*-pinocamphone content. The convulsant effects of hyssop oil were first researched in 1891. Doses of 2.5 mg/kg were injected into dogs, producing almost immediate epileptiform seizures [431]. In 1945, similar tests, injecting hyssop oil at 1–2 ml, revealed a two-phase response. During the first phase, blood pressure went down, breathing became rapid, and random clonic movements appeared. The second phase was characterised by hypertension, rapid heartbeat and numerous clonic contractions. The rapid onset of both phases was noted [353].

Tests in rats found that convulsions appeared for hyssop oil, at dose levels over 0.13 g/kg, and for sage oil, at dose levels over 0.5 g/kg [234, 325]. These figures indicate that hyssop is more powerfully neurotoxic than sage in rats. Intraperitoneal pinocamphone was found to be convulsant and lethal to rats above 0.05 ml/kg, with the comparative dose for thujone being 0.2 ml/kg [234, 325]. It has been established that camphor, thujone, pinocamphone, hyssop oil and sage oil are all potentially convulsant and neurotoxic to laboratory animals [432].

The amounts of hyssop oil used in these animal tests were generally high. However, there are three reported cases of low-dose hyssop oil ingestion resulting in epileptiform convulsions [140]. The first case was a 6-year-old whose mother frequently gave him 2–3 drops of hyssop oil for his asthma. During one severe attack, he was given 'half a coffee spoon' (perhaps 15–20 drops?) shortly after which he suffered a convulsion. He fully recovered after 3 days in hospital.

The second case, an 18-year-old girl, drank 30 drops of hyssop oil to treat a cold. 1 hour later she suddenly lost consciousness for 10 minutes, during which she suffered generalised contractions and bit her tongue. In the third case a 26-year-old woman took 10 drops of hyssop oil on each of 2 consecutive days, and suffered a seizure on the second day [234].

Camphor

The potential for camphor to produce convulsions in humans is well known. Before electrical stimulation was used, injections of camphor were given as a treatment for depression and schizophrenia because of its ability to provoke convulsions [203]. Thankfully, this use rapidly became obsolete. One statistical analysis reveals that 45% of cases of symptomatic camphor poisoning in children under 5 suffered convulsions [433].

A 3-year-old girl suffered a generalised convulsion 2 hours after ingesting 0.7 g of camphor [257]. A 22-year-old female experienced two violent epileptiform convulsions following the ingestion of 12 g of camphor [275]. Three further cases of camphor ingestion which caused convulsions are discussed on page 106.

Other oils

There is one report of thuja oil causing convulsions [234]. A 50-year-old woman took 20 drops twice a day for 5 days. (She was advised to take it by a 'naturologist', but did not follow instructions to dilute the oil to 1% before taking the drops.) 30 minutes after her tenth dose she suffered a tonic seizure and fell, fracturing her skull [234].

Convulsions were induced in an adult who accidentally ingested one 'swallow' of sage oil [140]. Presumably these were caused by either the camphor and/or the thujone content. The essential oil of *Artemisia caerulescens* (rich in camphor and *alpha*-thujone) produces epileptiform convulsions in various test animals, notably cats [139]. Calamus oil may be convulsant because of its *beta*-asarone content [351].[10]

Pennyroyal oil has been a rare cause of epileptiform seizures when taken in an attempt to produce abortion [137]. Both pulegone and peppermint oil have been found to produce microscopic dose-related lesions in the brains of rats when fed to them in the diet [142, 143]. The doses causing lesions would be equivalent to the consumption of 3–6 g of peppermint oil by an adult. The lower dose level, producing no lesions, is equivalent to 0.65 ml in an adult. A proprietary menthol-containing oil has been reported as producing incoordination, confusion and delirium when 5 ml of the product (35.5% peppermint oil) was inhaled over a long time period [141].

Peppermint oil has been known to produce convulsions and ataxia, with paralysis, loss of reflexes and very slowed breathing in rats [337]. However, the doses were extremely high (0.5–2 ml/kg, i.p.). This toxicity may be due to the pulegone and/or menthone content; pulegone causes histopathological changes in the white matter of the cerebellum above 80 mg/kg [329]. Menthone produces similar cerebellar changes, but the lowest dosage level used was 0.2 g/kg/day [327]. There is no indication that peppermint oil can produce these toxic effects when used in aromatherapy, but some caution is required regarding oral dosage.

There are reports of nasal preparations containing menthol or cineole causing apnoea and instant collapse in infants following instillation into the nose [5, 250].

The risks

In the only recorded cases where seizures have been induced by essential oils, most of them apparently in non-epileptics, the oils were taken orally. Hyssop oil most definitely represents a serious risk, and has the potential to cause epileptiform seizures at oral dosage levels. Balsamite oil, camphor chemotype, may present a similar level of risk, due to its very high camphor content. Other oils, such as Dalmatian sage and wormwood, are probably only a potential risk if taken orally by people with a low convulsive threshold.

The majority of people with epilepsy take medication to suppress their fits. It is unlikely that this group would be any more at risk than the general population. However, there is a potential risk to those with a low convulsive threshold, i.e. a potential to have epilepsy which has not yet declared itself. This group could be as high as 5% of the population [490]. Therefore it would be prudent to avoid oral dosage of all the oils listed under 'A' and 'B' in Table 6.4. Most of these oils have already been classed as toxic. It is unlikely that the oils listed in 'C' present any risk when used externally, but they should be used with caution at oral dosage levels.

Because of the lower levels used, external use is much less of a risk. A 1967 review of olfactory stimulation of seizures in epileptics concluded that, 'Reflex epilepsy caused by scent in the narrowest sense of the word is very uncommon. It depends on the degree of oversensitivity of the individual to olfactory stimuli' (our translation from the original German) [440]. It is very unlikely that the external use of essential oils would lead to epileptiform seizures, but oils most likely to do so are those listed under 'A'.

People who are prone to epilepsy may have idiosyncratic reactions to essential oils (and to odours in general). This makes the prediction of adverse effects difficult. Epileptics should therefore exercise caution with essential oils, especially orally, if they suspect that they might react badly. It would be prudent for those with a strong family history of epilepsy to be cautious. The same would apply to those who may be predisposed to epilepsy, for instance people who had seizures some time ago, and are now off medication. Anyone with a fever is also more prone to convulsions.

It is worth noting that the commonest place where people with epilepsy die as a result of a

Table 6.4 Potentially convulsant essential oils, and the components responsible for this effect

Oil	Convulsant component	Percentage of component in oil

A: Do not use at all in anyone suspected of being vulnerable to seizures, or in anyone with a fever.

Oil	Convulsant component	Percentage
Annual wormwood	Artemisia ketone	35–63
Armoise	Thujone	35
	Camphor	30
Artemisia arborescens	Camphor	12–18
	Thujone	30–45
Artemisia caerulescens	Camphor }	major
	Thujone }	components
Balsamite (camphor CT)	Camphor	72–91
Buchu *(B. crenulata)*	Pulegone	50
Camphor (white)	Camphor	35–45
Ho leaf (camphor/safrole CT)	Camphor	42
Hyssop	Pinocamphone	40
	Iso-pinocamphone	35
Lanyana	Thujone	65
Lavender cotton	Artemisia ketone	10–45
Pennyroyal	Pulegone	55–95
Sage (Dalmatian)	Camphor	26
	Thujone	50
Tansy	Thujone	66–81
Thuja	Thujone	39–80
Western red cedar	Thujone	85
Wormwood	Thujone	34–71

B: Avoid orally in anyone suspected of being vulnerable to seizures, or in anyone with a fever. Use with caution externally.

Oil	Component	%
Lavandula stoechas	Camphor	15–30
Sage (Spanish)	Camphor	11–36

C: Probably safe to use externally. Caution required orally in anyone suspected of being vulnerable to seizures, or anyone with a fever.

Oil	Component	%
Cornmint	Menthone	15–30
Lavandin	Camphor	5–15
Peppermint	Menthone?	19
Rosemary	Camphor	10–20
Rue	2-undecanone?	31–49
Spike lavender	Camphor	10–20
Yarrow (camphor CT)	Camphor	10–20

Those suspected of being vulnerable to seizures include people who have had epilepsy in the past, anyone with a strong family history of epilepsy, and anyone with a fever. People with epilepsy-suppressant medication may not be especially vulnerable. Many of the oils listed here are contraindicated in aromatherapy due to general toxicity.

seizure is the bath [490]. Even though the risk from the external use of oils is very low indeed, it would be prudent for anyone who is at risk to avoid using any of the listed oils in the bath.

Essential oils most strongly implicated as being contraindicated in epilepsy and fever are those rich in artemisia ketone, pinocamphone, camphor, thujone or pulegone. It is notable that each one of these components is a ketone. Some consider ketones in general to be highly stimulant to the CNS, and therefore a risk to those prone to epilepsy [34]. However, we have no reason to believe that essential oils rich in other ketones present any danger in epilepsy. While it may be true that all convulsants found in essential oils are ketones, it does not necessarily follow that all ketones found in essential oils are convulsants.

Psychotropic effects

A psychotropic (= psychoactive) substance is one which affects the brain in such a way as to alter mood and behaviour. Many reports of essential oils with CNS activity relate to nutmeg and its constituents [321]. A few of these reports describe hallucinogenic effects. A hallucination is defined as 'The subjective experience of a perception, in the absence of the corresponding stimulus in the external real world' [534].

Nutmeg

Nutmeg has apparently been used as a hallucinogen since the time of the Crusades [233]. There are early reports of its psychotropic action:

The botanist Löbel provided the first written report (in 1576) of its (nutmeg's) inebriating properties, and he claimed that 'a pregnant English lady who, having eaten ten or twelve nutmegs, became deliriously inebriated'. Johannes Purkinje , famous for his numerous discoveries in cellular biology, ate three nutmegs in 1829, and claimed that the resultant intoxication was similar to that which he had already experienced with cannabis. [535]

According to one clinician, nutmeg intoxication can produce 'CNS excitation, hallucinations (especially visual) and color distortions along with feelings of unreality and depersonalization' [233].

Two early reports of poisoning from ingestion of a whole ground nutmeg (5–6 g) in order to correct irregular menstruation both describe a feeble pulse, cold skin, irregular respiration, and a state of collapse [448, 449]. Both cases recovered fully.

A woman who ingested 18 g of ground nutmeg to induce menstruation became disoriented, with episodes of screaming and thrashing of limbs. On admission to hospital she was in a semi-stupor, and remained in this phase for 12 hours. She then began to experience periods of wild excitement, with loud screaming. She fully recovered after 5 days [266]. Both this last case, and another (two nutmegs ingested to treat a boil on the neck) [231] experienced periods of acute anxiety with feelings of impending death.

In none of these last four cases was the individual anticipating anything but a therapeutic effect. They experienced psychotropic effects which were clearly unexpected. One study set out to prove the hypothesis that the reported psychotropic effects of nutmeg were entirely due to suggestion. In a placebo-controlled trial, no effects in human subjects were reported after dosing with 6 g of ground nutmeg [146]. It has been suggested that this amount may not be sufficient to produce psychotropic effects [150]. This may be true, and psychotropic effects from nutmeg may be very dose-sensitive. The nutmeg used might have been old, or an atypical sample. Another possible explanation for the lack of effects is the influence of the intention of the experimenters on the trial.

Two male students each ingested about 14 g of powdered nutmeg, in order to induce intoxication. 5 hours later, both became aware of a detached mental state, described as 'unreal' or 'dreamlike'. One of them became very agitated, and talked incoherently. Medical examination disclosed two people in acute distress, but with no hallucinations. One described a sense of impending doom. The sense of unreality persisted for 3 days; both later described the experience as unpleasant and frightening [262].

Ground nutmeg has been abused by prison inmates, who report that it positively affects mood,

sociability, perception and libido. In one study 10 inmates were each given between 31 g and 47 g of ground nutmeg, and were asked to report its effects. Nine of them experienced an elevated, lofty feeling, accompanied at times by drowsiness, and at other times by excitement. Effects were compared by the inmates to those of either cannabis (the majority), heroin or alcohol [136].

None of the above cases describes or mentions hallucinations. According to one review, hallucinations from nutmeg ingestion are infrequent, but 'definitely do occur with many individuals' [150]. Two cases of nutmeg intoxication describe distortion of reality [536]. In the first case, a 17-year-old girl took 25 g of ground nutmeg, in order to experience its narcotic action. 3 hours after ingestion she experienced visual perception disturbances; everything seemed far away and dreamlike, time was distorted, and people seemed very small (micropsia). A few hours later she slept for 40 hours. When she awoke she was euphoric, bought things she could not afford and danced in the streets. The following day she began to experience anxiety and feelings of unreality; she was admitted to hospital for a week, and then discharged. In the second case, a few hours after a 22-year-old woman took 15 g of ground nutmeg she experienced micropsia, was completely oblivious to the people around her, and experienced music very intensely.

Only one case of unequivocal hallucinations could be found. A 19-year-old girl took 30 g of ground nutmeg. After 4 hours she felt cold and shivery, and 6–8 hours later she was vomiting severely. She saw faces and the room appeared distorted, with flashing lights and loud music. Time appeared to stand still and she felt vibrations and twitches in her limbs. When she closed her eyes she saw lights, black creatures, red eyes and felt sucked into the ground. Her mood was one of elation. She was taken to hospital, and quickly fell into a sound sleep. For the next week she felt confused and could not concentrate. She commented that she would not take nutmeg again [537].

Two components of nutmeg oil, myristicin (2–10%) and elemicin (0–3.2%) have been reported to be the principal psychotropic ingredients [369]. A metabolic pathway has been proposed which could convert myristicin and elemicin to TMA (3,4, 5-trimethoxyamphetamine) [147, 227] or MMDA (3-methoxy-4,5-methylenedioxyamphetamine) [227], both known hallucinogens. Further work has confirmed that myristicin and elemicin are potentially convertible to chemicals similar to hallucinogens [148]. Myristicin is itself very similar in structure to MMDA, an ancestor of the currently fashionable and dangerous drug, Ecstasy [138]. There is no reason to believe, however, that nutmeg is as dangerous as Ecstasy and there is no evidence that myristicyn is in fact converted in vivo to either TMA or MMDA.

Myristicin has been shown to be an inhibitor of an enzyme important in eliminating certain of the body's excess neurotransmitters—monoamine oxidase. This may partly explain nutmeg's euphoriant effects [367, 371]. Furthermore, myristicin increases the levels of the neurotransmitter serotonin in rat brain [369]. If this occurs in humans, a psychotropic consequence is almost certain.

Nutmeg, deprived of its essential oil, seems to have no psychotropic effect on either animals or humans [321, 368]. This would seem to confirm that the psychotropic active ingredients are in the essential oil. However, myristicin alone, taken orally by humans in doses of 400 mg, (equivalent to around 40 g of ground nutmeg) appears to have no psychotropic effect [368]. Myristicin and elemicin have been reported to be insufficiently potent to explain the psychotropic effects of nutmeg [369].

One of us (RT) ingested 1 ml of nutmeg oil orally with no perceptible effects, and 1.5 ml (equivalent to about 15 g of ground nutmeg) which resulted in a euphoric mood elevation lasting some 36 hours. Because of the first experience (after taking 1 ml) he was not expecting any effect at all after taking 1.5 ml, and did not realise for perhaps half an hour that anything unusual was happening. The effects began some 15 minutes after taking the oil in a glass of milk. His breath still smelt very strongly of nutmeg 8 hours after taking it.

He described the experience as a mild mood elevation, with waves of euphoria. It was not intense, perhaps equivalent to 1 or 2 units of alcohol, but it lasted much longer. There was a very slight feeling that reality had been distorted, and he was definitely feeling 'high'.

There were no obvious negative effects other than feeling unusually tired the morning after, having only slept for 4 hours. He was able to function the next day in his office, while still feeling the effects, but was much more prone to laughter than normal. He particularly noticed this when talking on the telephone about a very important business matter to a solicitor he had never spoken to before. He was quite relieved, the following day, when he realised that the 'nutmeg euphoria' had ended.

Three other individuals, A, B and C, who each ingested 1.5 ml of nutmeg oil from the same source, experienced variable effects. 'A' reported feeling slightly euphoric for about 20 minutes shortly after taking the oil but noticed no other effects at all. 'B' took the oil in the morning, and initially felt heavy-headed and nauseous. The nausea disappeared after eating breakfast but was soon replaced by a feeling of restlessness, and a slight feeling of discomfort in the stomach. 2 hours after taking the oil 'B' reported feeling 'definitely affected' and this increased when she attended an aerobics class. While physically active, she felt 'happy' and noticed sweating more than normal, with a high pulse rate. When not active she felt 'tired, heavy and apathetic but restless'. This continued throughout the evening, but she fell asleep at 1 a.m. without any problem. She still felt affected the next morning but not so intensely. By the evening the effect had faded away but she had a restless sleep that night.

'C' took the oil in the early evening and described a feeling of levity, which began about 20 minutes after taking the oil. She laughed at everything for about 30 minutes. She described feelings of unreality, heightened awareness, and clarity of thought and emotion, which more or less continued into the following day. The only negative effect reported was a brief 'panic attack' the next morning.

One of us (RT) experienced no psychotropic effects from ingestion of 1 ml of parsleyseed oil with a myristicin content of 35%. This ineffective dose of myristicin was four times as high as the amount in the nutmeg oil which did produce an effect (after ingesting 1.5 ml). This would seem to confirm that myristicin alone is not sufficient to produce psychotropic effects, at least at these dosage levels. One or more other components of the essential oil probably act synergistically with myristicin and elemicin to produce the psychotropic effect. Other oils rich in myristicin (parsleyseed, parsley leaf, parsnip) may not be psychotropic. Mace oil may have similar effects to nutmeg oil as the two oils are chemically similar.

Current knowledge strongly suggests that ground nutmeg is moderately to strongly psychotropic when taken in relatively high doses, and may occasionally cause hallucinations. There are few reports of psychotropic effects from the essential oil, but those few suggest that the effects may be less intense than those from ground nutmeg. There are no reports of nutmeg oil causing hallucinations. Nutmeg oil seems likely to demonstrate moderate psychotropic effects when taken in oral doses, but non-oral dosage is unlikely to have any such effects.

Cannabis

The substances which imbue cannabis with its psychotropic and hallucinogenic effects are the tetrahydrocannabinols [153]. These are present in quantity in the glandular hairs of the plant, where the essential oil originates [458]. Relatively small amounts (1–2%) of tetrahydrocannabinols find their way into the distilled oil [457, 458]. We must assume that the essential oil is potentially psychotropic, although it is probable that little or no effect would result from the amounts used in non-oral aromatherapy. It is difficult to predict whether or not a psychotropic effect is likely from oral doses. Cannabis oil is not commercially available.

Thujone

In 1915, France banned the production of *absinthe* containing wormwood oil. It was claimed, with some justification, that the oil acted as a narcotic in higher doses, and was habit-forming. It was,

and is still, believed that the thujone in wormwood oil was largely or solely responsible for these effects [439]. Oral thujone is lethal, convulsive and psychotropic in mice at 0.25 g/kg [521]. It has been suggested that thujone and *delta*-9-tetrahydrocannabinol, the most active ingredient in cannabis, interact with a common receptor in the CNS and so have similar psychotropic effects [201].

Anethole

Trans-anethole has demonstrated psychotropic activity in mice at levels over 0.3 g/kg (equivalent to human ingestion of around 20 ml) [149]. Such high doses clearly have no relevance to aromatherapy.

Table 6.5 lists potentially psychoactive components of essential oils, together with the oils in which each is commonly found. The amounts of essential oil used externally in aromatherapy are unlikely to cause a psychotropic effect, but it would be prudent to avoid oral dosing with both mace and nutmeg oils, unless a psychotropic effect is desired. The thujone-rich oils are already contraindicated due to their toxicity, and cannabis oil is not commercially available.

Table 6.5 Components with possible psychotropic effects and the essential oils they are commonly found in

Component	Oil
Myristicin + unknown synergist	Mace, nutmeg
Thujone	Armoise, *Artemisia arborescens*, lanyana, sage (Dalmatian), tansy, thuja, Western red cedar, wormwood
Tetrahydrocannabinols	Cannabis

Those with clinical psychological conditions, a history of psychotic episodes, or a history of use of cannabis and/or psychedelic drugs should be carefully monitored for their reactions to aromatherapy.

Sedative effects

Melissa [212] and lavender [459] oils have been found experimentally to have marked sedative

---

Box 6.3 Essential oil constituents which can cause central nervous depression (sedation)

Trans-anethole	Geraniol
Beta-asarone	Linalool
Benzaldehyde	Linalyl acetate
Citronellol	Methyleugenol
Elemicin	Myristicin
Iso-elemicin	*Iso*-myristicin
Eugenol	Nerol
Eugenol methyl ether	Nerolidol
Farnesol	Terpineol

Box 6.4 Sedative essential oils

Almond (bitter)	Neroli
Asafoetida	Nutmeg
Calamus	Rose
Chamomile (German)	Rosewood
Clary sage	Savory
Elemi	Taget
Galbanum	Valerian
Lavender	Violet leaf[A]
Marjoram	Yarrow
Melissa	

---

activity. Other tests have shown perilla oil to be highly sedative after oral administration, probably because of its perillaldehyde content [330]. The essential oil constituents listed in Box 6.3 seem to be producers of CNS depression [208, 212].

When tests were performed using realistic amounts of essential oils, given to experimental animals by inhalation, those which most consistently produced a depressant effect on the CNS were valerian, asafoetida, lavender, rose, galbanum and violet 'extract' [204].

It has been shown that valerian oil has a direct effect on the CNS; it appears that the most important sedative constituents of the oil are: valeranone, valerenic acid, valerenal, and kesseglycol diacetate [219]. These are not all found in all types of valerian oil. Both nutmeg oil [472] and myristicin [208] are markedly sedative.

Very high doses of benzaldehyde and benzyl acetate can lead to depression and coma, but these effects would not be encountered in aromatherapy, where much smaller amounts are used [339, 345]. CNS depression has been reported after eugenol, *iso*-eugenol [339], anisole, estragole, anethole, safrole, *iso*-safrole, dihydrosafrole, clary sage, savory and marjoram when given to experimental animals in very high doses [207].

Linalyl acetate (<> lavender) is markedly narcotic, whether given by injection [210, 213, 220] orally [221] or by inhalation [443]. Linalool (<> rosewood, lavender) is sedative in experimental animals [213]. Acute toxicity from linalool presents as ataxia, and narcosis [328].

German chamomile [218] and yarrow oils both produce sedation and a fall in body temperature when given in large doses to experimental animals [206].

Essential oils with sedative properties are listed in Box 6.4.

Sedatives in general are sometimes contra-indicated for people who are either depressed or lethargic. However, there is no really strong case for advocating such a contraindication in aromatherapy at the present time. When essential oils are used externally, the fact that they are also fragrances means that, whatever their effect on the CNS, they engage the conscious mind through the sense of smell, often resulting in a more positive and aroused state. Clary sage, for example, although a sedative oil, also has a reputation as one of the primary euphoric, or antidepressant resources in aromatherapy [162]. Clearly this argument would not apply to anosmic individuals, or to oral administration.

Practitioners may wish to take the above information into account in individual cases, and carefully note responses to particular essential oils.

THE ENDOCRINE SYSTEM

The endocrine system consists of a diverse group of ductless glands whose function is to synthesise and store chemicals called hormones, and then to release them directly into the blood supply. Hormones may be steroids, polypeptides or amino acid derivatives. Once in the bloodstream, hormones are recognised by their specific target tissues, which then respond in a characteristic fashion. Together with the nervous system, which partly controls its functioning, the endocrine system constitutes the internal communication mechanism of the body.

N-propyl disulphide (<> onion), methyl disulphide and allyl disulphide (<> garlic) all inhibit iodine metabolism in rats at low concentrations [358, 359]. This effect is probably only significant when essential oils are taken orally by people with thyroid disease or by people living in low-iodine areas where there is no use of iodised salt (rare). Allyl disulphide is the most active compound in this group of anti-thyroid essential oil constituents. There is no indication that *Allium* consumption by humans decreases thyroid function [363, 364].

There is evidence that *trans*-anethole has oestrogen-like properties, although its oestrogenic activity is many times weaker than that of oestrogen [360, 361]. There is some discussion about whether the weak oestrogenic agent is *trans*-anethole, or a polymer of *trans*-anethole [361]. (A polymer is a long chain of, in this case, anethole molecules.) In any case, anethole, fennel oil and anise oil have all demonstrated oestrogenic activity in rodents [360].

In rats, the administration of 185 mg/kg/day of citral for 3 months produced benign prostatic hyperplasia [507] which may be testosterone-dependent [508]. Other studies have suggested that the effect occurs via competition with steroid receptors and that citral has an oestrogenic effect in causing vaginal hyperplasia in rats [510]. In all the above studies citral was applied topically, in amounts equivalent to around 10 ml in a human. This is at least 15 times the amount used topically in aromatherapy, in an essential oil with 75% citral, but is very close to an oral dosage level. It is not known whether citral-rich oils will have a hormonal effect as used in aromatherapy nor, if they do, whether that effect will be oestrogenic or androgenic.

Table 6.6 lists essential oils that may have hormonal activity together with the active component of each oil.

Table 6.6 Essential oils with possible hormonal activity

Oil	Component	Percentage of component in oil
Anise	Anethole	80–90
Backhousia citriodora	Citral	90
Eucalyptus staigeriana	Citral	16–40
Fennel	Anethole	52–86
Lemongrass	Citral	75
May chang	Citral	75
Melissa	Citral	35–55
Star anise	Anethole	75–90
Verbena	Citral	33

The anethole-rich oils should be used with caution orally in people with oestrogen-dependent cancers (these include some breast cancers) and in endometriosis, pregnancy and breast-feeding. The citral-rich oils should be used with caution orally in benign prostatic hyperplasia.

It has been claimed that hop oil has oestrogen-like properties [34]. However, investigations of oestrogenic properties in the herb have been inconclusive [121] and there appears to be no evidence that the essential oil has an oestrogenic effect.

Chronic toxicity from the administration of either safrole or sassafras oil includes cellular changes to the adrenal, thyroid and pituitary glands, and to the testes or ovaries of the rat [343].

OTHER SYSTEMS

Citral can cause a rise in ocular tension, which would not be good in cases of glaucoma. In tests a very low daily oral dose (2–5 µg) produced an increase in ocular pressure within 2 weeks in monkeys [348]. It would be prudent, therefore, to avoid oral use of citral-rich essential oils in cases of glaucoma.

With the exception of the most toxic oils, adverse reactions in other body systems, from any route of administration, seem to be very rare. It may be that this reflects a lack of easy notification and registration of toxic effects, or that problems are so low-level that they are not generally noticed.

Both the skin and the reproductive system have whole chapters devoted to them.

SUMMARY

• In general, undiluted essential oils should not be applied to mucous membrane (mouth, vagina, rectum) and should not be applied to or very near the eyes.

• Given in high doses, some essential oil components can deplete glutathione in the liver, causing paracetamol-like poisoning. Other possible effects include depleting the enzyme cytochrome P_{450}, and producing DNA abnormalities.

• Coumarin is hepatotoxic in some animals and is restricted by several regulatory agencies. However, recent evidence demonstrates that it is not hepatotoxic in humans.

• Certain essential oils should not be taken orally by those with alcoholism, any kind of liver disease, and those taking paracetamol (= acetaminophen).

• Menthol-rich oils (peppermint, cornmint) should not be taken orally by people with G6PD deficiency.

• Apart from apiol-rich oils, there is no evidence that essential oils can cause damage to healthy kidneys, at least in humans at the doses in which they are given in aromatherapy.

• There is a case for reducing oral dosage in kidney disease. Oral administration of anti-coagulant oils might be best avoided in kidney disease.

• There do not appear to be any essential oils which, as used in aromatherapy, produce or exacerbate hypertension or hypotension.

• Cornmint and peppermint oils should probably be avoided altogether by anyone with cardiac fibrillation.

• Onion oil, garlic oil and eugenol-rich oils should not be taken orally by those with blood clotting problems.

• Perilla is the only essential oil suspected of respiratory toxicity.

• Certain potentially convulsant essential oils would be best avoided orally (or in some cases altogether) in those suspected of being vulnerable to epilepsy, and in cases of fever.

• In spite of a body of opinion to the contrary, essential oils rich in ketones do not appear to be

necessarily dangerous in epilepsy, nor to be hazardous in general. Some are, some are not.

• Although myristicin is believed to be psychotropic, it appears that it requires a synergist, present in nutmeg oil (and possibly mace oil) to take full effect. Other myristicin-rich oils, such as parsleyseed, parsley leaf and parsnip, are probably not psychotropic.

• The amounts of essential oil used externally in aromatherapy are unlikely to cause a psychotropic effect, but it would be prudent to avoid oral dosing with both mace and nutmeg oils, unless a psychotropic effect is desired.

• Those with clinical psychological conditions, a history of psychotic episodes, or a history of use of psychedelic drugs should be carefully monitored for their reactions to aromatherapy.

• Sedative essential oils are not contraindicated in depression or lethargy. However, individual cases should be monitored for their response to sedative oils.

• Essential oils rich in anethole should be used with caution orally in people with oestrogen-dependent cancers, and in endometriosis, pregnancy and breast-feeding.

• Essential oils rich in citral should be used with caution orally in benign prostatic hyperplasia.

• Citral-rich essential oils should not be taken orally in cases of glaucoma.

Notes

1. Because the liver receives blood which carries material straight from the intestines, and because it is the main organ of detoxification, it receives an enormous load of toxic, or potentially toxic chemicals. The liver therefore has an amazing capacity to regenerate after injury but is extremely vulnerable to chemical injury. Liver damage typically presents as general malaise with or without jaundice. The liver often brings some of the damage upon itself by the biochemical methods it uses to detoxify molecules. This is because its detoxification mechanisms often result in the temporary production of highly reactive molecules called electrophiles. Electrophiles are on the look-out for a suitable molecule to attack because they have an unpaired electron which needs to combine with some other available electron. Unfortunately, most biological molecules, including proteins and DNA, provide suitable electrons and are quickly damaged by any electrophile which comes into contact with them.

2. The most notorious pharmaceutical drug that can cause liver damage by inducing glutathione depletion is

paracetamol (= acetaminophen), which is metabolised by the liver to a highly reactive electrophilic intermediate. Once the liver's stores of glutathione have been exhausted mopping up the electrophile, cellular damage ensues as liver proteins are attacked. This is why a paracetamol overdose is likely to be fatal.

3. Essential oil of bitter almond has been found to inhibit liver aldehyde dehydrogenase; both bitter almond and anise oils also inhibit liver alcohol dehydrogenase [23]. Both enzymes are essential for the metabolism of alcohol. However, this effect was obtained with dosage levels of 1.6 g/kg/day and 3.2 g/kg/day, and so the data have no relevance to aromatherapy.

4. The metabolite of estragole which actually seems to bind to DNA and which is directly responsible for carcinogenesis is 1′-hydroxyestragole [123]. 1′-hydroxy-2′,3′-dehydroestragole is a synthetic relative of 1′-hydroxyestragole and is much more carcinogenic. It is not known whether this dehydro- compound is made in vivo. Anethole, an analogue of estragole, is much less toxic [361].

5. Some substances which are metabolised by microsomal enzymes actually cause the liver to produce more enzyme, thus speeding up their own inactivation. Alcohol and barbiturates are well-known examples and the phenomenon explains why, for instance, habitual drinkers can take more alcohol before becoming drunk than can occasional drinkers.

6. Jaundice-like symptoms have also been seen after oral administration of anisole, estragole and *iso*-safrole [339], all of which have been shown to be hepatotoxic in a number of species [346]. Microscopic liver lesions in rats have been seen after oral administration of eugenol, safrole, *iso*-safrole, anisaldehyde, anethole and estragole [339], as well as from menthone [327]. Methyl salicylate has been shown to be hepatotoxic to dogs, producing abnormally large liver cells and livers when chronically fed at low levels [341].

7. The functional unit of the kidney is the nephron, a microscopic arrangement for filtration of the blood. In health, humans have greater kidney capacity than they need, as evidenced by the fact that it is possible to lose a kidney and remain perfectly healthy. However, once overall kidney function falls below a certain minimum level, toxic substances begin to build up in the bloodstream.

8. In humans, there are three interconnected circulatory systems which contain blood. The pulmonary circulation contains unoxygenated blood being pumped from the right side of the heart to the lungs in order to be oxygenated, and returning from the lungs in an oxygenated state to the left side of the heart. The systemic circulation pumps oxygenated blood from the left side of the heart around the body and collects deoxygenated blood from the tissues, returning it the right side. The portal circulation carries blood directly from the digestive organs to the liver, so its content of amino acids, fats and toxins can be processed before being released into the general circulation.

9. The blood–brain barrier is not a physical structure but represents the relative lack of permeability in cerebral blood capillaries when compared with peripheral capillaries and the presence of supporting cells through which molecules must pass if they are to reach the brain's nerve tissue. The effect is that of a screen, protecting the brain from a wide range of potentially toxic chemicals. Substances bound to plasma proteins and those with a very low fat-solubility do not readily pass through the barrier into the brain. Conversely, highly fat-soluble compounds do.

10. In in vitro studies, hyssop, sage, camphor, and thuja oils [144] as well as 1,8-cineole [145] and calamus oil [132] have all been shown to inhibit the respiration of rat brain [203, 357]. Thujone apparently produces seizures in the rat at doses well below the minimum lethal dose, whereas the convulsant dose for camphor is nearly fatal [133]. Most *Artemisia* species, as well as sage, tansy and thuja oils contain thujone [201, 354].

7

The skin

SKIN REACTIONS

The basic structure of the skin has already been outlined (see p. 27). The skin is of prime importance when considering the toxic effects of essential oils used in aromatherapy. The most common method of bringing healing oils into contact with the body is via the skin.[1]

Skin reactions to essential oils take three forms, which are not always given the same 'labels'. In this text they are referred to as:

- Irritation
- Sensitisation
- Phototoxicity.

As is generally true in toxicity, skin reactions are 'dose-dependent' and vary according to the concentration of oil applied to the skin. It is potentially dangerous to put undiluted essential oils on to damaged, diseased or inflamed skin. Under these circumstances the skin condition may be worsened, and larger amounts of oil than normal will be absorbed. Sensitisation reactions are also more likely to occur.

Compared to most other types of toxicity, skin reactions can vary considerably from one individual to another, and so are especially difficult to predict.

Irritation and sensitisation

There is sometimes confusion about the difference between these two types of skin reaction. They are quite distinct, and tend to be initiated by different essential oil constituents.

Irritation (= primary irritation) in the skin is produced by a 'primary irritant', such as a corrosive chemical. Primary irritants act on the first exposure, the reaction is rapid, and severity depends on the concentration of the irritant. The direct skin damage is usually of greater importance than the inflammatory response. Examples of essential oils producing severe primary irritation include horseradish and mustard.

Sensitisation is a type of allergic reaction. It occurs on first exposure to a substance, but on this occasion, the noticeable effect on the skin will be slight or absent. However, subsequent exposure to the same material, or to a similar one with which there is cross-sensitisation, produces a severe inflammatory reaction brought about by cells of the immune system (T-lymphocytes). The severity of this subsequent response can seem out of proportion to the concentration of substance present, and in some cases very unpleasant reactions can be evoked by minute quantities.

A relatively small group of chemicals found in essential oils are potential sensitisers, and sensitivity reactions to them are very idiosyncratic (i.e. they vary from person to person). Aldehydes, such as cinnamaldehyde, have been most commonly implicated, as have lactones, such as those found in costus and elecampane oils. Sensitisation occurs when the sensitising chemical in an essential oil has bound to a protein in the dermis. Consequently, sensitisation cannot occur until the sensitising chemical has been transdermally absorbed. (This, of course, is easier when the skin is damaged.)

PATCH TESTING

In patch testing, some of the material to be tested is put on to a patch of skin, either of a volunteer or an experimental animal, typically a rabbit. The skin may be intact, or it may be abraded in order to increase penetration and contact with the cells of the immune system. The substance may be applied neat, in petrolatum, or in alcoholic solution.

In testing for irritation, the substance is typically applied full-strength for 24 hours, probably under a covered patch if volatile. It may also be tested for longer (48 hours) at a lower concentration in petrolatum.

In testing for sensitisation, the substance is typically given repeatedly at a concentration of 2% in petrolatum to groups of volunteers until a maximal response is obtained. This is called a maximation test.

Animal skin and human skin do not absorb materials similarly and they react differently to irritants and sensitisers. However, there is sufficient similarity between animal and human skin for general conclusions to be drawn about the skin hazards associated with the more dangerous substances. Less hazardous substances probably need to be tested on humans for reliable data to be obtained.

Even if safety guidelines are carefully followed, aromatherapists are likely to encounter idiosyncratic skin reactions at some time. If there is any reason for doubt, either because the therapist is using an oil which might produce a reaction, or because there is reason to believe that the client may be particularly sensitive to essential oils (e.g. perfume allergy) patch testing may be considered worthwhile.

How to patch test

To test for irritation, apply the essential oil, at double the concentration you plan to use it at, to the inside of the forearm for 48 hours. Apply 2 drops of oil to the inside of a plaster (without antiseptic). Repeat a second time in order to test for sensitisation. Testing for sensitisation is important if a sensitivity (allergy) is suspected.

If there is a positive result (i.e. there *was* irritation) any of the following may be present: redness, itching, swelling, and (very rarely) blistering. If an adverse reaction does occur, either from a patch test, an accident, or aromatherapy, the following guidelines may be useful.

How to deal with adverse skin reactions

- Washing the skin gently with unperfumed soap will remove most of the oils on the surface of the skin.

- Exposing the skin to the air (but not to strong sunlight) will encourage evaporation of some of the essential oil.
- Application of a mild corticosteroid cream would be the standard medical approach.
- Essential oils of yarrow and German chamomile have been known to counter sensitisation to cinnamaldehyde (anecdotal). Either oil *may* be useful in other instances, but should be dispersed, preferably in a suitable cream/lotion/gel. There is evidence that *d*-limonene can reduce the degree of sensitisation to aldehyde-rich oils if applied at the same time [422]. A limonene-rich oil, such as lemon, *might* therefore help reduce the severity of a reaction after the event.

IRRITATION

Skin irritation is a difficult subject because data in the literature are sometimes inconsistent and because people tend to have idiosyncratic reactions to chemical substances, including essential oils. These two reasons are undoubtedly related.

The essential oils listed as potential irritants are therefore those which most consistently provoke dermal irritation in certain concentrations, although this is no guarantee that they will in a particular individual. At the same time, oils regarded as non-irritant may very occasionally cause irritation in someone with sensitive skin, especially if used in a not very dilute concentration.

Many skin irritation tests for essential oils have been carried out on rabbits. While these tests may provide useful information, there are oils which have been shown to be irritating to rabbit skin but which seem to have no such effect on human skin; an example of this would be sandalwood oil [420].

In Box 7.1, the essential oils which have been tested are classed into five categories. The severely irritant oils should be avoided altogether for dermal application. The oils followed by a '?' have not been tested. They have been placed in the category suggested by their chemical composition. However, it is possible that, when these oils are tested, they may turn out to have a different irritation rating. The ratings are the result of testing on both animals (usually rabbits) and humans.

Case examples

Cinnamon oil (type unspecified) caused severe burns in an 11-year-old boy after a vial broke in his trouser pocket and the oil remained in contact with his skin for 48 hours [392]. Cinnamon oil (type unspecified) has recently received attention in the USA as a substance of abuse by adolescents. They either suck the oil on toothpicks or repeatedly sniff impregnated tissue paper or absorbent cotton wrapped around impregnated toothpicks [409]. This apparently produces a sensation of warmth and facial flushing. Contact with the skin is frequently associated with a burning sensation, and occasionally with blistering [393].

Spillage of clove oil on to the skin has caused transient irritation followed by apparent permanent anaesthesia and loss of the ability to sweat by the affected area. The skin remained sensitive to deep pressure only [403]. Linalool has been reported as having mildly irritant effects upon the skin, producing an eczematous reaction in some people [328].

SENSITISATION (ALLERGIC SKIN REACTION)

A skin allergy is an immune reaction which usually manifests as a rash. This reaction is sometimes referred to as allergic urticaria, or urticarial hypersensitivity. The allergen must first penetrate the skin, and the immune reaction results in the release of histamine and other irritant substances from mast cells and basophils, causing the rash. It is believed by researchers that the molecules of most allergy-causing agents are too small to be recognised by the immune system, and must first be bound together with skin protein to stimulate an immune response [468].

The allergen may be a drug, a cosmetic, a household or industrial chemical, or an essential oil constituent. Sensitisation is, to an extent, unpredictable, as some individuals will be sensitive to a potential allergen and some will not.

Box 7.1   Grading of skin irritancy

The oils were all tested on human subjects at concentrations between 1% and 30%, depending on the maximum use level of each oil in fragrances.

A: Severely irritant

Horseradish	Mustard

B: Strongly irritant

Garlic ?	Massoia	Onion ?	Pine (dwarf) (only if oxidised)	Terebinth (only if oxidised)

C: Moderately irritant

Abies alba (cones)	Cinnamon leaf	Laurel	Sage (Dalmatian)	Thyme
Ajowan ?	Clove bud	*Ocimum gratissimum*	Sassafras	Thyme (wild) ?
Almond (bitter) FFPA*	Clove leaf	Oregano	Savory (summer)	Verbena
Betel leaf ?	Clove stem	Parsley leaf	Savory (winter)	Wintergreen
Birch (sweet)	Fennel (sweet)	Parsleyseed	Spruce hemlock	
Cade (rectified)	Fig leaf[A]	Pimento berry ?	Taget	
Cassia	Fir needle (Siberian)	Pimento leaf	Tarragon	
Cinnamon bark	Hyacinth[A]	Rue	Tejpat ?	

*FFPA = free from prussic acid

D: Very mildly irritant

Abies alba (needles)	Chamomile (German)	Geranium (Bourbon)	Mace	Rosemary ?
Ale	Chamomile (Roman)	Geranium (Maroc)	Mandarin (expressed)	Sandalwood
Almond (bitter)	Citronella	Ginger	Marjoram (Spanish)?	Savin ?
Angelica root	Clary sage (French)	Grapefruit	Mastic[A]	Spearmint
Angelica seed	Copaiba	Guaiacwood	May chang	Spike lavender
Anise	Coriander	Hibawood	Mimosa[A]	Spruce
Armoise	Costus (*but a severe*	Ho leaf	Myrtle	Star anise
Basil ?	*sensitiser*)	Honeysuckle[A]	Narcissus[A]	Tangelo
Bay (W. Indian)	Cubeb	Hay[A]	Nutmeg	Tangerine
Benzoin ?	Cumin	Immortelle	Orange (bitter)	Tansy
Bergamot	Cypress	Immortelle[A]	Orange (sweet)	Tea Tree
Bergamot mint	Davana	Jonquil[A]	Orange flower[A]	Thuja
Birch (sweet)	Dill seed	Juniper	Palmarosa	Tobacco leaf[A]
Birch tar	Eau de brouts[A]	Karo karoundé[A]	Patchouli	Tonka[A]
Boldo	Elecampane (*but a*	Labdanum	Pennyroyal (European)	Turmeric
Buchu ?	*severe sensitiser*)	Lavandin[A]	Pennyroyal (N. Ameri-	Valerian ?
Cabreuva	Eucalyptus ?	Lavandin	can) ?	Verbena[A]
Camphor (white)	*Eucalyptus citriodora*	Lavender	Pepper (black)	Vetiver
Camphor (yellow)	Elemi	Lavender[A]	Perilla	Wormseed
Cananga	Fennel (bitter)	Lemon	Peru balsam	Wormwood
Caraway	Fenugreek[A]	Lemon leaf	Peruvian pepper	Yarrow ?
Carrot seed	Flouve	Lemongrass ?	Petitgrain (Bigarade)	Ylang-ylang
Cedarwood (Atlas)	Frankincense[A]	Lime (distilled)	Petitgrain (Paraguay)	
Cedarwood (Texas)	Galbanum	Linaloe	Rose absolute	
Cedarwood (Virginian)	Geranium (Algerian)	Lovage	Rose otto	

E: Non-irritant

Abies alba	Cardamon	Dill weed	Marjoram (sweet)	Pine (Scotch)
Ambrette seed	Cascarilla	Fir needle (Canadian)	Myrrh	Rosewood
Broom[A]	Celery seed	Gurjun balsam	Myrrh[A]	Sage (Spanish)
Cajeput	Clary sage (Russian)	Hyssop	Neroli	Snakeroot
Calamus	Cornmint	Jasmine[A]	Phoenician juniper	Violet leaf[A]
Camphor (brown)	Deertongue[A]	Labdanum[A]	Orris[A]	

Allergic urticaria is certainly the most common reason for unpredictable sensitivity to essential oils. Its frequency of occurrence suggests that aromatherapists should check whether clients have noticed any rashes when previously using essential oils or perfumes.

Urticarial hypersensitivity has been reported for d-limonene, *alpha*- and *beta*-pinene [84] and for menthol [85]. Such reactions are very rare and generally occur where there is a history of skin sensitivity. Hydroperoxides formed on prolonged storage of certain essential oils may be responsible for allergic reactions [38]. (This is one of the principal reasons for ensuring that essential oils used in aromatherapy have not been oxidised.) Oils of terebinth and dwarf pine have produced eczematogenic effects; this is thought to be due to oxidised *delta*-3-carene (a component of these oils) and may also apply to other pine oils, when oxidised [74, 84, 422].

Perhaps the most notorious essential oil causing sensitisation is costus. It was formerly much used in perfumery, and has been shown in patch tests to be the cause of numerous cases of contact dermatitis [102]. The agents in the oil which have so far been implicated are the sesquiterpene lactones costunolide, costuslactone and dehydrocostuslactone.

Maximation tests using costus oil at a concentration of 4% in petrolatum have produced 25 sensitisation reactions out of 25 volunteers. Of these, eight had such severe reactions that they were unable to continue with the procedure [420].[2]

Similar in severity to costus is elecampane oil, which elicits severe skin reactions at the same concentrations as costus and contains similar chemicals. A major component of elecampane oil, alantolactone, is a potent dermal sensitiser. Not surprisingly, those sensitised to costus cross-react to elecampane, and vice versa [476].

There are many reports of contact dermatitis or allergy from laurel leaf oil [83, 388, 460, 461, 471]. There is potential for confusion between 'laurel oil' (produced by solvent extraction of the leaves) and 'laurel essential oil' (produced by steam distillation of the leaves) [471]. The RIFM monograph on laurel leaf oil reports that three different samples produced no sensitisation reactions when tested on 25, 25 and 49 volunteers [422].[3]

Some laurel leaf oils are thought to contain small amounts of costunolide [388]. However, it is likely that costunolide is not the allergen responsible for most reactions to this essential oil [471]. It is not clear from the literature whether or not laurel leaf oil is likely to cause sensitisation problems. This may be because some laurel leaf oils contain sensitising agents and others do not.

Garlic can cause sensitisation reactions which may be due to its diallyl disulphide content [362]. This, garlic oil's major component, has been shown to be a more probable allergen than other, related compounds in garlic oil [362].[4]

Citral is able to induce sensitisation reactions, and this effect can be reduced by the co-presence of d-limonene [104], which appears to be present in all the essential oils that contain significant amounts of citral, such as lemongrass, melissa, verbena, may chang and *Backhousia citriodora*. Sensitisation to citral-rich oils is therefore hardly ever encountered in aromatherapy, with one notable exception. Verbena oil is a strong potential sensitiser; it is not clear which of its components is responsible [429].

Sensitisation is seen with cinnamaldehyde, which some workers feel can cause contact dermatitis [75]. Sensitisation to cinnamaldehyde can be reduced by the co-presence of d-limonene or eugenol. In 10 people who developed urticaria after cinnamaldehyde had been applied to the skin, 6 had a greatly diminished reaction when it was applied combined with eugenol [408]. Interestingly, *iso*-eugenol is both an irritant and a sensitiser, although it is unlikely to cause problems unless clove oil is applied undiluted to the skin [421, 430].

Tea absolute has been found to cause sensitisation in dilutions as low as 0.001% in guinea pigs; however, it has not been tested on humans (private communication from RIFM). Peppermint may occasionally produce contact dermatitis [38] as may *Zanthoxylum budrunga* oil [223].

Sandalwood, thyme and guaiacwood oils only give reactions in people found to be sensitive to balsams. It has been commented that cinnamaldehyde may be the culprit in many of the

Box 7.2 Grading of skin sensitisation

A: Severe

Do not use on skin or mucous membrane.

Costus	Elecampane	Tea[A]	Verbena

B: Strong

Best avoided on both skin and mucous membrane. Do not use unless diluted to suitable concentration—percentages shown are maximums to avoid sensitisation. Do not use at all on hypersensitive, diseased or damaged skin, or on infants.

Cassia (< 0.1%)	Fennel (bitter, oxidised)	Garlic (patch test first	Oakmoss[A] (< 0.6%)	Verbena[A] (< 0.1%)
Cinnamon bark(< 0.1%)	Fig leaf[A]	at 0.1%)	Treemoss[A] (< 0.6%)	

Also oxidised oils, especially pine and spruce oils.

C: Slight risk

Anise	Laurel leaf	Melissa	Perilla	Star anise
Catnep	Lavender[A]	Myrrh[A]	Pine (dwarf)	Ylang-ylang
Citronella	Lemongrass	Onion	Pine (Scotch)	
Khella	May chang			

D: Negligible risk

Almost certainly will not cause skin sensitisation.

Abies alba (cones)	Cedarwood (Virginian)	Galbanum	Marjoram (Spanish)	Rosemary
Abies alba (needles)	Chamomile (German)	Geranium (Algerian)	Marjoram (sweet)	Rosewood
Ale	Chamomile (Roman)	Geranium (Moroccan)	Mastic[A]	Rue
Almond (bitter)	Celery seed	Geranium (Réunion)	Methyl salicylate	Sage (Dalmatian)
Almond (bitter) FFPA	Cinnamon leaf	Ginger	Mimosa[A]	Sage (Spanish)
Ambrette seed	Clary sage (French)	Grapefruit	Myrrh	Sandalwood
Angelica root	Clary sage (Russian)	Guaiacwood	Myrrh[A]	Sassafras
Angelica seed	Clove bud	Gurjun balsam	Myrtle	Sassafras (Brazilian)
Armoise	Clove leaf	Hay[A]	Narcissus[A]	Savory (summer)
Basil	Clove stem	Hibawood	Neroli	Snakeroot
Bay (W. Indian)	Copaiba	Ho leaf	Nutmeg	Spearmint
Benzoin	Coriander	Honeysuckle[A]	Orange (bitter)	Spike lavender
Bergamot	Cornmint	Hyacinth[A]	Orange (sweet)	Spruce
Bergamot mint	Cubeb seed	Hyssop	Orange flower[A]	Star anise
Birch (sweet)	Cumin	Immortelle	Oregano	Taget
Birch tar	Cypress	Immortelle[A]	Orris[A]	Tangelo
Boldo leaf	Davana	Jasmine[A]	Palmarosa	Tangerine
Broom[A]	Deertongue[A]	Jonquil[A]	Parsley leaf	Tansy
Cabreuva	Dill seed	Juniper	Parsleyseed	Tarragon
Cade	Dill weed	Karo karoundé[A]	Patchouli	Tea tree
Cajeput	Eau de brouts[A]	Labdanum[A]	Pennyroyal (European)	Terebinth
Calamus	Elemi	Labdanum (cistus)	Pepper (black)	Thuja
Camphor (brown)	Eucalyptus globulus	Lavandin	Peru balsam	Thyme
Camphor (yellow)	Eucalyptus citriodora	Lavandin[A]	Peruvian pepper	Tobacco leaf[A]
Camphor (white)	Fennel (bitter, un-	Lavender	Petitgrain (Bigarade)	Tonka[A]
Cananga	oxidised)	Lemon	Petitgrain (Paraguay)	Turmeric
Caraway	Fennel (sweet)	Lemon leaf	Phoenician juniper	Vetiver
Cardamon	Fenugreek[A]	Lime (distilled)	Pimento leaf	Violet leaf[A]
Carrot seed	Fir needle (Canada)	Linaloe	Rose[A] (French)	Wormseed
Cascarilla	Fir needle (Siberia)	Lovage	Rose (Bulgarian)	Wormwood
Cedarwood (Atlas)	Flouve	Mace	Rose (Moroccan)	
Cedarwood (Texas)	Frankincense[A]	Mandarin	Rose (Turkish)	

Possible risk of sensitisation

These oils have not been tested for sensitisation. Their composition suggests that they just might cause sensitisation, but the risk is very small.

Combava (leaf)	Hinoki (root)	St John's Wort	*Eucalyptus staigeriana*

Box 7.3 Contraindications and cautions for use of oils on hypersensitive or diseased or damaged skin

A: Avoid
Due to their irritant and/or sensitisation potential the following essential oils should be avoided altogether on hypersensitive, damaged or diseased skin, such as in eczema, dermatitis or psoriasis. Some of these oils are also contraindicated for other reasons.

Cassia	Elecampane	Massoia	Tea[A]	Verbena[A]
Cinnamon bark	Garlic	Mustard	Treemoss[A]	
Costus	Horseradish	Oakmoss[A]	Verbena	

Also any oxidised oils, in particular the following:

| Abies alba (cones) | Abies alba (needles) | Fennel (bitter) | Pine oils generally | Terebinth (= yarmor) |

B: Use only with caution
Due to their irritant and/or sensitisation potential the following essential oils should be used with caution on hypersensitive, damaged or diseased skin, such as in eczema, dermatitis or psoriasis. Some individuals will react to oils not listed here. In case of doubt patch testing is recommended

| Boldo | Clove bud | Clove stem | Onion | Wormseed |
| Cade | Clove leaf | Laurel | Star anise | |

sensitivity reactions seen. Star anise oil has been the cause of relatively frequent sensitivity reactions (in 5% of patients in a related test). People who reacted to star anise frequently showed sensitivity to *trans*-anethole, *alpha*-pinene, *d*-limonene and safrole [101].

Taking into account all the available information, Box 7.2 is offered as a guide to the risk of skin sensitisation. Box 7.3 gives contraindications and cautions for the use of oils on skin that is hypersensitive, damaged or diseased.

PHOTOTOXICITY

An excessive reaction to sunlight can be induced by certain chemicals, particularly if they are applied to the skin. In phototoxicity, the reaction occurs only if the sensitising agent is present, and results in rapid tanning. Substances known to be phototoxic include many of the topical antifungal and antiseptic preparations added to soaps, toiletries and deodorants; psoralens in suntan preparations and some plants; cosmetic dyes; coal-tar preparations.

The most common phototoxic agents are the psoralens, or furanocoumarins. These are polycyclic molecules whose structure gives them the ability to absorb ultraviolet (UV) photons, store them for a while, and then release them in a burst on to the skin (Fig. 7.1) [82]. Phototoxic components are present only in a few essential oils, and in relatively small amounts, normally less than 2%. However, even at this level, and

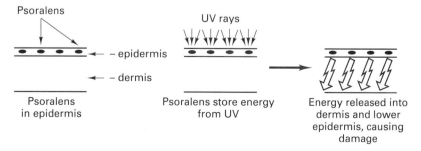

Figure 7.1 Phototoxic agent absorbing UV light and then releasing it in a burst on to the skin.

even if the essential oil is diluted to, say, 2%, they are often capable of producing phototoxic effects if the skin is then exposed to sunlight. If such essential oils are used undiluted, and/or if the skin is exposed to concentrated UV light (sunbeds or strong sunlight) very severe phototoxic effects can result.

Studies in the 1950s and 60s revealed that a reaction took place between furanocoumarins (such as bergapten) and the skin, in the presence of UV light [491, 513, 514, 515]. In 1970, problems with a tanning lotion containing bergamot oil were reported in France [80]. In 1972, Urbach & Forbes reported that severe phototoxic effects were experienced when humans were treated with expressed bergamot oil and simulated sunlight [cited in 419]. Other studies have confirmed the phototoxicity of bergamot oil [485, 486, 492].

It has been found that the bergapten content of bergamot oil produces abnormally dark pigmentation and reddening of the surrounding skin after exposure to an ultraviolet lamp. This condition is known as berloque dermatitis, or bergapten dermatitis, and was first described in 1916 [483]. The patches of darkened skin characteristic of this condition can remain for many years.

A woman was treated for minor burns after a 20-minute session on a sunbed, taken immediately after a sauna bath with lemon oil [517]. A few drops of the oil were placed in a pot in the sauna room, not on the woman's skin. The burns were to one arm and leg.

In a second, more serious case severe, full-thickness burns were sustained following a 20-minute session on a sunbed, which followed bergamot oil application (private communication). In this instance, a few drops of undiluted bergamot oil were rubbed on both arms and both legs; 15 minutes later, the woman showered, and went on to the sunbed. (The shower was an attempt to remove any bergamot oil much of which, by that time, had crossed the epidermis.) Burns steadily developed during the following 48 hours, at which time she was admitted to hospital and remained there for 7 days. The skin on her arms and legs had a roasted appearance, and some blisters were 10 cm in diameter. At the time of writing she had not yet fully recovered.

There is a whole family of psoralens, of which the best known is bergapten (Fig. 7.2) found primarily in bergamot oil. Both bergapten and xanthotoxin (<> rue) are phototoxic [485]. Bergamottin (<> bergamot + lemon) is also phototoxic [487]. In addition to bergapten (0.15–0.87%) bergamot oil also contains citropten (0.01–0.75%), bergaptol (0.2%), bergamottin (1.1–2.2%) and other furanocoumarins, all of which are phototoxic [79, 216]. Bergapten is also present in lemon oil, as are eight other phototoxic constituents including oxypeucedanin. Lime oil and bitter orange oil also contain oxypeucedanin [77].[5]

The psoralens and coumarins are relatively non-volatile molecules and, in the case of citrus oils, tend to be found in expressed (cold-pressed) oils, but not in distilled oils. It is also possible to remove (or, more correctly, considerably reduce) the bergapten in bergamot oil. The resulting oil is known as bergaptenless bergamot, or bergamot FCF (furanocoumarin-free). This bergamot oil is

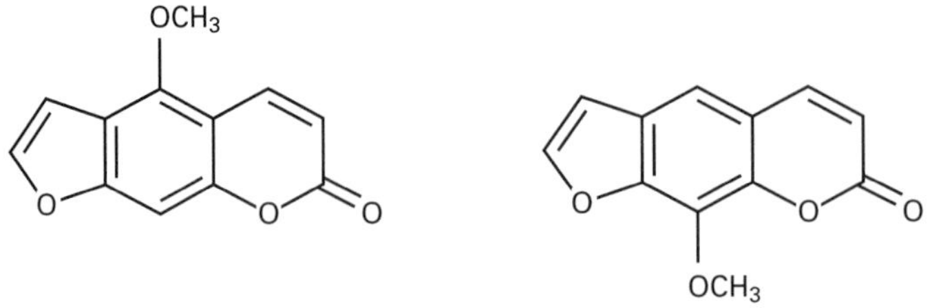

Figure 7.2 Molecular structure of bergapten and xanthotoxin.

regarded as having an inferior fragrance to the oil in its natural state [484].

It is clear from the essential oil monographs published by RIFM, and from the consequent International Fragrance Research Association (IFRA) guidelines, that the risk of phototoxicity depends, not on the mere presence of phototoxic furanocoumarins, but on the type and amount present. Some essential oils contain several phototoxic furanocoumarins, which probably all contribute to the phototoxicity of the oils.

We are concerned not only with which essential oils are phototoxic, but also with their degree of phototoxicity. Taget oil, for instance, appears to be around 100 times more phototoxic than expressed grapefruit oil. Box 7.4 shows the

percentage dilution at which each phototoxic oil is considered safe to use on the skin, based on current IFRA guidelines. Any higher percentage is considered as carrying a risk. Table 7.1 shows how percentages convert to drops of essential oil in base.

The ABC categorisation is ours, and this is based on the amounts generally used in aromatherapy. For example, an oil mixture containing equal proportions of three essential oils (one of them bergamot) diluted to 3%, will contain 1% of bergamot oil, about double the recommended maximum. In this instance 'normal' amounts could very well be phototoxic. For the 'A' category oils, phototoxicity is very likely, and there is no safe dilution. For the 'C' category oils, there is little risk unless the oils are used in higher concentrations than normal in aromatherapy.

There is reason to suspect the following essential oils of phototoxicity, but it is not known whether or not they are phototoxic and, if so, at what level they could be safely used:

- Khella (oil contains furanocoumarins)
- *Skimmia laureola* (oil probably contains bergapten).

Box 7.4 Recommended maximum use levels to avoid phototoxic reactions

A: severe

It is recommended that these oils are not used on the skin, or in any aromatherapy products intended for retail.

Fig leaf absolute   Verbena oil[6]

B: strong

We recommend that treated skin should not be exposed to sunlight or UV lamps for at least 12 hours, if used at levels higher than those indicated. Retail products should not contain these oils at levels higher than those indicated.

Taget	0.05%
Verbena absolute	0.2%
Bergamot	0.4%
Cumin	0.4%
Lime (expressed)	0.7%
Angelica root	0.78%
Rue	0.78%
Opopanax	No level set

C: moderate

We recommend that treated skin should not be exposed to sunlight or UV lamps for at least 12 hours, if used at levels higher than those indicated. Retail products should not contain these oils at levels higher than those indicated.

Orange (bitter, expressed)	1.4%
Lemon (expressed)	2.0%
Grapefruit (expressed)	4.0%

Table 7.1 Calculating essential oil concentrations

ml	5	10	1 5	20	25	30	50
%							
0.1	0.1	0.2	0.3	0.4	0.5	0.6	1
0.4	0.4	0.8	1.2	1.6	2	2.4	4
0.5	0.5	1	1.5	2	2.5	3	5
0.7	0.7	1.4	2.1	2.8	3.5	4.2	7
0.8	0.8	1.6	2.4	3.2	4	4.8	8
1.0	1	2	3	4	5	6	10
1.4	1.4	2.8	4.2	5.6	7	8.4	14
1.5	1.5	3	4.5	6	7.5	9	15
2.0	2	4	6	8	10	12	20
2.5	2.5	5	7.5	10	12.5	15	25
3.0	3	6	9	12	15	18	30
4.0	4	8	12	16	20	24	40
5.0	5	10	15	20	25	30	50

These figures assume that 20 drops of essential oil = 1 ml. It is an easy reference guide for attaining particular dilutions. For instance, a 0.4% dilution of bergamot oil will be obtained by mixing 2 drops of bergamot oil in 25 ml of vehicle, or 4 drops in 50 ml.

Box 7.5 Citrus oils which are not phototoxic
Lemon (distilled)
Lime (distilled)
Mandarin (expressed)
Orange (sweet, expressed)
Tangelo (expressed)
Tangerine (expressed)

Box 7.6 Botanical grouping of many of the phototoxic essential oils

Rutaceae	**Apiaceae**
Bergamot	Angelica root
Fig leaf[A]	Cumin
Lemon (expressed)	Opopanax
Lime (expressed)	
Orange (bitter, expressed)	
Rue	

The presence of bergapten in *Skimmia laureola* oil is disputed, and the oil has not been tested for phototoxicity. On two occasions bergapten has been reported as being present in skimmia oil [216], while a 1992 report failed to detect any bergapten in the oil [473].

A number of the citrus oils are not phototoxic; they are listed in Box 7.5. No information could be found regarding the phototoxicity of combava, satsuma and yuzu, which are also citrus oils. Satsuma is very similar to mandarin, and yuzu is very similar to orange. It is interesting that most of the phototoxic oils are found in two botanical families – Rutaceae and Apiaceae (= Compositae) (see Box 7.6).

When there is no risk, or a reduced risk, of phototoxicity

There is no risk of phototoxicity if the oils are used in a product which is not applied to the skin, or which is washed off the skin, such as shampoo, bath preparations or soap.

There is no risk if non-phototoxic citrus oils are used, such as furanocoumarin-free bergamot, or distilled oils. However, distilled citrus oils tend to be used in flavourings rather than fragrances, and are not much used in aromatherapy, due to their relatively poor fragrance impact.

There is no risk if the parts of the skin to which the oils have been applied are covered in such a way as to prevent UV rays from reaching them. It should be assumed, for the sake of safety, that typical summer clothing (lightweight, lightly coloured) is UV permeable, whereas more

substantial clothing blocks out UV. Most fabrics commonly used to manufacture lightweight clothing only have a sun protection factor of between 5 and 15 [470].

The use of sunscreens, in a potentially phototoxic preparation, will reduce the risk of phototoxicity.

Common sense tells us that the risk of phototoxicity will diminish with time following application to the skin. In reality the intensity of the phototoxic response increases over the first hour following application, remains at a peak for the next hour, and then decreases over the following 8 hours (see Table 7.2). This is certainly true for bergamot oil [485] and the same timescale will probably hold for other citrus oils. The steady increase and then gradual decrease presumably mirrors the time taken for the furanocoumarins to reach the dermis, and then to pass beyond it.

As can be seen in Table 7.2, a 0.5% concentration of bergamot oil produced no phototoxic reactions, 1% was safe after 8 hours, and 2.5% was safe after 10 hours. Tests were carried out on human volunteers.

We would recommend that skin which has been treated with phototoxic oils at levels higher than those maximum use levels, should not be exposed to UV light for at least 12 hours. This gives a safety zone of 2 hours, and assumes a maximum concentration of 5% essential oil. A 15% or 20% concentration of bergamot oil can still produce a phototoxic reaction 12 hours after application [485]. A 0.5% concentration of xanthotoxin (very roughly equivalent to undiluted bergamot oil in terms of phototoxic potential) continued to produce phototoxic reactions for 36 hours, but ceased after 48 hours [485].

Table 7.2 The effect of the time interval between the application of varying concentrations of oil of bergamot and the subsequent irradiation of five subjects expressed as number producing a positive response

Concentration of oil of bergamot in ethanol	Number of positive responses at different time intervals (hours)								
	0	1/2	1	2	4	6	8	10	12
0.5%	0/5	0/5	0/5	0/5	0/5	0/5	0/5	0/5	0/5
1.0%	0/5	1/5	2/5	3/5	2/5	1/5	0/5	0/5	0/5
2.5%	0/5	3/5	4/5	3/5	2/5	1/5	1/5	0/5	0/5
5.0%	0/5	3/5	5/5	5/5	2/5	2/5	2/5	0/5	0/5
Total	0/20	7/20	11/20	11/20	6/20	4/20	3/20	0/20	0/20

Reproduced, by permission of the publishers, from Zaynoun S T, Johnson B E, Frain-Bell W 1977. A study of bergamot and its importance as a phototoxic agent. Contact Dermatitis 3: 225-239, © 1977 Munksgaard International Publishers Ltd., Copenhagen, Denmark.

Individual differences

In a study on 63 volunteers, differences in eye colour, age, sex and ability to tan did not significantly affect phototoxic responses to bergamot oil, but skin colour was a significant factor [485]. There was no statistical difference between those with skin described as fair, sallow and light brown. The average concentration of bergamot oil required to produce a phototoxic response in these individuals was 2.4%, compared to an average of 15% in those with skin colour described as dark brown or black [485]. A suntan gave light skin some extra protection.

When there is an increased risk of phototoxicity

There are two scenarios in which the risk of phototoxicity may be increased. Firstly, if several phototoxic oils are used together, the risk increases proportionally. For instance, if bergamot and cumin oils are both used in a product in equal proportions, the maximum safe percentage will be 0.2% for each, not 0.4% for each.

Secondly, if concentrated (deterpenated) citrus oils are used, the maximum percentage of essential oil should be reduced in proportion to the degree of concentration. Citrus oils generally contain a large amount of terpenes. Deterpenated (= deterpenised, terpeneless, or terpenefree) citrus oils therefore possess all their other components (including the furanocoumarins) in much larger amounts. So, a 10-times concentrated expressed lemon oil will have a maximum safe concentration of 0.2%, instead of 2%.

Photocarcinogenesis

Photocarcinogenesis is the initiation of cancer by UV light, or by a chemical in the presence of UV light. While there are many phototoxic compounds in some essential oils, only very few of these appear to be photocarcinogenic. There is reason to believe that bergapten is photocarcinogenic, and that therefore essential oils containing it (see Table 7.3) will also be. Tests in yeast, whose DNA is processed by the cell similarly to that of humans, have shown that bergapten is highly mutagenic in the presence of UV light, a property casting some

Table 7.3 Essential oils that contain bergapten

Oil	% bergapten
Bergamot	0.3–0.4
Lime (expressed)	0.1–0.3
Lemon (expressed)	0.15–0.25
Orange (bitter, expressed)	0.069–0.073
Grapefruit (expressed)	0.012–0.013
Rue	No reliable figures
Skimmia laureola (controversial)	No reliable figures
Taget	No reliable figures

suspicion on it as a potential carcinogen [482]. Citropten is also photomutagenic [516].

Bergamot oil, in the absence of UV light, is not carcinogenic [483]. Several studies have shown that bergamot oil has carcinogenic properties when applied to mouse skin which is then irradiated with UV light [483, 493, 494]. In the most recent study of this type, bergapten was applied to hairless mouse skin in concentrations of 5 p.p.m., 15 p.p.m. and 50 p.p.m.. 15 p.p.m. is the equivalent concentration of bergapten to a 0.4% dilution of a typical commercial bergamot oil, and this level is the maximum recommended by IFRA.

Even at 5 p.p.m., bergapten was shown to lead to skin tumour formation, sometimes benign and sometimes malignant, in the presence of UV light. It was established that bergapten is the only phototumorigenic component in bergamot oil, and that even low-concentration sunscreens can completely inhibit bergapten-enhanced phototumorigenesis [483].

In the study in question the bergapten/UV light procedure was carried out every weekday for 75 weeks, in many cases more than half the lifespan of the mice. The authors of the paper point out that mice are thought to be less capable of DNA repair than humans, and that mouse data cannot be used to predict absolute human risk [483].

There would appear to be little doubt that bergapten is a potential photocarcinogen. Paradoxically, it is also claimed that bergapten can help prevent sunlight-related skin cancer.

Tanned skin is known to be more protective against sunlight damage than non-tanned skin [483]. An initial study on pigs showed that bergamot oil, plus a sunscreen, provided very good protection against subsequent epidermal cell damage from UV light [495].

Two further studies, on humans, show that a tan gained with bergapten results in less DNA damage from subsequent UV radiation than a tan gained without bergapten [441, 442]. The levels of bergapten used, 15–45 p.p.m., are equivalent to those which cause photocarcinogenesis in mice, cited above. The most recent paper concludes: 'Ultimately, the question of the addition of bergamot oil (5-MOP) to sunscreens must be addressed in terms of risk–benefit analysis. We

present evidence that the judicious use of 5-MOP-containing sunscreens may confer benefit against solar DNA damage in people who seek a tan' [442].

Bergapten has similar phototumorigenic properties to xanthotoxin and psoralen [496]. All three furanocoumarins are found in rue oil [216]. No phototumorigenic data could be found on rue oil or on citrus oils other than bergamot. Although it appears that only a few of the phototoxic compounds found in these oils are photo-tumorigenic, virtually all of the phototoxic oils contain one or more of the three known photocarcinogenic compounds.

We believe that concern about photo-carcinogenesis is justified. Animal and other studies do suggest a possible human cancer risk greater than that associated with exposure to UV in the absence of these essential oils. It is strongly recommended that the IFRA guidelines for maximum use levels are adhered to. At the same time it is worth emphasising that, when applied with bergapten, sunscreens appear to offer considerable protection from photocarcinogenesis.

One 1990 review paper on bergamot oil made the following comment on the IFRA guideline for bergapten: 'Considering that (i) reported acute phototoxic side effects from perfumes are very rare, (ii) repeated exposure to sunlight just after perfume application is very unusual and (iii) there are no reports of increased incidence of epithelial tumours on skin sites that are habitually perfumed, the recommendation of IFRA appears to be very reasonable.' [484].

SUMMARY

• It is potentially dangerous to apply undiluted essential oils to damaged, diseased or inflamed skin. Under these circumstances the skin condition may be worsened, and larger amounts of oil than normal will be absorbed. Sensitisation reactions are also more likely to occur.

• Aromatherapists should check with new clients to find out whether they have ever experienced skin reactions to essential oils, perfumes or perfumed substances.

- Old, oxidised essential oils are more likely to produce skin reactions, particularly sensitisation.
- The most likely system to show toxic effects produced by essential oils is the skin. This is because of the possibility of idiosyncratic skin reactions, which are quite common both to natural and to synthetic chemicals.
- Patch testing is a useful way of predicting adverse skin reactions to essential oils.
- Skin which has been treated with phototoxic oils at levels higher than those recommended, should not be exposed to UV light (sunlight, sunbeds) for at least 12 hours. Lightweight, summer clothing is semi-permeable to UV rays.
- People with dark brown or black skin are still at risk, but can tolerate higher concentrations of phototoxic oils without risk. Tanned white skin also offers a degree of protection, and sunscreens offer considerable protection.
- If several phototoxic oils are used together, the risk increases proportionally.
- If concentrated (deterpenated) citrus oils are used, the maximum percentage should be reduced in proportion to the degree of concentration.
- Most phototoxic oils are potentially photo-carcinogenic. It is strongly recommended that the IFRA guidelines for maximum use levels are adhered to. The notes for phototoxicity (time lapse before exposure, skin colour, sunscreens, etc.) also apply to photocarcinogenesis.

Notes

1. It is possible that a skin reaction might be part of a healing process, or might be due to a contaminant in the oil. Either of these would be difficult to distinguish from a primary reaction to the oil. We would therefore recommend that in the event of any reaction the suspect essential oil be avoided. It is also possible that a skin reaction might be due to myelosuppression although there is no recorded case of an essential oil causing this.
2. It is possible to prepare costus oil so that it is free of toxic agents by using a polymer, aminoethyl polysty-rene. IFRA recommends that natural (untreated) costus oil is not used in fragrances [430].
3. Laurel leaf (*Laurus nobilis*) more commonly known as bay leaf, should not be confused with another essential oil known as 'bay leaf oil' (*Pimenta racemosa*). To avoid confusion *Laurus nobilis* oil is referred to as laurel leaf.
4. Diallyl disulphide is a hapten, a molecule too small to be recognised by the immune system unless it is bound to a larger molecule, such as a protein. Presumably diallyl disulphide combines with a protein before provoking an allergic reaction. This is the normal mechanism for sensitisation.
5. Bergamottin = 5-geranoxypsoralen
   Bergapten = 5-methoxypsoralen
   Bergaptol = 5-hydroxypsoralen
   Citropten = 5,7-dimethoxycoumarin
   Xanthotoxin = 8-methoxypsoralen
6. Verbena oil is not very powerfully phototoxic. Out of six samples tested, three were phototoxic when applied undiluted; one of these was not phototoxic at 12.5%, and another was not phototoxic at 50% [429]. IFRA's recommendation that verbena oil is not used in fragrances is presumably more due to its sensitisation potential.

8

Cancer

This chapter considers the questions:

- Is there any need for concern that essential oils could sometimes cause cancer?
- Are some oils potentially dangerous and best restricted, or is the risk so small that oils can be used in aromatherapy without giving any thought to cancer?

Photocarcinogenesis has already been covered in Chapter 7.

WHAT IS CANCER?

The body has complex mechanisms for regulating how and when cells divide. Control of cell division is dependent on good communication between cells: how else could a limb, for instance, 'know' when to stop growing? Chemicals which effect this intercellular communication interact with DNA in the cell nucleus, either causing it to replicate, thereby instructing the cell to divide (if tissue growth is needed) or causing the cell to stop dividing.[1]

Sometimes, the system of communication becomes deficient, and an abnormal amount of tissue develops, often as a tumour. In benign tumours, the excess tissue has well-defined borders and resembles the normal tissue from which it is derived; that is, it is well differentiated. This means, for example, that the tumour is only able to survive in particular environments, and so is unlikely to be able to start new growth in a different region of the body should a piece break free.

Malignant growth, on the other hand, produces new tissues whose borders are ill-defined and which therefore infiltrate surrounding tissues like a crab (cancer means 'crab'). The tumour is poorly differentiated, making it able to survive in different regions of the body, should part of it break free of the main mass. Most tumours lie somewhere between the extremes of a malignant and a benign tumour; a benign tumour may over time become malignant and, more rarely, the reverse may occur.[2]

Some tumours are undoubtedly caused at random in the population by the background of radiation reaching us from inside the earth and from outer space. Others may be set off by physical trauma; bone tumours are occasionally produced this way. Some people have a genetic predisposition to cancer; their developmental genes (oncogenes) are, for some reason, particularly likely to switch on inappropriately. However, we are concerned here with one specific means by which cancerous changes can be produced: chemical action.

It has been estimated that 70–90% of all human cancers can be attributed to environmental chemical causes [489]. Many of the known carcinogens to which humans are or have been exposed (diethylstilboestrol, benzene, asbestos, etc.) are products of the industrial revolution. It is often said that the great rise in cancer incidence in the civilised world this century is due to the fact that cancer is an age-related disease, and that people are living longer. There seems little doubt that this is true. However, we do not know how many as yet unidentified carcinogens, whether synthetic or natural, may be also contributing to human cancers today.

CANCER PRODUCTION BY CHEMICALS

It is thought that the chemical production of malignant tumours depends on two processes, initiation and promotion. Initiation is the production of an irreversible change in a cell which is necessary, but by itself insufficient, to cause malignancy. Initiating agents are believed to act either at the cell surface, thereby affecting intercellular communication, or by altering DNA (causing mutation). Promotion of a cancer is generally by non-specific irritants. These facilitate malignant growth by altering the cellular environment or by inhibiting DNA repair mechanisms. Some carcinogens are both initiators and promoters, and so are able to produce malignant changes by themselves.[3]

Some chemical carcinogens are so powerful that even one exposure at a dangerous level is very likely to result in cancer. Although it is always difficult to grade levels of risk in cancer, we do know that, in many of the studies, it requires daily exposure, over a long period of time (many months) to induce cancer in rodents with the essential oils and components discussed below.

Alkylating agents

Alkylating agents are frequently carcinogenic because they can bind irreversibly to DNA and protein. Safrole is a good example of an alkylating agent found in essential oils, notably sassafras [153, 309]. Other alkylating agents are used in the chemotherapy of cancer; that is, they are cytotoxic, but they are also carcinogenic, mutagenic, teratogenic and immunosuppressive.

TESTS FOR CARCINOGENICITY

These tests fall into two groups: in vitro and in vivo. A positive result in an in vitro test puts the tested substance under suspicion of being carcinogenic, but certainly does not prove that it is carcinogenic. A positive in vitro result would normally be followed by in vivo testing. In vitro tests include tests for mutagenicity, cytotoxicity and genotoxicity.

Mutagenicity

It is possible to perform preliminary tests for mutagenic activity (the ability to alter chromosomal DNA) using bacteria and fungi [294]. Many of the metabolites of alkenylbenzenes (a class of potentially carcinogenic chemicals, including safrole, and estragole) are mutagenic in bacteria; 1´-hydroxyestragole and 1´-hydroxysafrole are more strongly mutagenic than estragole or safrole [292].

Metabolites of safrole and eugenol are only weakly mutagenic; estragole itself is weakly mutagenic [294]. On the basis of in vitro tests on microorganisms, the alkenylbenzenes seem to be at worst very weak mutagens. However, all these chemicals, with the exception of anethole, estragole and *iso*-safrole, have given positive results in tests on microorganisms [295].

Tarragon oil has been shown to be mutagenic, with activity residing in the estragole fraction of the oil [281]. Peppermint, perilla, onion and cinnamon bark oils have given weakly positive results in tests on the ability of chemicals to interact with chromosomes in vitro [288]. There is no evidence of a hazardous action in humans.

Bergamot oil, bergapten and citropten are strongly mutagenic, but only in the presence of UV light [482, 516].

Genotoxicity

Substances which interfere with the structure and/ or function of cellular DNA are known as genotoxic (i.e. toxic to genes).[4] It is possible to look for such abnormal cellular changes using cultures of human or animal cells; rat hepatocytes were used in all the following studies. Safrole, estragole and methyleugenol are genotoxic [122, 289, 291] as is *beta*-asarone [498]. Two studies show myristicin and elemicin to have a low but possibly significant level of genotoxicity [122, 289]; a further study found elemicin to be significantly genotoxic, but not myristicin [445]. Anethole, *iso*-safrole, dill apiol and parsley apiol show zero to low levels of genotoxicity [122, 289, 291, 445]. Eugenol is totally devoid of genotoxicity [122, 289, 291].

Live animal tests

For obvious ethical reasons chemical carcinogens are not tested experimentally on humans. In some instances useful data are obtained from non-experimental exposure in humans, the most obvious example being cigarette smoke. No such data exist, however, for any essential oils or essential oil components. What we are left with are the in vitro type of tests described above, and tests on animals, usually rodents. The relevance of the relatively high doses used in animal tests to actual human exposure to these same substances has always been controversial.

In their 1987 paper, 'Ranking Possible Carcinogenic Hazards', Ames, Magaw & Gold state:

'Extrapolation from the results of rodent cancer tests done at high doses to effects on humans exposed to low doses is routinely attempted by regulatory agencies when formulating policies attempting to prevent future cancer. There is little sound scientific basis for this type of extrapolation, in part due to our lack of knowledge about mechanisms of cancer induction, and it is viewed with great unease by many epidemiologists and toxicologists [references given]. Nevertheless, to be prudent in regulatory policy, and in the absence of good human data (almost always the case), some reliance on animal cancer tests is unavoidable. [305]

There are arguments which support the predictive value of rodent carcinogenicity testing. Most of the chemicals that are carcinogenic to humans are carcinogenic to at least one, and in most cases to more than one, animal species; experimental evidence of carcinogenicity has in several cases preceded human observation and could have predicted it; there is a good correlation between animal and human target organ sites [489]. In general, rodent tests are probably good predictors of human carcinogenesis.

POSSIBLE CARCINOGENS
Safrole

Probably the most notorious essential oil constituent with carcinogenic potential is safrole. This material is the major component of sassafras oil (85–90%), and of both yellow and brown camphor oils. It is also present in an unknown quantity in cangerana oil, and in small amounts in a few other oils (see Table 8.3). Natural root beer, made with sassafras, used to be popular in the USA and is still made illicitly. It has been banned for a number of years because of its safrole content.

Safrole was itself banned as a food additive in the USA in 1961.

Safrole has a hepatocarcinogenic effect when given subcutaneously or by stomach tube to male mice during infancy [153]. When administered orally, safrole is a low-level hepatic carcinogen in the rat [320]. Oesophageal tumours have also been reported in rats given safrole in the diet [319]. Oral administration appears to require long periods of administration before cellular changes manifest. In one study, sassafras oil and safrole, given in the diet, produced no tumours after 22 months, but a large percentage of the same animals showed initial tumour development at 24 months [343]. Safrole produces kidney epithelial and liver tumours in the offspring of pregnant mice given it in the diet [161].

Data suggest that metabolites of safrole, *iso*-safrole and estragole [309] are possible carcinogens [153]. A metabolite of safrole, 1′-hydroxysafrole, is almost certainly the culprit [286] and is a much more potent hepatic carcinogen than safrole itself [32]. Other toxic metabolites may also be formed in vivo [59, 72, 287]. Safrole and 1′-hydroxysafrole can both cause hepatic cell enlargement in experimental animals [313]. There is some evidence that safrole may be able to activate a cancer-causing virus, the polyoma virus, at least in rats. It is not known whether safrole has a similar action in humans [344].

IFRA recommends that safrole as such should not be used as a fragrance ingredient. When safrole-containing essential oils are used, it recommends a maximum use level of 0.05% safrole in fragrance compounds (equivalent to 0.01% in the final product) [430]. Both the UK and EC 'standard permitted proportion' of safrole in food flavourings is 0.001 g/kg [455, 456].

Iso-safrole

Some sources regard *iso*-safrole as potentially carcinogenic while others do not. The weight of opinion is currently in favour if its being safe. The matter is only of academic interest, since *iso*-safrole does not occur in more than trace amounts in commercially available essential oils.

Estragole

Estragole (= methyl chavicol) is found in significant quantities in tarragon oil (70–87%), in most types of basil oil (commonly 20–70%, but both higher and lower levels are possible) and in *Ravensara anisata* (88%). It is found in smaller quantities (2–7%) in fennel oil, and at around 75% in chervil oil, which is not commercially available.

It has been shown that rodent livers can metabolise estragole to potentially carcinogenic compounds [62, 312, 478]. It is thought that these metabolites are potential carcinogens because of their ability to bind to DNA in vivo [122, 289] a process associated with an increased chance of genetic mutation [312].

The metabolite of estragole which actually binds to DNA and which is directly responsible for carcinogenesis is 1′-hydroxyestragole [123]. Both estragole and 1′-hydroxyestragole induced hepatic tumours on administration for 12 months in the diet of female mice [309]. Both compounds also caused significant increases in the hepatocellular carcinomas of male mice that were given them by subcutaneous injection at 1–22 days of age [312]. Estragole is not restricted by any regulatory agencies.

Methyleugenol

Methyleugenol is found in snakeroot oil (36-45%), *Melaleuca bracteata* (major component), Russian Tarragon (5-29%), French Tarragon (< 1.5%) and East Indian nutmeg oil (< 1.2%). It is also the major component of several essential oils which are not commercially available (see p. 194). Methyleugenol comes into the same category as safrole and estragole. It is genotoxic [122, 289, 291, 479] and is carcinogenic in rodents [309, 310]. Its 1′-hydroxy metabolite, 1′-hydroxymethyleugenol, is more strongly genotoxic [479] and more strongly carcinogenic [309] than methyleugenol.

Elemicin

Elemicin and *iso*-elemicin, both related to safrole, are metabolised by a pathway similar to that which metabolises safrole. There is the suspicion, therefore, that they may be weak hepatic carcinogens of the same general type as safrole [61]. Recent work shows that elemicin is genotoxic in rat hepatocytes, which goes some way to confirming these suspicions [445]. However, two earlier studies revealed only a low level of genotoxicity [122, 289] and one live rodent test found both elemicin and its metabolite, 1′-hydroxyelemicin to lack carcinogenesis [309]. Other work found that 1′-hydroxyelemicin possessed weak, but statistically significant hepatocarcinogenic activity [453]. There is currently insufficient data to regard elemicin as being carcinogenic.

Asarone

Calamus oil (containing 75.8% *beta*-asarone) was weakly carcinogenic to rodents, producing malignant duodenal tumours after 59 weeks of dietary administration [285]. *Beta*-asarone produced malignant liver tumours in rodents given four intraperitoneal injections; tumours were found on autopsy at 13 months [453]. One study failed to confirm the genotoxicity of *beta*-asarone and calamus oil in rat hepatocytes [454].

Recent work on rat hepatocytes suggests that *beta*-asarone's genotoxicity is due to a novel, but P_{450}-dependent activation mechanism [445]. It is possible that *beta*-asarone is the most active carcinogenic compound to be found in essential oils.

Beta-asarone is banned in the USA as a pharmaceutical ingredient, while the *alpha* isomer is permitted [6]. Both the UK and EC 'standard permitted proportion' of *beta*-asarone in food flavourings is 0.0001 g/kg [455, 456].

IFRA recommends that *alpha*- and *beta*-asarone should not be used as fragrance ingredients, and that the level of asarone in consumer products containing calamus oil should not exceed 0.01% [430]. Concordance with these guidelines would mean that calamus oil containing *beta*-asarone should not be used in aromatherapy. *Alpha*-asarone is not found in significant quantities in any commercially available essential oils.

Benzo[*a*]pyrene

Benzo[*a*]pyrene, a polynuclear hydrocarbon, is a well-known carcinogen [158]. It is present in unrectified cade oil at 8000 p.p.b. (= 8 p.p.m.) and in rectified cade oil at up to 20 p.p.b.; this last amount is less than that found in some foodstuffs [462]. Other potentially carcinogenic polynuclear hydrocarbons are also present in cade oil. Unrectified cade oil produced a worrying level of (potentially carcinogenic) DNA adducts in the skin of psoriasis patients receiving cade oil therapy [463].

IFRA recommends that unrectified cade oil should not be used as a fragrance ingredient. IFRA does permit the use of cade oil that has been rectified by fractional distillation in order to remove (considerably reduce) the undesirable polynuclear hydrocarbons [430]. (See note on birch tar oil, p 203.)

d-limonene

In tests using various undiluted citrus oils on mouse skin, orange, lime, lemon and grapefruit oils have all been found to produce tumours at the site of application [282, 283, 284, 315]. It is important to bear in mind that these tumours appear in mice who have been primed with a cancer initiator, and that the essential oils do not give rise to tumours on their own. Some of the tumours produced were malignant, and some were benign. One paper concludes that 'Anatomical differences and considerations of dosage render it unlikely that the citrus oils are a serious tumour-promoting hazard for man.' [316].

In searching for the component of citrus oils causing these tumours, suspicion fell on *d*-limonene, and these suspicions were eventually confirmed. However, further research revealed that it was not *d*-limonene itself which caused the problem, but chemicals formed by the oxidation of *d*-limonene, which is an unstable terpene [301]. The implication of this is that only old, oxidised citrus oils would be capable of causing any problems.

Paradoxically, more recent research has demonstrated that *d*-limonene itself is antitumoral, and prevents malignant tumours in rodents primed with cancer initiators [410, 411, 412]. The early tests, it is assumed, must have all used oxidised citrus oils or oxidised limonene.

Although even oxidised citrus oils are very unlikely to present a hazard in aromatherapy, this research very much underlines the importance of using relatively fresh essential oils, which have not oxidised.[5]

Trans-anethole and citral

Trans-anethole has not been found to demonstrate any carcinogenic activity [309] nor has a metabolite 3-hydroxyanethole [309]. Another metabolite, anethole 1,2-epoxide, is non-genotoxic [504].

There is evidence that trans-anethole has oestrogen-like properties, although its oestrogenic activity is many times weaker than that of oestrogen [360, 361]. Citral is suspected of demonstrating oestrogenic activity when applied dermally to rats in doses some 15 times higher than those used externally in aromatherapy [510].

Certain cancers are known as oestrogen-dependent, because increased levels of oestrogen in the body can exacerbate them. Oestrogen-dependent cancers include, for example, certain breast cancers and some endometrial carcinomas in post-menopausal women.

Neither citral, trans-anethole, nor essential oils containing them, is restricted by any agency, and these materials are freely used in foodstuffs. Clearly they are not regarded as a hazard to the general population. However, in the light of the

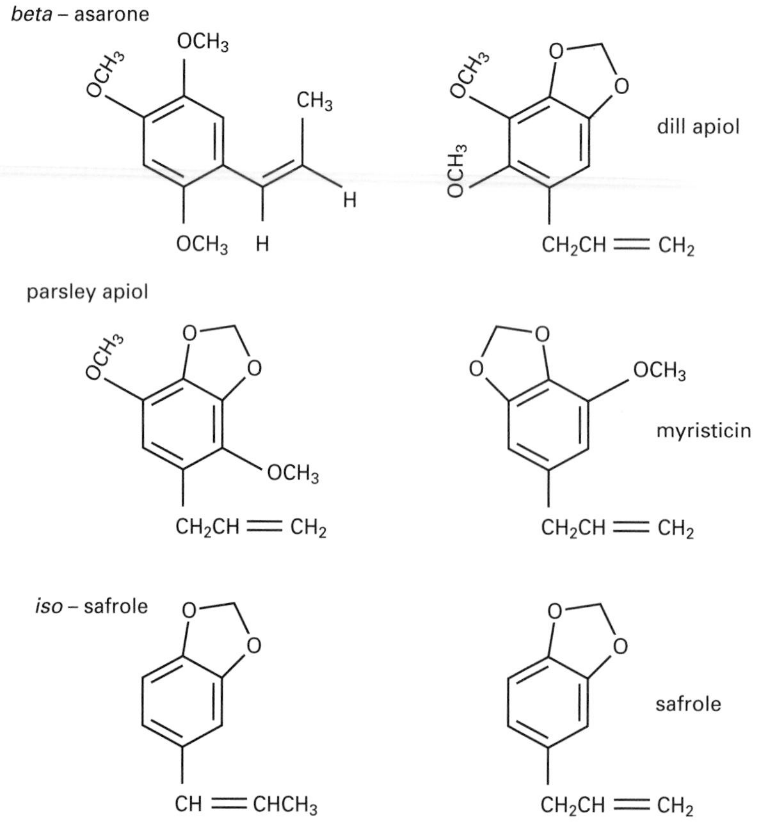

Figure 8.1A Molecular structure of some alkenylbenzenes.

above comments it might be prudent to avoid the oral use of anethole-rich oils in oestrogen-dependent cancers.

Other compounds

Eugenol, dill apiol, parsley apiol and myristicin have not been found to demonstrate any carcinogenic activity [309, 310]. This correlates well with their relative lack of genotoxic activity. While weakly mutagenic, cinnamaldehyde is not carcinogenic [453].

Discussion

In general there is a good correlation between those compounds demonstrating genotoxic activity in rat hepatocytes, and hepatocarcinogenic activity in rodents. Apart from benzo[*a*]pyrene, all of the essential oil compounds suspected of carcinogenicity are alkenylbenzenes. These include *beta*-asarone, safrole and estragole (see Fig. 8.1 A, B). Other, similar alkenylbenzenes include *trans*-anethole and myristicin. It might be tempting to suspect all of these substances of being carcinogenic, since they have very similar molecular structures. However, this does not seem to be the case.

It has been suggested by Caldwell that the simpler alkenylbenzenes (*beta*-asarone etc.; Table 8.1) are carcinogenic, while the more complex ones (*trans*-anethole etc.; Table 8.1) are not [445]. Most of the available evidence certainly supports this theory. Myristicin, for example, is not genotoxic [445], it shows high activity as an inducer of glutathione-*S*-transferase (this correlates with

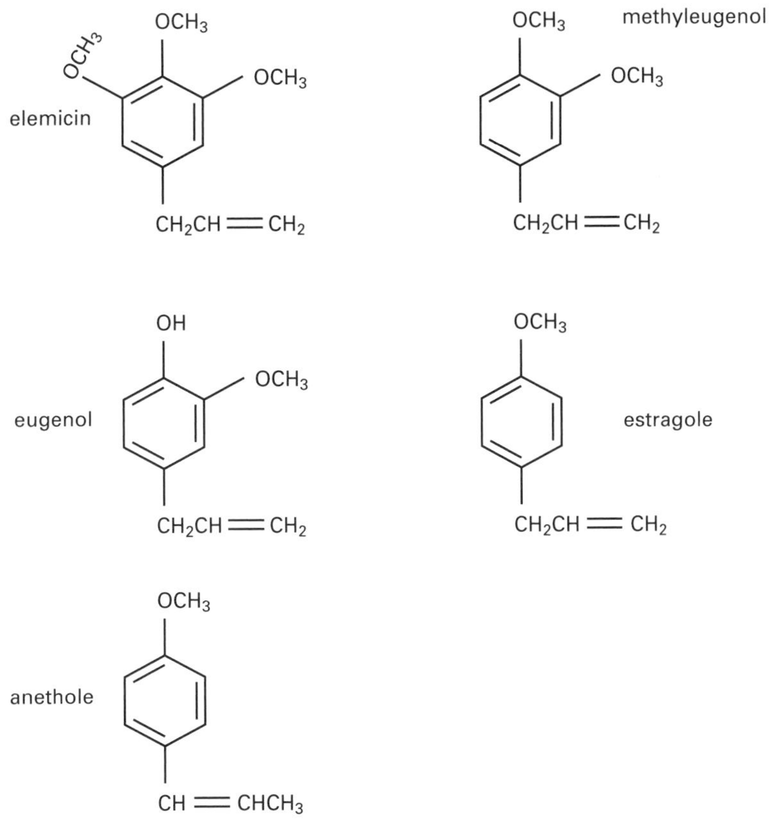

Figure 8.1B

Table 8.1 Carcinogenic and non-carcinogenic alkenylbenzenes

Component	Carcinogenic metabolite
Potentially carcinogenic	
Beta-asarone	Unknown
Estragole	1´–hydroxyestragole
Methyleugenol	1´–hydroxymethyleugenol
Safrole	1´–hydroxysafrole
Some uncertainty	
Elemicin	1´–hydroxyelemicin ?
Iso-safrole	1´–hydroxyiso-safrole ?
Not carcinogenic	
Trans-anethole	
Dill apiol	
Eugenol	
Myristicin	
Cinnamaldehyde	
Parsley apiol	
Methyl *iso*-eugenol	

tumour inhibition) [466] and it inhibits benzo[*a*]pyrene-induced tumours in mice [467].[6]

The reason for the difference between the two groups of compounds seems to be that the simpler compounds are able to undergo a particular metabolic change in vivo (1´-hydroxylation) while the more complex compounds are not. (The metabolites, such as 1´-hydroxyestragole, are the actual carcinogens.)

It is curious that safrole is restricted by regulatory agencies such as the US Food and Drug Administration (FDA), the EC and IFRA, and estragole is not. The metabolism and method of carcinogenesis of the two substances are remarkably similar, and in one study 1´-hydroxyestragole proved to be more strongly carcinogenic than 1´-hydroxysafrole [312]. The reasons for this discrepancy appear to be political and historical, rather than toxicological. The first study showing the carcinogenic potential of safrole was published in 1961, and in the same year the FDA withdrew it from the market [344]. Estragole did not come under suspicion until 1976 [312]. More than a dozen papers have since been published on estragole, but it remains unrestricted.

The 1976 paper commented that 'The potential risks from the use of estragole by humans are

presently difficult to determine, but they are probably small.' [312]. The metabolic activation of safrole and related naturally occurring alkenylbenzenes such as estragole is under study and there is evidence that the risk to humans is considerably less than that to some other species [311]. In 1981 Zangouras et al pointed out that:

The carcinogenic potential of a chemical is generally assessed from the results of life-time feeding studies in rodent species. However, extrapolation of the findings to man is frequently difficult for many reasons, and one of the more important of these is that human exposure generally occurs at much lower levels than the ones used in animal studies. The metabolism of many chemicals is dose-dependent and this is of particular significance when biotransformation produces a carcinogenic metabolite. At low doses a compound may be safely disposed of metabolically whereas at higher doses these pathways become saturated and alternative routes become increasingly employed which lead to the formation of toxic metabolites. [62]

This same paper showed that the amount of 1´-hydroxyestragole formed from estragole increased from 1% to 9% as the dosage given was increased from 0.05 mg/kg to 1000 mg/kg. A later study found that, at the same two dosage levels, the percentage of 1´-hydroxyestragole formed rose from 1.3–13.7% in rat, and from 1.3–9.4% in mouse [306]. At lower dose levels estragole is detoxified much more efficiently by enzyme systems than at higher dose levels. It is important to note that, in humans, the amount of 1´-hydroxyestragole found in excreted urine is 0.3% of the amount of ingested estragole [30]. Caldwell has calculated that the amount of 1´-hydroxyestragole formed after normal dietary ingestion of estragole (in the form of basil, tarragon, fennel herbs or essential oils) is 13 000 000 times less than the doses required to produce liver tumours in rodents [479, 480]. Caldwell comments:

Estragole . . . causes hepatocellular carcinoma in rodents when given at very high doses of 0.5 g/kg/day. Estragole is regularly consumed in the human diet, but only at very low levels, around 1 mg/kg/day, i.e. some 500 000 times less than the estragole dose required to produce cancer in animals.

The carcinogenicity of estragole is due to a metabolite, 1´-hydroxyestragole. It has been shown that the proportion of a dose of estragole converted to this carcinogenic metabolite depends very much upon the size of the estragole dose given. At the very high dose administered to animals, 1´-hydroxyestragole is a major metabolite, accounting for 10% or more of the material administered. In contrast, at the very low doses which are present in the human diet, 1´-hydroxyestragole is a very minor metabolite indeed.

The risk of cancer from estragole is thus not related to the dose ingested, but rather to the amount of the carcinogenic metabolite formed. Viewed in this way, the safety margin for estragole in the human diet is not the 500 000 fold difference between the animal and human exposures, but the 13 000 000 fold difference between the amounts of the metabolite formed at these two extremes.

This argument has been cited by the WHO/FAO Joint Expert Committee on Food Additives as a case example of the use of metabolic principles in the safety assessment of natural food components and supports the safety of estragole in the diet. [308]

However, it is also important to bear in mind that the most common route of administration in aromatherapy is dermal, not oral. In the case of estragole, Caldwell comments:

It is important to appreciate that these principles apply only to the normal food use of herbs, spices and oils containing estragole. As yet we do not have the same framework for justifying the safety of such materials when applied to the skin. In particular, we do not know how much or how efficiently estragole is absorbed through the skin, or how much of the carcinogenic metabolite is formed when applied to the skin. Since the amounts of estragole applied to the skin in aromatherapy are as much as 50 mg (or 700 µg/kg) some 700 times greater than normally encountered in the diet, it seems prudent to avoid the use of estragole-containing oils until more knowledge is available. [308]

In terms of essential oils, we have a situation in which a few oils, and a few components, are weak carcinogens in rodents, when used at relatively high dosages. Rodent carcinogens are more likely than not to be carcinogens in humans. This is because the chemical nature of DNA, and its behaviour in replication and cell division, are almost identical between animal species. However, this only applies if exposure is at the same dosage levels.

It is important to remember that the normal dietary exposure to safrole and estragole is significantly lower than the amount encountered

in aromatherapy. We can be reasonably certain that the compounds in question will present a risk to humans, if exposure is at a sufficiently high level. At the same time these compounds will almost certainly present no risk if exposure is sufficiently low. Ultimately, then, it is a question of dosage and frequency.

ESSENTIAL OILS AND CANCER

There is no evidence that tumours in humans have ever been provoked by the use of essential oils. Nevertheless, a few essential oils do contain potentially carcinogenic substances, and there is concern that these oils may not be safe to use in aromatherapy.

It is reasonable to assume that essential oils containing high concentrations of carcinogenic compounds will act no differently to any other dilution of such compounds, i.e. they will be predictably carcinogenic. Tests using sassafras oil, for example, demonstrate a level of carcinogenesis similar to that of its major component, safrole [343]. However, the effect of those oils containing low concentrations of carcinogens is more difficult to predict, especially if other components in the same oil demonstrate anti-carcinogenic effects.

Nutmeg oil, for instance, contains up to 4% of safrole (carcinogenic) and up to 10% of myristicin (anti-carcinogenic). Apart from sassafras, none of the suspect oils have been tested. In the absence of such data it seems prudent to err on the side of caution.

There is little doubt that a single application of a potentially carcinogenic essential oil in an aromatherapy context presents no risk. On the other hand, aromatherapy oils or other products containing significant concentrations of, say, basil oil or sassafras oil might feasibly contribute to cancer if used on a regular basis.

Currently, there is no way of knowing what constitutes a 'regular basis'. Therefore we do feel justified in recommending that the use of certain essential oils should be restricted on the basis of the current evidence.

Setting safe levels

It is worth noting that some common foodstuffs contain tiny amounts of chemicals shown to be carcinogenic in animal tests. These include bacon, mushrooms, carrots, lettuce, white bread, apples, celery and peanut butter [413]. Safrole and *beta*-asarone are found in certain alcoholic drinks. None of these foodstuffs is regarded by regulating

Table 8.2 Maximum use levels recommended by regulatory agencies

	Beta-asarone	Safrole
EC maximum use levels for foodstuffs	0.0001 g/kg food	0.001 g/kg food
IFRA maximum use levels	0.01%	0.01%

Table 8.3 Essential oils to be avoided altogether in aromatherapy, due to their carcinogenic potential

Maximum concentration that can be used safely*	Oil	Maximum concentration likely to be found in commercial oils	Component
0.0%	Cade (unrectified)	8 p.p.m.	Benzo[a]pyrene
0.1%	Sassafras	90%	Safrole
0.1%	*Ravensara anisata*	88%	Estragole
0.1%	Basil (high estragole)	87%	Estragole
0.1%	Tarragon (French)	87%	Estragole
0.1%	Camphor (brown)	80%	Safrole
0.1%	Calamus (Indian)	80%	*Beta*-asarone
0.2%	Tarragon (Russian)	46%	Estragole and methyleugenol
0.2%	Snakeroot	45%	Methyleugenol
0.5%	Camphor (yellow)	20%	Safrole
?	*Melaleuca bracteata*	Major component	Methyleugenol

* Assuming a maximum use level of 0.1% for *beta*-asarone, methyleugenol safrole and estragole

Table 8.4 Essential oils which are safe to use in aromatherapy, at the appropriate external maximum use levels, but which should not be taken in oral dosages

Maximum concentration that can be used safely*	Oil	Maximum concentration likely to be found in commercial oils	Component
1.5%	Fennel	7%	Estragole
2.0%	Basil (low estragole)	5%	Estragole
2.0%	Ho leaf (camphor/safrole CT)	5%	Safrole
2.25%	Nutmeg (E. Indian)	4.5%	Safrole and methyleugenol
5.0%	Mace	2%	Safrole
10.0%	Cinnamon leaf	1%	Safrole
10.0%	Star anise	1%	Safrole

* Assuming a maximum use level of 0.1% for *beta*-asarone, methyleugenol, safrole and estragole

Note: Cangerana oil, which *is* commercially available, contains an as yet undetermined quantity of safrole.

agencies as being carcinogenic, since the amounts present in the foods of the chemicals in question are so small.

IFRA has set maximum use levels of 0.01% for safrole and *beta*-asarone in fragrances (Table 8.2); there is no maximum use level for estragole or methyleugenol. Taking into account all of the above arguments, particularly the significant difference between rodent and human in the amount of carcinogenic metabolite formed, it is our opinion that a 0.1% maximum external use level for asarone, estragole, methyleugenol and safrole is both safe and appropriate for aromatherapy use.

Except in the case of oestrogen-sensitive cancers, we do not believe that the few potentially carcinogenic essential oils present any more danger to people with cancer than they do to the general population. This is because carcinogens do not work in a way that would speed up the progression of an already established cancer.

Common sense tells us that giving potentially carcinogenic essential oils to a person with cancer is not prudent, and advice to this effect would be understandable. However, the authors are of the opinion that a contraindication of this nature would be unwarranted. Therefore we have not indicated that essential oils with low levels of potential carcinogens, such as nutmeg and fennel, should be avoided in people with cancer. There might, however, be a case for avoiding regular use of such oils in pre-cancerous states.

Potentially carcinogenic essential oils that should be avoided altogether in aromatherapy are listed in Table 8.3; those that are safe to use externally up to certain concentrations, but which should not be taken in oral doses, are listed in Table 8.4.

MASSAGE IN CANCER

Issues relating to the safety of massage do not come within the scope of this text. However, questions about massage in cancer are asked so frequently that we decided to include a brief section to discuss this important safety issue which, at the time of writing is not covered anywhere else.

Can massage spread cancer?

There is a common belief that massage might spread cancer cells from one part of the body to another, by stimulating lymph flow. There is no clinical evidence either that massage can spread cancer, or that it is safe to use in cancer. However, the considered opinion of everyone we have consulted is that *gentle* massage will not make any difference to the distribution of cancer cells in the body.

Common sense tells us that lymph flow will not be stimulated any more by *gentle* massage than it will by the muscular contraction caused by normal body movements. There is no reason to believe that gentle mechanical stimulation of lymph flow will cause cancer cells to spread which would not otherwise have done so. Gentle massage is probably no more dangerous in cancer than is gentle exercise. However, it would be prudent to avoid any kind of deep massage over or near lymph glands. (Tumour cells are, of course, not just spread through lymphatics but are also disseminated directly by the bloodstream.)

Is it safe to massage someone with cancer?

The majority of opinion is against the use of any kind of deep-tissue, or heavy massage, although some therapists do not agree with this. Certainly, deep massage over or close to a tumour site or to lymph glands is inadvisable. (There is evidence that the surgical handling of tumours, during resection, can disseminate tumour cells.) We would strongly caution against *deep* massage in all situations involving cancer, since the risks probably outweigh the benefits.

Any kind of massage should be completely avoided over areas of the body receiving radiation therapy, since the skin becomes especially fragile, and can break down. Massage should also be avoided on areas of skin cancer.

It is recommended that massage of someone with any kind of cancer is only undertaken by health professionals who have a very good understanding of both massage *and* cancer, or by people who have been shown exactly what to do by such health professionals.

SUMMARY

- The chemical production of malignant tumours depends on two processes, initiation and promotion.
- Alkylating agents are frequently carcinogenic because they can bind irreversibly to DNA and protein. Safrole is a good example of an alkylating agent found in essential oils.
- Mutagenic or genotoxic substances are those which interfere with the maintenance and replication of cellular DNA. Such activity is a possible, but not definite indication of carcinogenicity.
- There is no evidence that either essential oils or aromatherapy have ever caused cancer in humans.
- Substances which are carcinogenic in rodents are more likely than not to be carcinogens in humans. However, some account must be taken of dosage levels, since low dosages may not be dangerous.
- *d*-limonene possesses antitumoral properties. Oxidised *d*-limonene is potentially carcinogenic, but only after cells have been 'initiated' by another carcinogen, such as benzo[*a*]pyrene.
- Compounds which are likely to be carcinogenic in humans, if levels of exposure are sufficient, include *beta*-asarone, methylengenol, safrole and estragole.
- Essential oils with significant levels of these compounds should be avoided in aromatherapy. Those with low levels of these compounds are safe to use up to certain concentrations.
- Little is known about the risk of dermally applied potentially carcinogenic substances found in essential oils.
- It might be prudent for people with oestrogen-dependent cancers to avoid oral administration of essential oils rich in *trans*-anethole, which has weak oestrogen-like properties.
- It is very unlikely that gentle massage can cause cancer to spread through stimulation of lymph flow.

- Massage should be completely avoided over areas of the body receiving radiation therapy, over or close to tumour sites, and on areas of skin cancer.

Notes

1. Statistics published in 1985 show that cancer is responsible for 25% of all deaths in the UK. Paradoxically, cancers represent only 5% of hospital admissions and 1% of GP consultations. Clearly, many cases are not diagnosed or treated. Cancer is an old-age-related disease, and incidence is much lower in undeveloped countries where life expectancy is shorter. Taking the whole world, cancer accounts for around 8% of all deaths [60].
2. Malignant tumours may arise almost anywhere in the body. Those derived from epithelial tissue are called carcinomas, those derived from connective tissue or muscle, sarcomas. They may arise in tissues of the immune system: myeloid tumours in the bone marrow, lymphoid tumours in lymphoid tissue. Embryonic tissue may also, on occasion, become malignant. One of the great problems facing medicine is in understanding why such malignant changes should sometimes occur in previously normal tissues. There is undoubtedly no single reason.
3. Some of the most notorious chemical carcinogens are the polycyclic hydrocarbons, such as benzo[*a*]pyrene, found in cigarette smoke. Burned organic material very often contains these chemicals. They probably need to be converted to highly reactive intermediates (epoxides) by metabolic processes before cancer can result. Apart from cade oil, which is produced by burning during distillation (dry distillation) essential oils do not contain polycyclic hydrocarbons or other chemicals with similarly high carcinogenic potential.
4. The terms mutagenic and genotoxic are almost interchangeable, and both refer to cellular DNA damage. We have used these terms here to distinguish between tests carried out in bacteria or fungi (mutagenic) and tests carried out in live rat hepatocytes (genotoxic).
5. One of the cancer initiators found in cigarette smoke (benzo[*a*]pyrene) was used in some of the tests where subsequent application of citrus oils lead to tumour development. Factors in the citrus oils, now believed to be oxidised components of the oils, were thus acting as promoters of tumour development.
   There is a theoretical possibility that oxidised citrus oils could act as promoters in people already initiated by constituents of cigarette smoke. The actual risk to smokers from using oxidised citrus oils is vanishingly small, and is not currently regarded as a risk by any regulatory agencies. However, the studies highlight the importance of using fresh, unoxidised citrus oils, which present no risk at all.
6. Here we can see an example of a minor difference in the structure of a molecule making a major difference to its pharmacological effect. While estragole and safrole are potentially carcinogenic, myristicin has the opposite effect.

9

Reproduction

Few aspects of toxicology arouse such concern as the effects of chemical substances on the reproductive system and on the development of the fetus. Apparently avoidable disasters such as the stilboestrol and thalidomide scandals have made people aware that the reproductive process can be extremely sensitive to chemical agents, tolerating little in the way of chemical insult. Today about 20 drugs and chemicals have been proven to be teratogenic [476].[1]

There is naturally concern that some essential oils may not be safe if used during pregnancy, or may somehow affect fertility. This concern is highlighted by the fact that the great majority of those who engage in aromatherapy are women [450].

In both sexes, reproductive function is dependent upon sex hormones (oestrogens and progestogens in women; androgens, chiefly testosterone, in men). One concern is that some essential oil components may have hormone-like activity, and indeed this has been found for a very few terpenoids. There is then the possibility that hormone-like essential oil constituents could disturb sexual functioning by disrupting the normal, finely balanced ebb and flow of sex hormones in the circulation.

A second concern is that some essential oil ingredients could prove damaging to the development of the fetus; that is, they could be teratogenic, or that they could in some way disturb the normal outcome of the pregnancy, for instance by promoting the resorption of the early embryo. A third concern is that some essential oils could cause abortion. Pennyroyal and savin are probably most notorious in this regard.[2]

Studying the effects of drugs and other foreign materials during pregnancy is highly problematic [186]. Results of animal studies often correlate even more poorly with the human situation than in other fields of research. One offspring per pregnancy, the typical human result, is highly atypical of laboratory mammals, where there may be great variability in the toxic effects of administered substances within and between litters. Interpreting the results of animal toxicity studies is very difficult indeed.

Human and animal reproductive physiology are markedly different. Animal data on reproductive toxicity must therefore be extrapolated to the human situation with circumspection. In fact, thalidomide was tested on rats and mice, but did not cause the teratogenic effects later seen in humans. However, as in the case of cancer, toxicology testing in humans is not going to happen, for obvious ethical reasons, and considerable weight is normally given to tests in animals for reproductive toxicity.

This leaves us little choice but to take the attitude that substances which prove hazardous in pregnant animals are best avoided by pregnant humans. Case reports, such as those given below under 'Camphor' and 'Nutmeg and mace' are rare, but can provide useful information. Aromatherapists, and other practitioners who use essential oils, should be encouraged to report cases in which essential oils may have adversely affected pregnancy, as well as cases in which essential oils have apparently been safely used during pregnancy.

Reproductive toxicology includes the study of abnormal fertility, embryotoxicity and (later) fetotoxicity, perinatal and postnatal toxicity. These categories overlap, and substances may produce effects in only one or in several of them.[4]

FERTILITY
Sex-hormone-like activity and effects on fertility

Perhaps surprisingly, hormone-like activity is quite widespread in the plant kingdom. Potatoes,

beetroot, yeast and palm kernel oil all have very low-level oestrogen-like activity, apparently due to their content of oestrone [360].[3]

Trans-anethole is related to oestrone methyl ether, a chemical with oestrogenic properties, found in several plants. Polymerised anethole seems to stimulate lactation and bears some resemblance to diethylstilboestrol, a synthetic oestrogen [361]. However, the hormone-like action of *trans*-anethole, anise oil and fennel oil, demonstrated in rodents, is relatively weak [360] and anethole-rich essential oils are not considered to present any risk during pregnancy, unless taken in oral doses.

Oral dosing might not be advisable because the (weak) oestrogen-like properties of anethole may conceivably affect the progress of pregnancy or, more likely, may have an impact on the development of the fetus. Since this is highly speculative, we have indicated a caution, rather than a contraindication, for oral, anethole-rich oils in pregnancy.

It has been reported that citral (<> lemongrass) can impair reproductive performance in female rats by reducing the number of normal ovarian follicles [199]. The effect has been seen only after a series of intraperitoneal injections at a dose of 0.3 g/kg. This is equivalent to injecting around 25 ml of lemongrass oil into the abdomen every 4–5 days for 60 days. Clearly it has no bearing on the use of essential oils in aromatherapy.

East Indian nutmeg oil has been found to reduce fertility in male mice given the oil (Pecevski et al 1981). The dose-dependent effect occurred with oral doses between 60 and 400 mg/kg/day given 5 days per week for 8 successive weeks. The number of fertile mice was reduced from 95% (control) to 71% at the lowest dose, and to 32% at the highest. No chromosomal damage was found in these mice, but some of their male offspring did exhibit chromosomal damage. It is difficult to draw any firm conclusions from this research. The lower dosage used (equivalent to 4 g in an adult human) is a relatively high dosage and frequency, and effects on fertility are often species/strain-specific. However, this finding adds weight to the argument that East Indian nutmeg oil should not be taken in oral doses.

Methyl salicylate can decrease the litter size and number of live-born progeny in rats when given at levels above 3000 p.p.m. (0.3%) in the diet of pregnant animals [181]. In part, this is probably due to its action on kidney development in the fetus [129]. It seems that rather large quantities would need to be taken by a pregnant woman to produce fetal abnormality—of the order of 30 g at some point during early gestation [179, 180]. Wintergreen and sweet birch oils should be avoided in any case because methyl salicylate is toxic in a wider sense (see p. 194).

Essential oils and contraceptives

There are three types of oral contraceptive available. All contain a synthetic form of progesterone (for instance the progesterone-only pill) while others contain variable amounts of oestrogens as well (the combined and phased pills). The progesterone-only pill does not always suppress ovulation, and works mainly by making the cervical mucus impenetrable to sperm. It is somewhat less reliable than the oestrogen-containing types.

It seems very unlikely that oestrogenic essential oils could interfere with oral contraceptive control: any oestrogen-like effect of the oils would probably be far weaker than the pill's hormonal action, especially after dermal application of the oil.

Latex condoms can be weakened by both vegetable oils and essential oils. Corn oil, for instance, has been shown to cause a loss of up to 77% of a condom's strength after only 15 minutes [477]. Condoms are used either to avoid pregnancy, or to guard against spreading sexually transmissible disease. In either case it is important to avoid contact with essential or vegetable oils.

Essential oils and hormone replacement therapy (HRT)

It is unlikely that aromatherapy treatments could adversely affect hormone replacement therapy as any hormonal effect of the oil would almost certainly be considerably weaker than that of the HRT.

EMBRYOTOXICITY AND FETOTOXICITY

There are very few data on the distribution and fate of drugs within the human embryo because it is almost impossible to design safe experiments. It is often assumed that drug concentrations in the embryo reach similar levels to those in the mother's serum. This is likely to be wildly inaccurate for general application and the truth is that we simply do not know to what extent certain foreign substances circulating in the mother's bloodstream reach the developing child [186].

Crossing the placenta

The placenta acts as a barrier against both electrically neutral and positively charged molecules, leaving negatively charged molecules to cross into the fetus with comparative ease [200]. Exactly where this would leave essential oils regarding passage across the placenta is not clear, but small molecules all tend to pass into the fetal circulation relatively easily. As a broad approximation it is estimated that substances with a molecular weight of less than 1000 can cross the placenta [488]. Since all essential oil constituents have molecular weights well below 500 it would be prudent to assume that they are all capable of crossing the placenta.

The blood–brain barrier is somewhat underdeveloped at and before birth, increasing the likelihood that compounds which do cross the placenta will reach the fetal central nervous system [200]. Their effect on the fetal CNS is often much greater than on the mother's. This is because of the incomplete development of the fetal CNS and the relative ineffectiveness of the fetal blood–brain barrier [16]. The fetal CNS, because it is still growing, is more susceptible to damage by chemicals than is the adult CNS.

Crossing the placenta does not necessarily mean that there is a risk of toxicity to the fetus; this will depend on the toxicity and the plasma concentration of the compound. Fetal metabolic pathways are often immature. In some ways this puts the fetus at greater risk, but in other instances it may be protective; for example, in cases when the fetal liver is not yet capable of metabolising a compound into a more toxic one [511].

In rats, cineole appears to cross the placenta in sufficient quantity to affect the activity of fetal liver enzymes when given by subcutaneous injection to pregnant rats at a dose of 0.5 g/kg for 4 days [326]. Although this dose represents a vastly higher maternal blood level of cineole than would ever be encountered in aromatherapy, the study underlines the ability of some essential oil constituents to cross the placenta and reach the fetus. Eucalyptus oil (around 75% cineole) showed no embryotoxicity or fetotoxicity when tested on rodents (injected subcutaneously at 135 mg/kg on days 6 to 15 of gestation) and had no effect on birth weight or placental size [193].

Camphor is definitely able to cross the placenta. Following accidental ingestion of camphorated oil by a pregnant woman, camphor was found to be present in the body of her stillborn baby [241].

Camphor

There have been three cases of accidental camphorated oil ingestion by pregnant women which, taken together, constitute useful information. In the first case, reported in 1957, 45 ml of camphorated oil was ingested, while in hospital, during the third month of gestation. The woman vomited several times almost immediately, then had a convulsion and became unconscious. Further convulsions followed, as did therapy, including gastric lavage, and she recovered consciousness 1 hour later. A normal infant was eventually born [416].

In the second case a woman, who was 40 weeks pregnant, ingested 2 ounces of camphorated oil, also while in hospital. (In both cases the attending staff were not aware that camphorated oil was being taken.) Gastric lavage was initiated some 20 minutes later, and although the woman was severely intoxicated by the camphor, she recovered fully. Her baby was stillborn 36 hours later. To quote from the report: 'Several factors may have contributed to the death of this infant... Whether or not the camphor precipitated these difficulties is problematical'. It seems likely that the camphor at least contributed to the infant's death; camphor was detected in its liver, brain and kidneys [241].

In the third case, the baby was also a full-term infant. The mother ingested 2 ounces of camphorated oil, and had the first of three seizures 20 minutes later. She was admitted to hospital, and gastric lavage was performed. Spontaneous labour commenced the following morning, and her baby, smelling distinctly of camphor, was born without complication. The baby was closely examined and monitored for 3 days, and camphor was just detectable in its blood. The mother's blood, collected 24 hours after ingestion, contained large amounts of camphor [240].

The baby in the third case was some 1.5 kg heavier than the second case infant, and there were no complications. Otherwise the two cases are remarkably similar, except that one infant died and the other survived.

Since camphorated oil is 20% camphor, the amount of actual camphor ingested in the last two cases is approximately 12 ml. Referring to Table 3.1 (p. 25) we find that the maximum amount of essential oil likely to reach the bloodstream from external aromatherapy is 0.3 ml. Assuming that the quantity of camphor present in the oil is 20% (as in rosemary oil) the amount of camphor reaching the bloodstream will be no more than 0.06 ml. This is 200 times less than in the two cases of pregnancy.

If our suggested maximum concentration of essential oil for application by massage in pregnancy (2%) is assumed (see p. 110), this figure becomes 500 times less. If 12 ml of camphor is close to the smallest amount which can cause fetal death, since it caused no apparent damage to a surviving fetus, then over 200 times less than this can be regarded as safe, even for earlier stages of pregnancy, when the developing child will be much smaller than in the above cases. However, an oral dose of 2 ml would give a differential of only around 15 times for camphor oil—a much less comfortable margin, especially in early pregnancy and/or if taken for several consecutive days. Such a dose might not even be safe taken once only, if ingested during a time when the embryo or fetus was particularly sensitive to camphor.

Oral administration of camphor-rich essential oils should be avoided during pregnancy. It is our opinion that external use of camphor-rich oils such as rosemary is safe in pregnancy, but oils which contain very high levels of camphor should be avoided altogether.

Sabinyl acetate

Sabinyl acetate is found in essential oils of plectranthus (*Plectranthus fruticosus*) (> 60%), Spanish sage (*Salvia lavandulifolia*) (0.1–24%), and savin (*Juniperus sabina*) (20–53%). *Juniperus pfitzeriana* also contains sabinyl acetate (2–17%) but is not commercially available [497]. Oils rich in sabinyl acetate are probably the most dangerous ones in pregnancy.

Plectranthus oil is both embryotoxic and fetotoxic in rodents. It dramatically increased the rate of embryo resorption after oral administration, at doses equivalent to about 1 g on each of 2 days of a human pregnancy [197]. Plectranthus oil is strongly teratogenic (see below). It is not currently available commercially.

There are very worrying anecdotal reports of savin oil causing abortion, and being able to cross the placenta [159, 435]. Savin oil is embryotoxic, but not fetotoxic in rodents. Savin oil ingestion is associated with weight loss in pregnant animals [190].[5]

Spanish sage oil can contain up to 24% of sabinyl acetate, although levels are more commonly closer to 10%. A fraction of Spanish sage oil, containing 50% sabinyl acetate, 41% 1,8-cineole, and 5% camphor, caused dose-dependent maternal toxicity in rodents, and demonstrated a dose-dependent abortifacient effect, but was not fetotoxic [194]. Cineole is not teratogenic [193].

These data are most probably applicable to humans and suggest that plectranthus, savin and Spanish sage oils should be avoided altogether by pregnant women.

Other toxic oils

Safrole is able to cross the placenta and produces kidney epithelial and liver tumours in the offspring of pregnant mice given it in the diet [161]. Safrole is known to be carcinogenic, and safrole-rich oils should be avoided altogether in aromatherapy.

Thujone-rich essential oils should be avoided during pregnancy. Thujone can cause convulsions when taken by mouth [158] and is suspected of being particularly toxic to the CNS [14]. Relatively low doses of thujone have been shown to affect nervous tissue [133, 144]. This strongly suggests that it has the ability to cross the blood–brain barrier and to enter the CNS after absorption into the bloodstream.

In addition to safrole-rich and thujone-rich oils others, which are toxic in a general sense, should be avoided in pregnancy. These oils are not listed in Table 9.1 (p. 111), since they should be avoided altogether in aromatherapy. Because of its severe neurotoxicity it is recommended that hyssop oil rich in pinocamphone should be avoided altogether during pregnancy.

Nutmeg and mace

The administration of up to 0.56 g/kg of nutmeg oil to pregnant rodents for 10 consecutive days had no effect on nidation or on maternal or fetal survival; no teratogenic effect was observed [NTIS 1972, cited in 422]. A 30-year-old woman, at 30 weeks of gestation, ingested several cookies made with 7 g of ground nutmeg, instead of the recommended eighth of a teaspoon [474]. 4 hours later she experienced a sudden onset of palpitations, blurred vision, agitation and a sense of impending doom. The fetal heartbeat was 160–170 b.p.m., and returned to a normal 120–140 within 12 hours. The fetal response was attributed to the myristicin content of nutmeg oil, and its anticholinergic (i.e. stimulant) effect. It is assumed that myristicin readily crosses the placenta.

This case is not a sufficient basis to contraindicate nutmeg oil in pregnancy. Whatever applies to nutmeg oil also applies to mace oil, since the two are similar.

TERATOLOGY

This is the study of structural or functional abnormalities which arise during embryonic development. Teratology most often appears as congenital malformations which are apparent at birth, but can present in different ways. Major malformations are generally thought to occur in 2% of all live births, but the proportion increases if observations are continued beyond childhood and on into adulthood.

Studies have attempted to show correlations between molecular structure and teratogenicity for fragrance additives [184]. Although it is much easier to show such links than it is to explain why they exist, there do appear to be particular chemical structures especially associated with the ability to cause fetal abnormality. Unsaturation (the presence of one or more carbon–carbon double bonds) and carbonyl or aldehyde groups may enhance a molecule's ability to interact with lipid constituents of the embryo's cell membranes [198].

Plectranthus

Plectranthus is the only essential oil which is known to be strongly teratogenic. Its effects on mouse fetal development include kidney and heart defects, skeletal alterations and above all, anophthalmia (lack of eyes) [189]. Sabinyl acetate is thought to be responsible for these effects, and is present in higher concentrations in plectranthus oil than in any other essential oil. Plectranthus oil is not commercially available.

Citral

Citral, geranial and citronellal have produced malformations in chick embryos when injected into fertilised hen's eggs [182]. Of these compounds, citral is the most potent, and geranial the weakest. The authors suggest that the aldehydic structure of citral and citronellal may help make them teratogenic. However, work in rats has suggested that citral is non-fetotoxic [196]. This makes it most unlikely that either citral or citronellal present any hazard to humans during pregnancy. Experiments with chick embryos are of limited value for the risk assessment of humans.

Methyl salicylate

Salicylates, including methyl salicylate (<> wintergreen) have been implicated as teratogenic in mammals, but only in relatively high doses [417].

ABORTION

Approximately one fetus is aborted for every child born in Western countries [476]. Savin, tansy, juniper, pennyroyal and rue have all been considered abortifacient at one time or another [173]. Work using the isolated human uterus shows that the essential oils of these plants have no direct action on uterine muscle [174]. Furthermore, they do not tend to stimulate abortion by causing the death of the fetus [175, 178, 188]. However, the fact that these oils do not stimulate the isolated human uterus does not in itself prove that they are not abortifacient. It is possible that components of the essential oils are metabolised in vivo into more toxic ones. We know, for example, that pulegone is metabolised into menthofuran, a more toxic compound.

Pennyroyal

Pennyroyal has already been mentioned in the context of the effect that its pulegone content has on the liver. The plant has enjoyed folk status as an abortifacient from ancient times [171].[6]

In two separate cases, 7.5 ml of pennyroyal oil, taken orally, failed to induce miscarriage [168]. In a third case, 10 ml, taken orally, also failed [170]. A much more serious case was reported in *The Lancet* in 1955. It was not possible to determine how much oil was taken, but abortion did take place, and the mother died following massive urea leakage into the blood [169].

Pennyroyal oil, it seems, is not abortifacient, unless taken in such massive quantities that it causes acute hepatotoxicity in the mother. She miscarries only because she is so poisoned by pulegone that the pregnancy cannot be maintained [242]. Although not abortifacient, pennyroyal and other pulegone-rich oils should not be used in pregnancy, due to their hepatotoxicity. They are not listed in Table 9.1 (p. 111), since they should not be used in aromatherapy anyway.

Parsley

Preparations made from parsley have been used to procure abortion for a great many years, and are still in use today, notably in Italy. Apiol, a major component of most parsley leaf and seed oils, is generally held to be responsible for the abortifacient action of parsley, and is also used in its own right as an abortifacient. Data is often difficult to obtain, partly because of the legal implications of patients admitting to illegal abortion, and partly because in some cases death follows abortion, and there is no record of how much apiol was taken. Naturally, the cases which have been recorded tend to be the worst ones, in which medical intervention has been urgently sought.

D'Aprile (1928) is the most prolific source of case data. Out of five cases, all of whom were between 2 and 7 months pregnant, one aborted and later died, one did not abort but died, and three aborted and survived [530]. In the case which did not abort, the fetus was found to be dead. Post-abortive vaginal bleeding, sometimes profuse, is a feature of these cases. A cumulative effect is apparent, apiol being taken daily for between 3 and 8 days before either death or abortion ensued. One of the cases cited had traces of apiol in her urine 12 days after the last ingestion.

The lowest daily dose of apiol which induced abortion was 0.9 g taken for 8 consecutive days. This is approximately equivalent to 6 ml of parsley leaf oil, between 1.5 ml and 6 ml of parsleyseed oil, or 5 ml of Indian dill oil (See Table 6.2, p. 64). The inevitable conclusion is that all of these apiol-rich essential oils present a very high risk of abortion if taken in oral doses, and that external use would also seem inadvisable in pregnancy.

In animal studies, considerably higher dosages of apiol appear to be tolerated. In pregnant guinea pigs, abortion generally did not occur except at lethal doses, around 2 g [530]. In pregnant rabbits, abortion occurred at dosages of 5–14 g, with severe haemorrhage [531]. In both types of animal, the dosage is equivalent to approximately 100–200 g

in a human. This is some 20–40 times higher than the amount of apiol causing abortion in humans, and highlights the poor correlation between animals and humans in this area.

Sabinyl acetate

A fraction of Spanish sage oil, containing 50% sabinyl acetate, 41% 1,8-cineole, and 5% camphor, demonstrated a dose-dependent abortifacient effect in mice [194]. The abortifacient effect seen here was obtained even at very low dosage levels (0.015 g/kg, equivalent to 2 ml of Spanish sage oil with 24% sabinyl acetate) and is surely due to sabinyl acetate. If so, plectranthus and savin oils are probably also abortifacient.

Rue

There are isolated reports that rue oil has a direct action on the uterus, but the details of the research on which they are based are obscure [176]. Used in massive amounts on pregnant rabbits (12 ml/kg) and guinea pigs (50 ml/kg) rue oil, not surprisingly, was equally toxic to mother and fetus, causing widespread tissue damage and some fatalities [177]. (These dosage levels are equivalent to human ingestion of 800 ml and 3.2 litres!) Two pregnant guinea pigs, each fed 12 drops of rue oil, aborted; rue oil was found in the fetal tissue, and was considered toxic to it [335]. This is still a very high dosage, around 3 ml/kg (equivalent to human ingestion of around 200 ml).

There are anecdotal reports that rue oil can cause abortion in humans. In this respect the oil is similar to savin [436] which has a strong folk tradition as an abortifacient. It is possible that any abortifacient activity that rue and savin oil have is, like that of pennyroyal, due entirely to maternal toxicity.

The FDA consider that rue oil is safe to humans as currently used in food flavourings, 'but not necessarily under different conditions of use'; in particular, further teratological research is recommended [335]. Our recommendation is that rue oil should be used with caution, i.e. at low levels, or avoided altogether during pregnancy until further data become available.

Juniper

Although an ethanolic extract of juniper berries has demonstrated clear abortifacient effects [464], there is every reason to believe that the essential oil is not responsible (see juniper profile, p. 142). Historically, juniper has a reputation as an abortifacient. It is clear, however, that there has been confusion between juniper (*Juniperus communis*) and savin (*Juniperus sabina*) (see juniper profile).

Because juniper berries and savin oil are probably both abortifacient, it is easy to understand how the belief has arisen that juniper oil is also abortifacient. However, it is our opinion that this is an erroneous belief, and that juniper oil (*Juniperus communis*) presents no abortifacient risk (see juniper profile).

Emmenagogue oils

There is a popular belief among aromatherapists that emmenagoguic essential oils (those which stimulate menstruation) are unsafe during pregnancy, because they might lead to miscarriage. Since many essential oils have been labelled as emmenagogues in the popular literature this has led to essential oils such as clary sage, cypress, lavender, marjoram, peppermint and rose being flagged as dangerous in pregnancy [451]. Whether or not these oils are emmenagogues, there is no evidence that they are abortifacient in the amounts used in aromatherapy. If any of them do have an emmenagoguic action, the nature of it is such that it presents no danger in pregnancy.

Discussion

With the exception of apiol-rich and sabinyl-acetate-rich oils, there appears to be no clear evidence that any essential oils present an abortifacient risk, as they are used in aromatherapy. It is possible that, as in the case of juniper, some of the plants referred to above are abortifacient, but their essential oils are not. Rue

might present a risk, and this in turn might depend on chemotype. Pennyroyal does not appear to be abortifacient except in lethal dosages, and it is extremely unlikely that juniper oil represents a risk.

Miscarriage is very common, and women who use essential oils while they are pregnant and then miscarry may suspect that the essential oils contributed to or caused the miscarriage. In most cases no definite conclusion will be possible, but the overall risk from essential oils is very small indeed.

AROMATHERAPY IN PREGNANCY

As a general safety precaution during pregnancy, a maximum concentration of 2% essential oil is recommended for massage over large areas of skin. It is also recommended that essential oils are only administered orally, rectally or vaginally with great caution during pregnancy. The last two routes could deliver at least as much essential oil to the fetus as could oral administration.

The first 3 months?

The developing child is particularly sensitive to chemical insult during the first 3 months of pregnancy. However, the fetus remains vulnerable throughout pregnancy, and there is good evidence that different fetal systems are sensitive to different chemicals at specific times [5]. Such detailed data are not available for essential oils, and so we do not feel justified in recommending that oils be avoided only at specific stages of pregnancy, such as the first trimester. It is more prudent to recommend that those which are potentially dangerous be avoided throughout pregnancy.

Essential oils to avoid

Essential oils that should be avoided throughout pregnancy are listed in Table 9.1; oils that may be used with caution are listed in Table 9.2; and Table 9.3 lists oils that are safe for external use only.

Table 9.1 Essential oils which should be avoided altogether throughout pregnancy, and their toxic components: add to this list all the oils which are toxic in a general sense (see p. 230)

Oil	Component
Balsamite (camphor CT)	Camphor
Camphor (white)	Camphor
Ho leaf (camphor/safrole CT)	Camphor
Hyssop	Pinocamphone
Indian dill seed	Apiol
Juniperus pfitzeriana (NCA)	Sabinyl acetate
Parsley leaf	Apiol
Parsleyseed	Apiol
Plectranthus (NCA)	Sabinyl acetate
Sage (Spanish)	Sabinyl acetate
Savin	Sabinyl acetate

NCA = not commercially available

Oils which are safe to use

It is easier to list oils which are potentially dangerous than it is to list those which are safe during pregnancy, but this is a question which is often asked. The simple answer would be 'any oil in the book not listed as dangerous during pregnancy'. This would also be the most truthful answer, since there are no essential oils which have been experimentally proven safe in pregnancy in humans (the same applies to most other substances).

In order to satisfy those who want a short, positive list we have compiled one (Box 9.1). However, in no way does this imply that any essential oils not on the list, or included in Tables 9.1, 9.2 and 9.3, are unsafe during pregnancy.

Box 9.1 Examples of essential oils safe to use in pregnancy

Cardamon	Neroli
Chamomile (German)	Palmarosa
Chamomile (Roman)	Patchouli
Clary sage	Petitgrain
Coriander seed	Rose
Geranium	Rosewood
Ginger	Sandalwood
Lavender	

Table 9.2 Essential oils which should be used with caution throughout pregnancy due to suspect toxicity, and their toxic or suspect components

Oil	Component
Annual wormwood	Artemisia ketone
Buchu (*B. betulina*)	Pulegone
Cangerana	Safrole
Lavandula stoechas	Camphor
Lavender cotton	Artemisia ketone
Oakmoss[A]	Thujone
Perilla	Perilla ketone
Rue	?
Treemoss[A]	Thujone?

Table 9.3 Essential oils which are safe to use externally, but which should not be administered orally, rectally or vaginally during pregnancy

Oil	Component
Anise	Anethole
Fennel	Anethole
Lavandin	Camphor
Lavandula stoechas	Camphor
Rosemary	Camphor
Spike lavender	Camphor
Star anise	Anethole
Yarrow (camphor CT)	Camphor

Summary

- Anethole-rich oils have a weak oestrogen-like effect. Because of this they might be best avoided orally during pregnancy.
- Aromatherapy is very unlikely to have any unwanted effect on oral contraception or hormone replacement therapy.
- Certain essential oil components can cross the placenta and reach the embryo or fetus. The same is probably true of most, or even all essential oil components.
- Sabinyl acetate-rich oils and apiol-rich oils are probably the most dangerous ones in pregnancy.
- Essential oils which are sometimes described as emmenagogues are not necessarily abortifacient.

• There appears to be no clear evidence that any essential oils present an abortifacient risk, as far as external use is concerned in aromatherapy.

• It is more prudent to recommend that those oils which are potentially dangerous be avoided throughout pregnancy, than to specify the first trimester.

• For aromatherapy massage over large areas of skin, a maximum concentration of 2% essential oil is recommended during pregnancy.

• It would be prudent to administer essential oils with great caution orally, rectally and vaginally during pregnancy.

Notes

1. Stilboestrol is a drug which caused vaginal cancer in daughters born to women who had taken it. Thalidomide is a drug which causes gross limb deformities in the fetuses of those who take it. It is still prescribed in some parts of the world, and the problem continues, notably in South America.
2. Pennyroyal is not the only plant which has enjoyed a reputation as an abortifacient. In Mexico, cottonroot bark is used; in India, pulsatilla; and in the United States, rue and sage are mentioned in herbal medicine textbooks [172]. It is, however, not always easy to see how these plants have gained their reputations as abortifacients.
3. Other naturally occurring oestrogenic compounds are present in plants. Genistein is responsible for low fertility and abortion in grazing sheep; coumestrol occurs in alfalfa and clover; daidzein is present in soya beans. These substances are not very potent oestrogens at all, and do not appear to constitute a hazard to human health (neither do they appear in essential oils). They do, however, resemble coumarin, which may rarely appear in essential oils.
4. A background of spontaneous malformations exists which makes it difficult to recognise weak fetotoxic effects. An observed effect may only occur under special conditions and in some species, and be irrelevant to humans; there is no clearly defined border between gross abnormalities due to toxic substances (teratogens) and structural 'variation'.
5. The situation with savin is somewhat complicated by the fact that the oil is frequently substituted by oils from other *Juniperus* species, notably *J. thurifera* and *J. phoenica*, neither of which contains much sabinyl acetate [192]. Such essential oils are not teratogenic in mice [187].
6. Pulegone occurs in varying amounts in species of *Mentha* other than pennyroyal. Its has been reported in *Mentha piperita* (peppermint) and in *Mentha arvensis* (cornmint) as well as in *Mentha spicata* (spearmint) [163]. The pulegone content of peppermint oil, the most widely available mint oil, is rather variable, and appears to depend on the time of picking and the type of soil on which the plant is grown, as well as on other, more elusive factors [163].
   Although pulegone is found in peppermint and other mint oils, the levels (0.1–2%) are not regarded by other authorities as hazardous in pregnancy. We are not concerned that their use in aromatherapy presents a hazard in pregnancy.

10

Profiles

INFORMATION GIVEN IN THE PROFILES

Botanical name: Sometimes more than one botanical name is given. This is either because the oil comes from more than one botanical source, or because there are alternative botanical names for the same plant. In the latter case alternative names are in brackets.

Family: Here also there are sometimes alternative names for the same family, and these are given in brackets.

Oil from: The part of the plant the essential oil is distilled or expressed from. When the material is normally an absolute this reads 'absolute from'.

Notable constituents: Only important components in terms of either toxicity or quantity are given. The percentages are given as ranges where possible and where appropriate. These ranges are for the types of essential oil commercially available. Ranges above or below these limits will occasionally be found in essential oils which, however, are not produced on a commercial scale. The abbreviation 'tr.' = trace.

The primary source of information for compositional data is Brian Lawrence's *Essential Oils* monographs [214–217].

Acute oral LD$_{50}$: We have given figures here only when they are both available and relevant. > 5 g/kg indicates that the LD$_{50}$ of the oil is over 5 g/kg.

Hazards: Brief details are given here of why the oil may be hazardous. In descending order of severity, the descriptive terms used are: *severe, strong, moderate, mild.*

Contraindications: When the oil should be avoided. The method of administration is also stated. Oral contraindications apply to the oral administration of essential oils as used in medicinal aromatherapy, and assume a minimum daily dose of 0.5 ml. Oral contraindications are not intended to apply to the use of essential oils in any other context, such as in foods, fragrances or external use in aromatherapy.

Contraindications for dermal application assume a minimum use level of 1%. Concentrations of less than 1% may be safe to use in many instances; in fact if the concentration is low enough virtually all essential oils could be considered safe. However, concentrations below 1% are not considered to come into the parameters of aromatherapy practice.

IRV means inhalational, or rectal, or vaginal.

Cautions: These are potential hazards which do not warrant contraindications, either because the evidence is flimsy or because the hazard is not especially worrying. 'Anticoagulants' refers to anticoagulant drugs, such as aspirin, heparin and warfarin; SLE = systemic lupus erythematosus.

Maximum use level: For some of the profiled oils a recommended maximum use level for external use is given. This is usually based on already established guidelines, and is intended to avoid either skin reactions or carcinogenesis.

Toxicity data & recommendations: Here details are given of the results of toxicity tests, and of safety recommendations. IFRA guidelines are based on the assumption that fragrance compounds are used at a concentration of 20% of the final product. Sometimes IFRA give figures for the final product, and sometimes for fragrance compounds.

Comments: Any further useful information or observations are given here.

Compare: Essential oils with similar chemistry and/or hazards are mentioned here.

THE PROFILES

There are profiles of 95 essential oils, each of which is of interest from a toxicological standpoint, although many present only a low level of risk. Generally speaking, essential oils which might present hazards but which are not commercially available are not included (e.g. chervil). However a few of these, which are of particular interest, are included (e.g. plectranthus). Unless otherwise stated, all the profiled oils are commercially available at the time of writing. Please consult the Safety Index for an easy reference regarding the safety of over 300 essential oils.

Phototoxic oils

There are two scenarios in which the risk of phototoxicity may be increased. Firstly, if several phototoxic oils are used together, the risk increases proportionally. For instance, if bergamot and cumin oils are both used in a product in equal proportions, the maximum safe percentage will be 0.2% for each, not 0.4% for each.

Secondly, if concentrated (deterpenated) citrus oils are used, the maximum percentage of essential oil should be reduced in proportion to the degree of concentration. Citrus oils generally contain a large amount of terpenes. Deterpenated (= deterpenised, terpeneless, or terpenefree) citrus oils therefore possess all their other components (including the furanocoumarins) in much larger amounts. So, a 10-times concentrated expressed lemon oil will have a maximum safe concentration of 0.2%, instead of 2%.

Ajowan

Botanical name: *Trachyspermum ammi* (= *Carum copticum*)

Family: Apiaceae (= Umbelliferae)

Oil from: Seeds

Notable constituents:

Thymol	40–50%
Para-cymene	20–24%
Gamma-terpinene	18–20%
Carvacrol	5–7%

Acute oral LD$_{50}$: No data could be found

Hazards:
Dermal irritant* (moderate?)
Mucous membrane irritant* (severe)

*Assumed from thymol/carvacrol content

Cautions (dermal): Hypersensitive, diseased or damaged skin, children under 2 years of age

Cautions (IRV): Do not use at more than 1% concentration on mucous membrane

Comments: General care is required as data on ajowan oil are scarce. Its composition is similar to that of thyme oil, and a similar 'hazards' profile has been assumed. Ajowan oil is rarely found outside India, the Seychelles and the West Indies.

Compare: Oregano, savory, thyme

Almond, bitter (unrectified)

Botanical name: *Prunus amygdalus* var. *amara*

Family: Rosaceae

Oil from: Kernels

Notable constituents:

Benzaldehyde	95%
Hydrocyanic acid	2–4%

Acute oral LD$_{50}$: 0.96 g/kg [425]

Hazards: Toxic (severe)

Contraindications: Should not be used in therapy, either internally or externally

Toxicity data & recommendations: Hydrocyanic acid (= prussic acid, = hydrogen cyanide = HCN) has an LD$_{50}$ of 0.0001 g/kg, making it about 1000 times more toxic than boldo leaf oil, the most toxic essential oil. A single drop of a 10% solution of HCN would be lethal to an average adult.

Because of its HCN content this oil is not commercially available. Both the UK and EC 'standard permitted proportion' of HCN in food flavourings is 0.001 g/kg of food [455, 456].

Comments: HCN is not present in the nuts in their natural state. Prior to distillation, the nuts are comminuted and reduced to a press-cake. This is macerated in warm water for 12–24 hours, during which time the essential oil is formed by the decomposition of amygdalin, a naturally occurring glycoside (compare mustard oil). It is interesting that HCN has a similar odour to benzaldehyde, even though the two compounds are chemically unrelated. This makes it impossible to tell the rectified from the unrectified oil by smell.

Rectified bitter almond oil is referred to as bitter almond oil FFPA (free from prussic acid). This oil is commercially available, and is relatively non-toxic (acute oral LD$_{50}$ 1.49 g/kg) [425]. It is commonly used as a flavouring agent.

The benzaldehyde in bitter almond oil, with or without HCN, has a tendency to oxidise to benzoic acid [158]. This is not in itself harmful, but neither is it desirable in terms of aromatherapy, so oxidation should be guarded against.

Compare: Mustard

Angelica root

Botanical name: *Angelica archangelica*

Family: Apiaceae (= Umbelliferae)

Oil from: Root

Notable constituents:

Beta-phellandrene	16–24%
Alpha-phellandrene	8–20%
Furanocoumarins	1–3%

Hazards: Phototoxic (strong)

Contraindications (dermal): If applied to the skin at over maximum use level, skin must not be exposed to sunlight or sunbed rays for 12 hours

Maximum use level: 0.78%

Toxicity data & recommendations: In phototoxicity tests, distinct positive results were obtained with concentrations of 100%, 50%, 25%, 12.5%, 6.25% and 3.125%. A doubtful reaction was obtained with 1.56%, and a negative result with 0.78% [421]. IFRA recommends that, for application to areas of skin exposed to sunshine, angelica root oil be limited to a maximum of 3.9% in fragrance compound (equivalent to 0.78% in the final product) except for bath preparations, soaps and other products which are washed off the skin [430]. See note on combinations of phototoxic oils, page 114.

Comments: Phototoxicity is primarily due to the presence of bergapten. Angelica seed oil is also available, and is not phototoxic [420]. Angelica seed oil is less expensive than the root oil, and is often used as an adulterant in it.

Compare: Bergamot, cumin, lime, rue

Anise

Synonym: Aniseed

Botanical name: *Pimpinella anisum*

Family: Apiaceae (= Umbelliferae)

Oil from: Seeds

Notable constituents: *Trans*-anethole 80–90%

Acute oral LD$_{50}$: 2.25 g/kg [419]

Hazards:
  Oestrogen-like action (mild)
  Sensitiser (mild)

Cautions (oral): Alcoholism, breast-feeding, oestrogen-dependent cancers, liver disease, pregnancy, endometriosis, paracetamol.

Cautions (dermal): Hypersensitive, diseased or damaged skin, all children under 2 years of age

Toxicity data & recommendations: *Trans*-anethole and its derivatives are oestrogen-like in effect, as is anise oil [360, 361]. The effect is relatively weak, but it would be prudent to avoid oral dosing during pregnancy and breast-feeding in order to avoid influencing oestrogen levels. Endometriosis and oestrogen-dependent cancers could be adversely affected by increased oestrogenic activity (see pp. 95, 96).

Trans-anethole shows a dose-dependent cytotoxicity to rat liver cells in culture, and causes glutathione depletion [452]. The mechanism of its action is unknown, but it is not due to a DNA-damaging effect. A primary metabolite of *trans*-anethole, anethole 1,2-epoxide, also depletes glutathione [504]. Because of this effect, anethole-rich oils such as anise might be best avoided orally in the conditions listed above.

Anise oil produced no irritation when tested at 2% on human subjects; several cases of sensitivity to anise oil have been reported, due to the *trans*-anethole content; when tested at 2% it produced no sensitisation reactions [419]. One authority states that anise oil is not a primary irritant to normal skin [Harry 1948, cited in 419].

Comments: *Cis*-anethole, which exists in synthetic anethole (but not in essential oils) is considerably more toxic than *trans*-anethole.

Compare: Fennel, star anise

Annual wormwood

Botanical name: *Artemisia annua*

Family: Asteraceae (= Compositae)

Oil from: Herb

Notable constituents: Artemisia ketone 35–63%

Acute oral LD$_{50}$: No data could be found

Hazards:
Toxic?
Neurotoxic?

Contraindications (oral): Should not be taken in oral doses

Contraindications (all routes): Children up to 2 years of age

Cautions: Use with caution generally, due to uncertain toxicity; use with caution in epilepsy and pregnancy

Toxicity data & recommendations: It is historically assumed, from anecdotal material, that essential oils rich in artemisia ketone are toxic. However, no toxicity data for this compound could be found. In the circumstances it would be prudent to only use annual wormwood oil with extreme caution.

Comments: The majority of artemisia oils are thujone-rich, and consequently present some degree of toxicity (see armoise, *Artemisia arborescens*, lanyana, wormwood).

Compare: Lavender cotton, wormwood

Armoise

Botanical name: *Artemisia herba-alba*

Family: Asteraceae (= Compositae)

Oil from: Leaves and flowering tops

Notable constituents:
Thujones	35%
d-camphor	30%

Acute oral LD$_{50}$: 0.37 g/kg [421]

Hazards:
Toxic (severe)
Neurotoxic*

*Assumed from thujone content

Contraindications: Should not be used in therapy, either internally or externally

Toxicity data & recommendations: Thujone acute oral LD$_{50}$ is 0.21 g/kg; chronic administration leads to fatty degeneration of the liver [354]. Intraperitoneal thujone was found to be convulsant and lethal to rats above 0.2 ml/kg [234, 325]. Both the UK and EC 'standard permitted proportion' of *alpha-* and/or *beta-*thujone in food flavourings is 0.0005 g/kg of food [455, 456]. Armoise oil is non-irritant, non-sensitising, and non-phototoxic [421].

Comments: The source of armoise oil has previously been referred to as *Artemisia vulgaris* commonly known as mugwort [155]. It has now been established that the correct botanical origin is *Artemisia herba-alba* [216].
There are four chemotypes:
1. thujone/camphor (shown above)
2. camphor
3. *alpha*-thujone
4. *beta*-thujone.
Although content ratios for these are not clear, the matter is academic, since all commercial armoise oils contain substantial quantities of thujone. In some oils there may be a synergistic toxic effect between thujone and camphor, increasing the toxic effect of the oil.

Compare: *Artemisia arborescens*, lanyana, sage (Dalmatian), tansy, thuja, western red cedar, wormwood

Artemisia arborescens

Botanical name: *Artemisia arborescens*

Family: Asteraceae (= Compositae)

Oil from: Herb

Notable constituents:

Beta-thujone	30–45%
Camphor	12–18%

Acute oral LD$_{50}$: No data could be found

Hazards:
Toxic* (severe)
Neurotoxic*

*Assumed from thujone content

Contraindications: Should not be used in therapy, either internally or externally

Toxicity data & recommendations: Thujone acute oral LD$_{50}$ is 0.21 g/kg; chronic administration leads to fatty degeneration of the liver [354]. Intraperitoneal thujone was found to be convulsant and lethal to rats above 0.2 ml/kg [234, 325]. Both the UK and EC 'standard permitted proportion' of *alpha*- and/or *beta*-thujone in food flavourings is 0.0005 g/kg of food [455, 456].

Comments: No toxicity data on this essential oil could be found, and toxicity is assumed from its content of thujone. With very few exceptions, oils from the *Artemisia* genus are toxic. Oil of *Artemisia arborescens* is occasionally available.

Compare: Armoise, lanyana, sage (Dalmatian), tansy, thuja, western red cedar, wormwood

Balsamite (camphor chemotype)

Botanical name: *Chrysanthemum balsamita*

Family: Asteraceae (= Compositae)

Oil from: Herb

Notable constituents: Camphor 72–91%

Hazards:
Convulsant*
Neurotoxic*

*Assumed from camphor content

Contraindications (all routes): Epilepsy, fever, pregnancy, children under 2 years of age

Contraindications (oral): Should not be taken in oral doses

Cautions: Use with caution generally: neurotoxicity, due to high camphor content

Toxicity data & recommendations: Camphor is readily absorbed through skin and mucous membranes and is classified as very toxic [424]. Camphor minimum LD$_{50}$ is 1.7 g/kg in rats [377]. Camphor is thought to be considerably more toxic in humans than in rodents. There have been reports of instant collapse in infants following the local application of camphor to their nostrils [5]. The lethal dose of pure camphor in children is about 1 g [272]. The adult human lethal dose is probably 5–20 g. See page 51 for cases of poisoning from camphorated oil. Camphor readily causes epileptiform convulsions if taken in sufficient quantity [269, 272] (see p. 67). Balsamite (camphor CT) oil should therefore be used with caution, and should not be taken in oral doses.

Comments: Both camphor (72–91%) and carvone (50–80%) have been found as major components in different chemotypes of balsamite oil. Balsamite oils rich in camphor should only be used with great care in aromatherapy. Those rich in carvone are (like caraway oil) probably safe. Balsamite oils are occasionally offered commercially.

Compare: Camphor (white)

Basil

Synonym: Sweet basil

Botanical name: *Ocimum basilicum*

Family: Lamiaceae (= Labiatae)

Oil from: Leaves

High estragole (> 10%) basil

Notable constituents:

Estragole	74–87% (Comoro type)
Estragole	40–55% (French type)

Acute oral LD$_{50}$: 1.4 g/kg [419]

Hazards:
Carcinogenic*
Hepatotoxic*

*Assumed from estragole content

Contraindications: Should not be used in therapy, either internally or externally

Low estragole (< 5%) basil

Notable constituents:

Estragole	<5%
Linalool	major component

Acute oral LD$_{50}$: No data could be found

Contraindications (oral): Should not be taken in oral doses

Maximum use level: 2%

Toxicity data & recommendations: Estragole (= methyl chavicol) produces DNA abnormalities in the livers of experimental animals [122, 289]. It is carcinogenic in mice [309, 312] because it is metabolised, in vivo, to the carcinogenic compound 1'-hydroxyestragole [62, 123, 312, 478]. The same metabolic process is believed to take place in humans [30]. High doses are potentially carcinogenic, but very low doses are not, since they are readily detoxified [62, 306]. See page 98.

Estragole is not restricted by any regulatory agencies. It is recommended that basil oils with an estragole content of only 5% or less are used in aromatherapy, and a maximum use level of 2% is also recommended. One chemotype does exist which contains 5% or less, with linalool as the major component.

Comments: There are two other types of basil oil produced commercially. *Ocimum gratissimum* contains 21–90% eugenol, and a chemotype of *Ocimum basilicum* contains 45–60% methyl cinnamate.

Compare: Fennel, *Ravensara anisata*, tarragon

Bay (West Indian)

Botanical name: *Pimenta racemosa* (= *Pimenta acris*)

Family: Myrtaceae

Oil from: Leaves

Notable constituents: Eugenol 38–75%

Acute oral LD$_{50}$: 1.8 g/kg [419]

Hazards:
Mucous membrane irritant*
Hepatotoxic*
Inhibits blood clotting*

*Assumed from eugenol content

Cautions (oral): Alcoholism, anticoagulants, haemophilia, kidney disease, liver disease, paracetamol, prostatic cancer, SLE

Cautions (IRV): Do not use at more than 3% concentration on mucous membrane

Toxicity data & recommendations: Eugenol is a powerful inhibitor of platelet activity [370, 434]. Platelet activity is essential for blood clotting. This being the case, it would be prudent to avoid oral administration of West Indian bay oil in people whose blood clots only slowly, whether or not this is caused by hereditary haemophilia. Caution is recommended in anyone taking anticoagulant drugs, such as aspirin, heparin or warfarin.

Eugenol has been shown to cause liver damage in mice whose livers have been experimentally depleted of glutathione [400]. As in paracetamol poisoning, pretreatment with *N*-acetylcysteine prevents glutathione loss and hence cell death [401]. Although the oral doses of eugenol used were high (5 ml/kg) the data suggest that oral doses of eugenol-rich oils such as West Indian bay oil might be best avoided in those with impaired liver function and in anyone taking paracetamol (= acetaminophen).

West Indian bay oil was not irritating to human skin when tested at 10% [419].

Comments: Should not be confused with laurel leaf oil, which is also known as bay leaf.

Compare: Clove, laurel

Bergamot

Botanical name: *Citrus bergamia* (= *Citrus aurantium* subsp. *bergamia*)

Family: Rutaceae

Oil from: Fruit by expression

Notable constituents:

Linalyl acetate	36–45%
Limonene	28–32%
Linalool	11–22%
Bergapten	0.3–0.4%

Hazards:
Phototoxic (strong)
Photocarcinogenic

Contraindications (dermal): If applied to the skin at over maximum use level, skin must not be exposed to sunlight or sunbed rays for 12 hours.

Maximum use level: 0.4%

Toxicity data & recommendations: Phototoxicity is due to the presence of bergapten and some eight other furanocoumarins [216]. Several studies have shown that bergamot oil has carcinogenic properties when applied to mouse skin which is then irradiated with UV light [483, 493, 494]. This photocarcinogenicity is due to bergapten [483] (see p. 87). Bergamot oil, in the absence of UV light, is not carcinogenic; even low concentration sunscreens can completely inhibit bergapten-enhanced phototumorigenesis [483].

IFRA recommends that, for application to areas of skin exposed to sunshine, bergamot oil be limited to a maximum of 0.4% in the final product, except for bath preparations, soaps and other products which are washed off the skin [430]. See note on combinations of phototoxic oils (p. 114).

Comments: A treated oil, sometimes rectified by distillation, is obtainable as bergapten-free oil. This oil is also known as furanocoumarin-free bergamot, or bergamot FCF. Its odour is inferior to that of the untreated, cold pressed oil, but it is not phototoxic or photocarcinogenic.

Compare: Angelica, cumin, lime, rue

Betel

Botanical name: *Piper betle*

Family: Piperaceae

Oil from: Leaves

Notable constituents: Eugenol 28–90%

Hazards:
  Mucous membrane irritant* (moderate/strong)
  Hepatotoxic*
  Inhibits blood clotting*

*Assumed from eugenol content

Cautions (oral): Alcoholism, anticoagulants, haemophilia, kidney disease, liver disease, paracetamol, prostatic cancer, SLE

Cautions (IRV): Do not use at more than 3% concentration on mucous membrane

Toxicity data & recommendations: Eugenol is a powerful inhibitor of platelet activity [370, 434]. Platelet activity is essential for blood clotting. This being the case, it would be prudent to avoid oral administration of betel leaf oil in people whose blood clots only slowly, whether or not this is caused by hereditary haemophilia. Caution is recommended in anyone taking anticoagulant drugs, such as aspirin, heparin or warfarin.

Eugenol has been shown to cause liver damage in mice whose livers have been experimentally depleted of glutathione [400]. As in paracetamol poisoning, pretreatment with *N*-acetylcysteine prevents glutathione loss and hence cell death [401]. Although the oral doses of eugenol used were high (5 ml/kg) the data suggest that oral doses of eugenol-rich oils such as betel leaf might be best avoided in those with impaired liver function and in anyone taking paracetamol (= acetaminophen).

Comments: Betel leaf oils are now being offered commercially. There are numerous cultivars, many of which contain eugenol as a major component; these are the ones generally offered commercially. Some other cultivars contain up to 45% of safrole, and therefore should be regarded as potentially carcinogenic.

Compare: Clove

Birch (sweet)

Synonyms: Mahogany or cherry or southern birch

Botanical name: *Betula lenta*

Family: Ericaceae

Oil from: Bark

Notable constituents: Methyl salicylate 98%

Acute oral LD$_{50}$: 1.7 g/kg [425]

Hazards: Toxic (moderate/severe)

Contraindications: Should not be used in therapy, either internally or externally; do not use if taking anticoagulants

Toxicity data & recommendations: Numerous cases of methyl salicylate poisoning have been reported, with a 50–60% mortality rate; 4–8 ml is considered a lethal dose for a child [424]. Methyl salicylate could be three to five times more toxic in humans than in rodents (see p. 50). In the years 1926, 1928 and 1939–1943, 427 deaths occurred in the USA from methyl salicylate poisoning [237].

Common signs of methyl salicylate poisoning are: CNS excitation; rapid breathing; fever; high blood pressure; convulsions; coma. Death results from respiratory failure after a period of unconsciousness [424]; 0.5 ml of methyl salicylate is approximately equivalent to a dose of 21 aspirins.

Methyl salicylate can be absorbed transdermally in sufficient quantities to cause poisoning in humans [100]. Topically applied methyl salicylate can potentiate the anticoagulant effect of warfarin, causing side-effects such as internal haemorrhage [521]. A similar interaction is possible, but by no means certain, with other anticoagulants such as aspirin and heparin. Many liniments contain methyl salicylate.

Comments: There have been sufficient cases of poisoning by methyl salicylate or by oils containing it that it would be prudent to avoid use of this oil in aromatherapy. Virtually all 'sweet birch oils' are in fact synthetic methyl salicylate.

Compare: Wintergreen

Boldo

Botanical name: *Peumus boldus*

Family: Monimiaceae

Oil from: Dried leaves

Notable constituents:

Ascaridole	16%
Para-cymene	28%

Acute oral LD$_{50}$: 0.13 g/kg [426]

Hazards:
Toxic (severe)
Neurotoxic

Contraindications: Should not be used in therapy, either internally or externally

Toxicity data & recommendations: Boldo oil produces convulsions in rats at oral doses of 0.07 g/kg [426].

Comments: One of the most toxic essential oils. The toxicity of both boldo and wormseed oils is undoubtedly due to their content of ascaridole.

Compare: Wormseed

Buchu

Botanical name: *Barosma betulina* (= *Agathosma betulina*), *Barosma crenulata* (= *Agathosma crenulata*)

Family: Rutaceae

Oil from: Leaves

Notable constituents: [391]

	B. betulina	B. crenulata
d-pulegone	3%	50%
Iso-pulegone	4%	10%
Diosphenol	9%	1%
Ψ-diosphenol	8%	1%
Iso-menthone	43%	22%
Menthone	17%	6%

Acute oral LD_{50}: No data could be found

Hazards: Toxic—assumed from pulegone content (severe) (*B. crenulata*)

Contraindications (*B. betulina*) (all routes): Children up to 2 years of age

Cautions (*B. betulina*): Use with caution due to uncertain toxicity. Pregnancy.

Contraindications (*B. crenulata*): Should not be used in therapy, either internally or externally

Toxicity data & recommendations: *B. crenulata* is presumed to be toxic because of its content of pulegone, and is likely to be similarly hazardous to pennyroyal. Pulegone is toxic to the liver because it is metabolised to epoxides. It is especially toxic by mouth; acute oral LD_{50} is 0.5 g/kg in rats [424]. Both the UK and EC 'standard permitted proportion' of pulegone in food flavourings is 0.025 g/kg of food [455, 456].

Comments: Very few data exist on the biological effects of buchu oils. The *British Herbal Compendium* gives pregnancy as a contraindication for buchu leaves [121]. It would be prudent to avoid both types of buchu oil in aromatherapy, especially since confusion between the two types is likely.

Compare: Pennyroyal, plectranthus, savin

Cade

Synonym: Juniper tar

Botanical name: *Juniperus oxycedrus*

Family: Cupressaceae (= Coniferae)

Oil from: Destructive distillation of the wood

Notable constituents:
  Beta-cadinene (major component)
  Para-cresol
  Guaiacol
  Benzo[*a*]pyrene

Hazards:
  Carcinogenic (unrectified oil)
  Irritant moderate (rectified oil)

Contraindications (unrectified oil): Should not be used in therapy, either internally or externally

Cautions (dermal, rectified oil): Hypersensitive, diseased or damaged skin, all children under 2 years of age

Cautions (IRV, rectified oil): Do not use at more than 1% concentration on mucous membrane

Toxicity data & recommendations: Benzo[*a*]-pyrene, a polynuclear hydrocarbon, is a well-known carcinogen [158]. It is present in unrectified cade oil at 8,000 parts per billion (= 8 p.p.m.) and in rectified cade oil at up to 20 p.p.b.; this last amount is less than that found in some foodstuffs [462]. Cade oil produced a worrying level of (potentially carcinogenic) DNA adducts in the skin of psoriasis patients receiving cade oil therapy [463].

The RIFM monograph on rectified cade oil finds it to be non-toxic, non-irritant, and non-sensising [421]. A more recent study found that a 3% concentration of rectified cade oil produced two mild irritation reactions in 25 human subjects; the unrectified oil, tested at 3%, produced one reaction in the same group [462].

IFRA recommends that unrectified cade oil should not be used as a fragrance ingredient. IFRA does permit the use of cade oil rectified by fractional distillation, in order to remove the undesirable polynuclear hydrocarbons [430].

Comments: Cade is the name of the oil—the plant is known as 'prickly juniper'. Cade oil is generally prepared by dry distillation, without water or steam. This process, also known as destructive, or empyreumatic distillation, causes the wood to burn as distillation takes place. The process of burning causes the formation of the polynuclear hydrocarbons found in the oil.

Burning organic material, such as (juniper) wood or (tobacco) leaves generally produces carcinogenic polynuclear hydrocarbons. Benzo[a]pyrene is one of the carcinogenic compounds found in cigarette smoke. Since unrectified cade oil is potentially carcinogenic and the rectified oil is not, it is clearly important to distinguish between the two. See note on birch tar oil, page 203.

Compare: Bay leaf

Calamus

Synonym: Sweet flag

Botanical name: *Acorus calamus* var. *angustatus*

Family: Araceae

Oil from: Rhizome

Notable constituents: *Beta*-asarone 45–80%

Acute oral LD$_{50}$: 0.84 g/kg [423]

Hazards:
Toxic (severe)
Carcinogenic
Hepatotoxic

Contraindications: Should not be used in therapy, either internally or externally

Toxicity data & recommendations: Calamus oil acute toxicity presents as tremor and convulsions [338]; severe liver and kidney damage are found on investigation [414]. Chronically, small daily quantities of calamus oil produce growth retardation and increase the likelihood of developing duodenal-ulcer-related malignant tumours in rats after 59 weeks' oral dosing (500–5000 p.p.m.) [285]. Chronic oral dosing at 0.78 g/kg produces tremors and weight loss with macroscopic liver changes, slight myocardial degeneration, and growth depression [336]. *Acorus calamus* is one of 30 herbs listed as unsafe by the FDA [164].

Beta-asarone produced malignant liver tumours in rodents given four intraperitoneal injections; tumours were found on autopsy at 13 months [453]. Recent work on rat hepatocytes suggests that asarone's genotoxicity is due to a novel, but P$_{450}$-dependent activation mechanism [445].

Beta-asarone is banned in the USA as a pharmaceutical ingredient, while the *alpha* isomer is permitted [6]. Both the UK and EC 'standard permitted proportion' of *beta*-asarone in food flavourings is 0.0001 g/kg of food [455, 456]. IFRA recommends that *beta*-asarone should not be used as a fragrance ingredient, and that its level in consumer products containing calamus oil should not exceed 0.01% [430].

Comments: The essential oil described above is the tetraploid form, and is the type of calamus oil generally offered commercially, originating in India. There is also the diploid form found in North America and Siberia (*Acorus calamus* var. *americanus*), and the triploid form found in Eurasia (*Acorus calamus* var. *calamus*). The triploid form contains 8–19% *beta*-asarone, and 23–32% shyobunones. The *beta*-asarone content of this oil is sufficient to make it unsafe for use in aromatherapy.

The diploid form contains little or no detectable *beta*-asarone, and 13–45% shyobunones. The toxicology of this type of oil and its major components, shyobunones, is unknown, although there is no reason to suspect toxicity. If offered commercially, it would probably be safe to use in aromatherapy.

Compare: Basil, camphor (brown and yellow), sassafras, tarragon

Camphor (white)

Botanical name: *Cinnamomum camphora* (= *Laurus camphora*)

Family: Lauraceae

Oil from: Crude camphor, present in the tree

Note: Camphor is the name of a raw material, the name of its essential oil, and the name of a chemical component found in this, and other essential oils. Crude camphor contains around 50% of camphor the chemical.

Camphor oil is separated into four distinct 'essential oils' by fractional distillation. These are known as white, brown, yellow and blue camphor oils. None of these, therefore, can be classed as true essential oils.

Notable constituents:

Camphor	30–50%
Cineole	50%

Acute oral LD$_{50}$: 5.1 ml/kg [419]

Hazards:
Convulsant*
Neurotoxic*

*Assumed from camphor content

Contraindications (all routes): Epilepsy, fever, pregnancy, children under 2 years of age

Toxicity data & recommendations: Camphor the chemical is readily absorbed through skin and mucous membranes and is classified as very toxic [424]. Camphor minimum LD$_{50}$ is 1.7 g/kg in rats [377]. Camphor is thought to be considerably more toxic in humans than in rodents. There have been reports of instant collapse in infants following the local application of camphor to their nostrils [5]. One teaspoon of camphorated oil, equivalent to about 1 ml of camphor, was lethal to a 16-month-old child and to a 19-month-old child [244, 247]. The adult human lethal dose is probably 5–20 g [38]. See page 50 for cases of poisoning from camphorated oil. Camphor readily causes epileptiform convulsions if taken in sufficient quantity [269, 272] (see p. 67). White camphor oil is non-irritant and non-sensitising [419].

Comments: White camphor oil is the lowest boiling point fraction (lightest) and does not contain safrole, which is present in brown and yellow camphor oils (see below). It is the type normally sold as camphor oil for use in aromatherapy. 'Camphorated oil' is 20% of camphor in cottonseed oil.

Compare: Balsamite (camphor CT), camphor (brown and yellow), ho leaf (camphor/safrole CT), hyssop.

Camphor (brown and yellow)

Botanical name: *Cinnamomum camphora* (= *Laurus camphora*)

Family: Lauraceae

Oil from: Crude camphor, present in the tree

Note: See previous profile for comments on camphor and camphor oil.

Notable constituents:

Brown camphor oil: Safrole 80%

Yellow camphor oil:
Safrole   10–20%
Camphor   15–23%

Acute oral LD$_{50}$:
Brown camphor: 2.5 ml/kg [422]
Yellow camphor: 3.73 g/kg [421]

Hazards:
Carcinogenic*
Hepatotoxic*

*Assumed from safrole content

Contraindications: Should not be used in therapy, either internally or externally

Toxicity data & recommendations: The minimum lethal oral dose of safrole in the rabbit is 1.0 g/kg [38]. Safrole is highly toxic, especially on chronic dosing; 95% lethal to rats at 10 000 p.p.m. after 19 days' dosing [154]. Safrole has a hepatocarcinogenic effect when given subcutaneously or by stomach tube to male mice during infancy [153]. When administered orally, safrole is a low-level hepatic carcinogen in the rat [320]. Oesophageal tumours have also been reported in rats given safrole in the diet [319]. Sassafras oil and safrole, given in the diet, produced no tumours after 22 months, but a large percentage of the same animals showed initial tumour development at 24 months [343].

A metabolite of safrole, 1'-hydroxysafrole, is a much more potent hepatic carcinogen than safrole itself [32, 153, 286]. Other toxic metabolites may also be formed in vivo [59, 72, 287]. Safrole and 1'-hydroxysafrole can both cause hepatic cell enlargement in experimental animals [313]. There is some evidence that safrole may be able to activate a cancer-causing virus, the polyoma virus, at least in rats. It is not known whether safrole has a similar action in humans [344].

Safrole was banned as a food additive in the USA in 1961.

IFRA recommends that safrole should not be used as a fragrance ingredient. It recommends a maximum use level of 0.05% in fragrance compounds for safrole (equivalent to 0.01% in the final product), when safrole-containing essential oils are used [430]. Both the UK and EC 'standard permitted proportion' of safrole in food flavourings is 0.001 g/kg of food [455, 456].

Comments: Brown and yellow camphor oils are assumed carcinogenic, due to their high content of safrole. Neither of these oils is normally available commercially, as they are commonly used for the isolation of safrole or cineole. Brown camphor oil is sometimes sold as 'Chinese sassafras oil'.

Compare: Basil, calamus, camphor (white), ho leaf (camphor/safrole CT), sassafras, tarragon

Cangerana

Botanical name: *Cabralea cangerana* (= *Cabralea canjerana*)

Family: Meliaceae

Oil from: Wood

Notable constituents:
Caryophyllene (major component)
Safrole

Acute oral LD$_{50}$: No data could be found

Hazards: Carcinogenic?

Cautions: Use with caution due to uncertain toxicity

Toxicity data & recommendations: Cangerana oil contains an undetermined quantity of the carcinogenic compound safrole. See previous profile for toxicity data on safrole.

Comments: Cangerana oil is commercially available.

Compare: Camphor (brown and yellow), sassafras

Cassia

Synonyms: Chinese or false cinnamon

Botanical name: *Cinnamomum cassia* (= *Cinnamomum aromaticum*)

Family: Lauraceae

Oil from: Leaves and twigs

Notable constituents: Cinnamaldehyde 75–90%

Acute oral LD$_{50}$: 2.8 ml/kg [421]

Hazards:
 Dermal sensitiser (strong)
 Mucous membrane irritant (moderate)

Contraindications: Should not be used in therapy, either internally or externally

Toxicity data & recommendations: In a maximation test cassia oil, at a 4% dilution, produced positive reactions in 2 out of 25 test subjects [421]. IFRA recommends that cassia oil should not be used as a fragrance ingredient at a level over 1% in a fragrance compound (equivalent to 0.2% in the final product) due to its sensitising potential [430].

 Sensitisation is also seen with cinnamaldehyde, which some workers feel can cause contact dermatitis [75]. Skin sensitisation from cinnamaldehyde seems to be reduced by *d*-limonene and by eugenol. In 10 people who developed urticaria after cinnamaldehyde had been applied to the skin, 6 had a greatly diminished reaction when it was applied combined with eugenol [408]. Because of the glutathione-depleting action of cinnamaldehyde [475], oral cassia oil might be best avoided in alcoholism and liver disease and if taking paracetamol.

Comments: An oil which is such a strong skin sensitiser is unlikely to be well tolerated if taken internally, and it would be wise to avoid it altogether. If cassia were used externally, a maximum use level of 0.1% would be recommended. Oral use might be feasible, but only if the oil is well tolerated.

Compare: Cinnamon bark, oakmoss[A], treemoss[A], verbena[A]

Cinnamon bark

Botanical name: *Cinnamomum verum* (= *Cinnamomum zeylanicum*)

Family: Lauraceae

Oil from: Dried inner bark of young trees

Notable constituents:
 Cinnamaldehyde 55–75%
 Eugenol 5–18%

Acute oral LD$_{50}$: 3.4 ml/kg [421]

Hazards:
 Dermal irritant (moderate)
 Dermal sensitiser (strong)
 Mucous membrane irritant (moderate)

Contraindications: Should not be used in therapy, either internally or externally

Toxicity data & recommendations: In two sensitisation tests, cinnamon bark oil, diluted to 8%, produced positive reactions in 18 out of 25 and 20 out of 25 test subjects [421]. In another test, the lowest concentration causing positive reactions in patch tests was 0.01% [382].

 IFRA recommends that cinnamon bark oil should not be used as a fragrance ingredient at a level over 1% in a fragrance compound (equivalent to 0.2% in the final product) due to its sensitising potential [430]. As a recommended maximum use level, this percentage seems high.

 Sensitisation is also seen with cinnamaldehyde, which some workers feel can cause contact dermatitis [75]. Skin sensitisation from cinnamaldehyde seems to be reduced by *d*-limonene and by eugenol. In 10 people who developed urticaria after cinnamaldehyde had been applied to the skin, 6 had a greatly diminished reaction when it was applied combined with eugenol [408]. Cinnamon bark oil contains eugenol.

 Undiluted cinnamon oil (type unspecified) has caused severe burns in an 11-year-old boy after remaining in contact with the skin for 48 hours, after a vial broke in his trouser pocket [392]. Contact with the skin by undiluted cinnamon oil (type unspecified) is frequently associated with a burning sensation, and occasional blistering [393].

Because of the glutathione-depleting action of cinnamaldehyde [475], oral cinnamon bark might be best avoided in alcoholism and liver disease and if taking paracetamol.

In mice, cinnamon bark oil has been shown to cause variable changes in blood pressure, and to cause oedema after intraperitoneal dosing at 100 mg/kg [334]. This dosage level is very much higher than the amount which would be absorbed from any method of therapeutic application. Cinnamon oil (type unknown) caused serious, non-fatal poisoning in a 7-year-old boy, who drank 60 ml (see p. 51).

Comments: An oil which is such a strong skin sensitiser is unlikely to be well tolerated if taken internally, and it would be wise to avoid it altogether. If cinnamon bark were used externally, a maximum use level of 0.1% would be recommended. Oral use might be feasible, but only if the oil is well tolerated.

Cinnamon bark oil should not be confused with cinnamon leaf oil, which is eugenol-rich, and is not a sensitiser. Cassia oil is sometimes sold as cinnamon bark oil. There is an unusual safrole chemotype of *Cinnamomum verum*, which contains 11% of safrole in the bark oil, and 52% of safrole in the leaf oil [384]. These safrole-containing oils would be potentially carcinogenic.

Compare: Cassia, cinnamon leaf, costus, elecampane, oakmoss[A], treemoss[A], verbena

Cinnamon leaf

Botanical name: *Cinnamomum verum* (= *Cinnamomum zeylanicum*)

Family: Lauraceae

Oil from: Leaves

Notable constituents:

Eugenol	70–90%
Safrole	< 1%

Acute oral LD$_{50}$: 2.65 g/kg [421]

Hazards:
Dermal irritant (mild/moderate)
Mucous membrane irritant (moderate)
Hepatotoxic*
Inhibits blood clotting*

*Assumed from eugenol content

Cautions (oral): Alcoholism, anticoagulants, haemophilia, kidney disease, liver disease, paracetamol, prostatic cancer, SLE

Cautions (IRV): Do not use at more than 3% concentration on mucous membrane

Toxicity data & recommendations: Eugenol is a powerful inhibitor of platelet activity [370, 434]. Platelet activity is essential for blood clotting. This being the case, it would be prudent to avoid oral administration of cinnamon leaf oil in people whose blood clots only slowly, whether or not this is caused by hereditary haemophilia. Caution is recommended in anyone taking anticoagulant drugs, such as aspirin, heparin or warfarin.

Eugenol has been shown to cause liver damage in mice whose livers have been experimentally depleted of glutathione [400]. As in paracetamol poisoning, pretreatment with *N*-acetylcysteine prevents glutathione loss and hence cell death [401]. Although the oral doses of eugenol used were high (5 ml/kg) the data suggest that oral doses of eugenol-rich oils such as cinnamon leaf oil might be best avoided in those with impaired liver function and in anyone taking paracetamol (= acetaminophen).

Full-strength cinnamon leaf oil was moderately irritating to mouse skin, and strongly irritating to rabbit skin; tested at 10%, cinnamon leaf oil produced no irritation in a closed-patch test on human subjects; [421].

Comments: The level of safrole present in cinnamon leaf oil is not considered to represent a hazard in aromatherapy except in oral dosages.

Compare: Clove, cinnamon bark

Clove

Botanical name: *Syzygium aromaticum* (= *Eugenia caryophyllata* = *Eugenia aromatica*)

Family: Myrtaceae

Oil from: Dried flower buds, or leaves, or stems

Note: Because they are so similar in terms of chemical composition and toxicology, these three distinct essential oils are treated here as one.

Notable constituents:
Eugenol	70–95%
Iso-eugenol	0.14–0.23%

Acute oral LD$_{50}$
Buds: 2.65 g/kg [421]
Leaves: 1.37 g/kg [424]
Stems: 2.02 g/kg [421]

Hazards:
Dermal irritant (moderate)
Mucous membrane irritant (bud, moderate; leaf and stem, strong)
Hepatotoxic*
Inhibits blood clotting*

*Assumed from eugenol content

Cautions (oral): Alcoholism, anticoagulants, haemophilia, kidney disease, liver disease, paracetamol, prostatic cancer, SLE

Cautions (dermal): Hypersensitive, diseased or damaged skin, all children under 2 years of age

Cautions (IRV, leaf, stem): Do not use at more than 1% concentration on mucous membrane

Cautions (IRV, bud): Do not use at more than 3% concentration on mucous membrane

Toxicity data & recommendations: Eugenol is a powerful inhibitor of platelet activity [370, 434]. Platelet activity is essential for blood clotting. This being the case, it would be prudent to avoid oral administration of clove oil in people whose blood clots only slowly, whether or not this is caused by hereditary haemophilia. Caution is recommended in anyone taking anticoagulant drugs, such as aspirin, heparin or warfarin.

Eugenol has been shown to cause liver damage in mice whose livers have been experimentally depleted of glutathione [400]. As in paracetamol poisoning, pretreatment with *N*-acetylcysteine prevents glutathione loss and hence cell death [401]. Although the oral doses of eugenol used were high (5 ml/kg), the data suggest that oral doses of eugenol-rich oils such as clove oil might be best avoided in those with impaired liver function and in anyone taking paracetamol (= acetaminophen). See page 51 for detail on cases of human poisoning from clove oil ingestion.

Clove oil is a strong mucous membrane irritant. Tests using eugenol at 5% on the tongues of dogs for 5 minutes caused erythema and occasionally ulcers [421]. Spillage of clove oil on to the skin has caused transient irritation followed by apparent permanent anaesthesia and loss of the ability to sweat by the affected area. The skin remained sensitive to deep pressure only [403].

In closed-patch tests clove oil caused primary irritation in 2 out of 25 normal subjects when applied at 20%, and evoked no reaction when tested at 2% on 30 normal subjects [421]. *Iso*-eugenol is both an irritant and a sensitiser [421, 430]. IFRA recommends a maximum use level for *iso*-eugenol of 0.2% in products which will come into contact with the skin [430]. None of the three types of clove oil produced a positive reaction in maximation tests [421, 424].

Comments: Clove oil appears to be no more than a moderate irritant to human skin, and is very unlikely to cause sensitisation problems unless it is applied undiluted.

Compare: Cinnamon bark, pimento

Cornmint (dementholised)

Synonym: Japanese mint

Botanical name: *Mentha arvensis* var. *piperascens*

Family: Lamiaceae (= Labiatae)

Oil from: Leaves

Notable constituents:

l-menthol	35–50%
Menthone	15–30%
d-pulegone	0.2–5%

Acute oral LD$_{50}$: 1.24 g/kg [421]

Hazards:
Neurotoxic?
Mucous membrane irritant (moderate)

Contraindications (oral): G6PD deficiency

Contraindications (all routes): Cardiac fibrillation

Cautions (oral): Epilepsy, fever. Dose levels over 1 ml per 24 hours in an adult not recommended

Cautions (IRV): Do not use at more than 3% concentration on mucous membrane

Toxicity data & recommendations: A proprietary menthol-containing oil has been reported as producing incoordination, confusion and delirium when 5 ml of the product (35.5% peppermint oil) was inhaled over a long time period [141]. Pulegone causes histopathological changes in the white matter of the cerebellum above 80 mg/kg [329]. Menthone produces similar cerebellar changes, but the lowest dosage level used was 0.2 g/kg/day [327]. There is no indication that cornmint oil can produce these toxic effects when used externally in aromatherapy, but some caution is required regarding oral dosage.

There have been reports of apnoea and instant collapse in infants following the local application of menthol to their nostrils [5].

Menthol has been shown to provoke severe neonatal jaundice in babies with a deficiency of the enzyme glucose-6-phosphate dehydrogenase (G6PD). This is a fairly common inheritable enzyme deficiency, particularly in Chinese, West Africans, and in people of Mediterranean or Middle Eastern origin [109]. Usually, menthol is detoxified by a metabolic pathway involving G6PD. When babies deficient in this enzyme were given a menthol-containing dressing for their umbilical stumps, menthol was found to build up in their bodies [109]. Peppermint oil should be avoided orally by males with a deficiency of this enzyme. Such people can be recognised because they will characteristically have had abnormal blood reactions to at least one of the following drugs, or will have been advised to avoid them: antimalarials; sulphonamides (antimicrobial); chloramphenicol (antibiotic); streptomycin (antibiotic); aspirin.

Large doses of menthol (above 0.2 g/kg) can produce long-term change in the appearance of hepatocytes when given orally to rats [329]. Menthone caused some liver toxicity at high dosage levels (over 0.2 g/kg/day for 28 days) [327]. Other studies have concluded that menthone is not hepatotoxic [111, 117]. It is unlikely to cause problems at dosage levels used in aromatherapy.

Menthol has been shown to dilate capillaries in the nasal mucosa upon inhalation in some individuals; this effect correlates with its reported ability to dilate systemic blood vessels after intravenous administration [222]. Mentholated cigarettes and peppermint confectionery have been responsible for several instances of cardiac fibrillation in patients prone to the condition who are being maintained on quinidine, a stabiliser of heart rhythm [236]. Bradycardia (slowing of heartbeat) has been reported in a person addicted to menthol cigarettes [388]. It is recommended that cornmint oil should be avoided altogether in cases of cardiac fibrillation.

Cornmint oil is non-irritant, non-sensitising, and non-phototoxic [421]. Natural cornmint oil, before dementholisation, may present more problems than the dementholised oil, due to its higher menthol content.

Comments: Natural cornmint oil contains 70–90% of menthol, but this oil is rarely seen on the marketplace. The normal article of commerce is 'dementholised' by freezing, a process which removes about 50% of the menthol. Cornmint oil is a primary source for menthol. Cornmint oil is more toxic than peppermint oil, perhaps due to its higher pulegone content. Cornmint oil is sometimes used as an adulterant of, or even a substitute for, peppermint oil.

Compare: Peppermint

Costus

Botanical name: *Saussurea costus* (= *Saussurea lappa* = *Auklandia costus*)

Family: Asteraceae (= Compositae)

Oil from: Dried root

Notable constituents: Costuslactones (major components)

Acute oral LD$_{50}$: 3.4 g/kg [420]

Hazards: Dermal sensitiser (severe)

Contraindications: Should not be used in therapy, either internally or externally

Toxicity data & recommendations: In a maximation test at 4%, costus oil produced 25 sensitisation reactions in 25 volunteers; in a maximation test at 2%, it produced 16 sensitisation reactions out of 26 people tested [420]. Dermal sensitisation is due to the lactone content of the oil; costus oil is non-irritant and non-phototoxic [420].

IFRA recommends that costus oil should not be used as a fragrance ingredient, due to its sensitising potential, unless the particular costus oil being used has been shown not to have sensitising potential [430].

Comments: An oil which is such a powerful skin sensitiser is unlikely to be well tolerated if taken internally, and it would be wise to avoid it altogether.

Compare: Elecampane, verbena

Cumin

Botanical name: *Cuminum cyminum*

Family: Apiaceae (= Umbelliferae)

Oil from: Fruit (seeds)

Notable constituents: Cuminaldehyde 20–40%

Hazards: Phototoxic (strong)

Contraindications (dermal): If applied to the skin at over maximum use level, skin must not be exposed to sunlight or sunbed rays for 12 hours

Maximum use level: 0.4%

Toxicity data & recommendations: Distinct phototoxic effects have been reported for cumin oil, but none for cuminaldehyde [420]. IFRA recommends that, for application to areas of skin exposed to sunshine, cumin oil be limited to a maximum of 2% in fragrance compound (equivalent to 0.4% in the final product) except for bath preparations, soaps and other products which are washed off the skin [430]. See note on combinations of phototoxic oils (p. 114).

Compare: Angelica, bergamot, lime, rue

Elecampane

Synonym: Alantroot

Botanical name: *Inula helenium*

Family: Asteraceae (= Compositae)

Oil from: Dried roots

Notable constituents:

Alantolactone	52%
Iso-alantolactone	33% [518]

Acute oral LD$_{50}$: No data could be found

Hazards: Dermal sensitiser (severe)

Contraindications: Should not be used in therapy, either internally or externally

Toxicity data & recommendations: A maximation test, using elecampane oil at 4%, elicited 'extremely severe allergic reactions' in 23 out of 25 volunteers [422]. Dermal sensitisation is due to the lactone content of the oil.

IFRA recommends that elecampane oil should not be used as a fragrance ingredient, due to its sensitising potential [430].

Comments: An oil which is such a powerful skin sensitiser is unlikely to be well tolerated if taken internally, and it would be wise to avoid it altogether. Elecampane oil has been frequently employed as an adulterant of costus oil. Elecampane oil should not be confused with *Inula graveolens*; both oils have been referred to as 'inula oil'.

Compare: Costus, verbena

Elemi

Botanical name: *Canarium luzonicum* (= *Canarium commune*)

Family: Burseraceae

Oil from: Gum

Notable constituents:

d-limonene	27–54%
Elemicin	3–12%

Acute oral LD$_{50}$: 3.4 g/kg [422]

Hazards: Carcinogenic?

Cautions: Use with caution generally, due to uncertainty regarding carcinogenesis; oral dosage over 1 ml per 24 hours in an adult not recommended

Toxicity data & recommendations: Two studies have revealed a low to moderate level of genotoxicity for elemicin [122, 289]. Recent research found it to be significantly genotoxic [445]. One study found both elemicin and its metabolite, 1'-hydroxyelemicin to be non-carcinogenic [309]. Other work found that 1'-hydroxyelemicin possessed a weak, but statistically significant carcinogenic activity [453]. Elemicin is suspect because its chemical structure has much in common with safrole and estragole, which are carcinogenic. There is insufficient evidence to contraindicate elemi oil. Elemi oil would be safe to use at a maximum concentration of 1% if elemicin was carcinogenic. *d*-limonene is antitumoral, and prevents malignant tumours in rodents primed with cancer initiators [410, 411, 412].

Comments: It is feasible that the *d*-limonene content of elemi oil might counter the potentially carcinogenic effect of elemicin. *d*-limonene loses its antitumoral properties on ageing. The chances of elemi oil causing serious problems in aromatherapy seem remote.

Compare: Parsley leaf

Fennel (sweet and bitter)

Botanical name: *Foeniculum vulgare*

Family: Apiaceae (= Umbelliferae)

Oil from: Seeds or herb (bitter fennel)

Notable constituents:

Trans-anethole	52–86%
Fenchone	1–14%
Estragole	2–7%

Hazards:
  Oestrogen-like action (mild)
  Dermal sensitiser (mild?)
  Potentially carcinogenic*

*Assumed from estragole content

Contraindications (oral): Should not be taken in oral doses

Maximum use level: 1.5%

Toxicity data & recommendations: *Trans*-anethole and its derivatives are oestrogen-like in effect, as is anise oil [360, 361]. The effect is relatively weak, but oral dosing might be best avoided in endometriosis, prostatic hyperplasia and oestrogen-dependent cancers.

Trans-anethole shows a dose-dependent cytotoxicity to rat liver cells in culture, and causes glutathione depletion [452]. The mechanism of its action is unknown, but it is not due to a DNA-damaging effect. A primary metabolite of *trans*-anethole, anethole 1,2-epoxide, also depletes glutathione [504]. Because of this effect, anethole-rich oils such as fennel might be best avoided orally in alcoholism, liver disease, or if also taking paracetamol (= acetaminophen).

Estragole is potentially carcinogenic, depending on dosage (see pp. 94, 98). Due to the level of estragole present in fennel oil, a maximum use level of 1.5% is recommended. Oral dosing is not recommended.

Undiluted fennel oil was severely irritating to mouse skin, and moderately irritating to rabbit skin. It produced no irritation when tested at 4% on human subjects, and no sensitisation reactions in a maximation test at 4% [420]. It is believed that *trans*-anethole can enhance dermatitis.

Comments: *Cis*-anethole, which exists in synthetic anethole, is considerably more toxic than *trans*-anethole.

Compare: Anise, basil, star anise

Fig leaf^A

Botanical name: *Ficus carica*

Family: Urticaceae

Absolute from: Leaves

Notable constituents:
Methyl salicylate
Germacrene D
Furanocoumarins?

Hazards:
Phototoxic (severe)
Dermal sensitiser (moderate)
Dermal irritant (moderate)

Contraindications: Should not be used in therapy, either internally or externally

Toxicity data & recommendations: Fig leaf absolute produced strong phototoxic effects on hairless mice at a 12.5% dilution; a 0.001% dilution still produced phototoxic reactions in three out of six mice [426]. In two maximation tests, using fig leaf absolute at a 5% dilution on 25 and 28 volunteers, there were two sensitisation reactions in each group [426].
IFRA recommends that fig leaf absolute should not be used as a fragrance ingredient, due to its sensitising, and extreme phototoxic potential [430].

Comments: Fig leaf absolute is typical of materials which are produced for the fragrance industry, but are unlikely to be used in aromatherapy.

Compare: Verbena

Garlic

Botanical name: *Allium sativum*

Family: Alliaceae

Oil from: Bulb

Notable constituents:
Diallyl disulphide (major component)
Diallyl thiosulphinate (= allicin)
Diallyl trisulphide
Allylpropyl disulphide
Allyl disulphide
Methyl disulphide
Methyl allyl trisulphide

Hazards:
Inhibits iodine metabolism
Inhibits blood clotting
Dermal irritant (strong?)
Dermal sensitiser (moderate/severe?)

Contraindications (oral): Thyroid disease

Cautions (oral): Anticoagulants, haemophilia, kidney disease, liver disease, prostatic cancer, SLE

Contraindications (dermal): Hypersensitive, diseased or damaged skin, children under 2 years of age

Cautions (dermal): Use only with very great caution on skin

Cautions (IRV): Do not use on mucous membrane

Toxicity data & recommendations: Many of the volatile components of garlic, including methyl disulphide, allyl disulphide and propyl disulphide, inhibit iodine metabolism in rats at low concentrations [358, 359]. The effect is probably only significant for oral administration in low-iodine areas and where there is no use of iodised salt (rare). Allyl disulphide is the most active compound in this group of anti-thyroid essential oil constituents. There is no indication that *Allium* consumption by humans decreases thyroid function [363, 364].
Garlic oil demonstrates anti-platelet activity [363, 365, 366, 372]. Platelet activity is essential for blood clotting. The active component is believed

to be methyl allyl trisulphide [373]; tests with the synthetic compound showed it to be active at a very low concentration in human platelet-rich plasma [363]. Garlic oil should be avoided orally in people whose blood clots only slowly, whether or not this is caused by hereditary haemophilia. Ingesting four cloves of fresh garlic inhibited platelet aggregation [373]. Caution is recommended in anyone taking anticoagulant drugs, such as aspirin, heparin or warfarin. Garlic oil (18 mg) has a hypotensive effect in humans after oral administration [366].

Garlic and garlic oil have been frequently reported as causing allergic dermatitis [388]. Local irritant effects of garlic oil have occurred after application to the skin or rectal mucosa, with some severe cases reported in infants [388]. Three allergens in garlic oil have been identified: diallyl disulphide, allicin and allylpropyl disulphide [362].

Since safe levels for dermal application are not known, patch testing is recommended before application of garlic oil to the skin. Dilutions in the region of 0.1% or less might be reasonable for the purposes of testing.

Comments: Safe levels for application of garlic oil to the skin are not known. It is a definite potential allergen, and some irritancy is strongly suspected. Other *Allium* oils rich in sulphur compounds, such as leek, may present similar problems to garlic oil. Sulphur-rich oils such as garlic are not used in fragrances, hence the lack of RIFM data; they are more commonly administered internally than externally in aromatherapy.

Compare: Massoia, onion

Grapefruit

Botanical name: *Citrus paradisi*

Family: Rutaceae

Oil from: Peel by expression

Notable constituents:

d-limonene	90%
Bergapten	0.012–0.013%

Hazards: Phototoxic (moderate)

Contraindications (dermal): If applied to the skin at over maximum use level skin must not be exposed to sunlight or sunbed rays for 12 hours

Maximum use level: 4%

Toxicity data & recommendations: IFRA recommends that, for application to areas of skin exposed to sunshine, expressed grapefruit oil be limited to a maximum of 4% in the final product except for bath preparations, soaps and other products which are washed off the skin [430]. See note on combinations of phototoxic oils (p. 114). Expressed grapefruit oil is non-toxic, non-irritant and non-sensitising [420].

Comments: Phototoxicity is due to bergapten, and up to nine other furanocoumarins found in the oil [216]. It is unlikely that grapefruit oil would be used at over 4% in aromatherapy.

Compare: Lemon, orange

Ho leaf (camphor/safrole chemotype)

Synonym: Shiu leaf

Botanical name: *Cinnamomum camphora*

Family: Lauraceae

Oil from: Leaves

Notable constituents:

Camphor	42%
Linalool	15%
Safrole	5%

Acute oral LD$_{50}$: No data could be found

Hazards:
Carcinogenic*
Neurotoxic**

*Assumed from safrole content

**Assumed from camphor content

Contraindications (oral): Should not be taken in oral doses

Contraindications (all routes): Epilepsy, fever, pregnancy, children under 2 years of age

Maximum use level: 2%

Toxicity data & recommendations: Camphor the chemical is readily absorbed through the skin and mucous membranes and is classified as very toxic [424]. Camphor minimum LD$_{50}$ is 1.7 g/kg in rats [377]. Camphor is thought to be considerably more toxic in humans than in rodents. One teaspoonful of camphorated oil, equivalent to about 1 ml of camphor, was lethal to a 16-month-old child and a 19-month-old child [244, 247]. The adult human lethal dose is probably 5–20 g [38]. Camphor readily causes epileptiform convulsions if taken in sufficient quantity [269, 272] (see p. 67). Ho leaf (camphor/safrole CT) should therefore be used with caution even in non-oral doses.

The oral minimum lethal dose of safrole in the rabbit is 1.0 g/kg [38]. Safrole is highly toxic on chronic dosing; it was 95% lethal to rats at 10 000 p.p.m. after 19 days' dosing [154]. Safrole has a hepatocarcinogenic effect when given subcutaneously or by stomach tube to male mice during infancy [153]. When administered orally, safrole is a low-level hepatic carcinogen in the rat [320]. Oesophageal tumours have also been reported in rats given safrole in the diet [319]. Sassafras oil and safrole, given in the diet, produced no tumours after 22 months, but a large percentage of the same animals showed initial tumour development at 24 months [343].

A metabolite of safrole, 1′-hydroxysafrole, is a much more potent hepatic carcinogen than safrole itself [32, 153, 286]. Other toxic metabolites may also be formed in vivo [59, 72, 287]. Safrole and 1′-hydroxysafrole can both cause hepatic cell enlargement in experimental animals [313]. There is some evidence that safrole may be able to activate a cancer-causing virus, the polyoma virus, at least in rats. It is not known whether safrole has a similar action in humans [344].

Safrole was banned as a food additive in the USA in 1961. IFRA recommends that safrole should not be used as a fragrance ingredient. It recommends a maximum use level of 0.05% in fragrance compounds for safrole (equivalent to 0.01% in the final product) when safrole-containing essential oils are used [430]. Both the UK and EC 'standard permitted proportion' of safrole in food flavourings is 0.001 g/kg of food [455, 456].

Comments: This chemotype of ho leaf oil [512] combines the problems of a moderately high camphor content (compare camphor (white)) with those of a low safrole content (compare low-estragole basil). There is also a linalool chemotype of ho leaf which contains around 80% linalool [214]. To the best of our knowledge this chemotype contains little camphor and no safrole, but this is unconfirmed at the time of writing, If so, this oil would be completely safe to use in aromatherapy. Ho leaf oil is sometimes considered a more environmentally friendly alternative to rosewood oil. Ho wood oil also exists, consists primarily of linalool, and has none of the safety problems associated with ho leaf (camphor/safrole CT).

Compare: Basil, camphor (brown and yellow), camphor (white), sassafras

Horseradish

Botanical name: *Cochlearia armoracia* (= *Armoracia rusticana* = *Armoracia lapathifolia*)

Family: Cruciferae

Oil from: Root

Notable constituents:

Allyl isothiocyanate	50%
Phenylethyl isothiocyanate	45%

Acute oral LD$_{50}$: No data could be found

Hazards:
Toxic (severe)*
Dermal irritant (severe)*
Mucous membrane irritant (severe)*

*Assumed from allyl isothiocyanate content

Contraindications: Should not be used in therapy, either internally or externally

Toxicity data & recommendations: Allyl isothiocyanate is extremely toxic, and a violent irritant to mucous membranes and skin. Acute oral LD$_{50}$ of a 10% solution in corn oil was 0.34 g/kg in rats, producing jaundice [338]. IFRA recommends that allyl isothiocyanate should not be used as a fragrance ingredient or in fragrance ingredients [430].

Comments: One of the most hazardous of all essential oils. The composition of horseradish oil is similar to that of mustard oil.

Compare: Mustard

Hyssop

Botanical name: *Hyssopus officinalis*

Family: Lamiaceae (= Labiatae)

Oil from: Leaves and flowering tops

Notable constituents:

Pinocamphone	40%
Iso-pinocamphone	30%

Acute oral LD$_{50}$: 1.4 ml/kg [424]

Hazards:
Toxic (moderate)
Neurotoxic

Contraindications (all routes): Epilepsy, fever, pregnancy, children under 2 years of age

Contraindications (oral): Should not be taken in oral doses

Toxicity data & recommendations: Intra-peritoneal pinocamphone was found to be convulsant and lethal to rats above 0.05 ml/kg [234, 325]. Tests in rats found that convulsions appeared for hyssop oil at dose levels over 0.13 g/kg [234, 325]. The convulsions were of the same type as those seen in a 6-year-old girl given a few drops of hyssop oil by her mother, and in a 26-year-old woman who had taken 10 drops of hyssop oil on each of 2 consecutive days [234]. See page 67 for further details on convulsions caused by hyssop.

Due to its neurotoxicity it would be prudent to avoid hyssop oil during pregnancy.

Comments: In spite of its moderate oral LD$_{50}$ hyssop should be regarded as a dangerous oil because of its potential to cause convulsions, especially if taken orally. Chemotypes of hyssop oil with little or no pinocamphone or *iso*-pinocamphone (such as the oil from *Hyssopus officinalis* var. *decumbens*) may be relatively safe to use.

Compare: Camphor (white), rue

Indian Dill

Botanical name: *Anethum sowa*

Family: Apiaceae (= Umbelliferae)

Oil from: Seeds

Notable constituents: Dill apiol 20–30%

Acute oral LD$_{50}$: No data could be found

Hazards:
   Abortifacient*
   Hepatotoxic*
   Nephrotoxic*

*Assumed from apiol content

Contraindications (all routes): Pregnancy

Contraindications (oral): Should not be taken in oral doses

Cautions (all routes): Kidney disease, liver disease

Toxicity data & recommendations: Both apiol and various preparations of parsley have been used for many years to procure illegal abortion in Italy [500, 502, 503]. The lowest daily dose of apiol which induced abortion was 0.9 g, taken for 8 consecutive days. This daily dose is approximately equivalent to 5 ml of Indian dill oil. The inevitable conclusion is that Indian dill seed oil presents a very high risk of abortion if taken in oral doses, and that external use would also seem inadvisable in pregnancy.

There are several recorded cases of fatal poisoning from apiol ingestion [501, 502, 503]. The lowest total dose of apiol causing death is 6.3 g (2.1 g/day for 3 days) [530]; the lowest fatal daily dose is 0.77 g, which was taken for 14 days [533]; the lowest single fatal dose is 8 g [500]. At least 19 g has been survived [530], but if 1 or 2 g of apiol can be fatal to an adult, it would seem prudent to completely avoid oral dosages of Indian dill seed oil.

Common effects of apiol poisoning are fever, severe abdominal pain, vaginal bleeding, vomiting and diarrhoea [501, 530]. Post-mortem examination invariably reveals considerable damage to both liver and kidney tissue, often with gastrointestinal inflammation and sometimes damage to heart tissue [501, 503, 533].

Comments: Indian dill seed oil should not be confused with European dill seed oil (*Anethum graveolens*) which contains a maximum of 1% apiol, and is not toxic.

Compare: Parsley leaf, parsleyseed

Juniper

Botanical name: *Juniperus communis*

Family: Cupressaceae (= Coniferae)

Oil from: Berries

Notable constituents:

Alpha-pinene	40%
Myrcene	12%
Beta-pinene	7%
Terpinen-4-ol	6%
Sabinene	5%
Limonene	3%

Acute oral LD$_{50}$: >5 g/kg [422]

Hazards: None

Contraindications: None

Toxicity data & recommendations: Juniper is the only profile included which the authors do not regard as hazardous. It is included because other sources have frequently flagged juniper as being contraindicated in both pregnancy and kidney disease [9, 389, 394, 395, 397, 398, 446]. A vigorous attempt was made to trace the source of these contraindications. There is evidence that juniper berries are abortifacient, but there is no evidence that juniper oil is responsible for this effect. No evidence could be found to support the contraindication of juniper oil in kidney disease.

Most of the above referenced sources refer to 'juniper' generically, without specifying a particular preparation. None of them refer to what seems to be the only scientific basis for the abortifacient activity of juniper. Ethanolic and acetone extracts of juniper berries have a significant antifertility effect in rats [464, 465]. An ethanolic extract of juniper berries was found to demonstrate both an early and a late abortifacient activity in rats [464].

An ethanolic extract would contain some essential oil, but there is no evidence that the essential oil is responsible for these effects. Since the oil constitutes only some 1.5% of the raw material, and since all the major components of the essential oil are apparently non-toxic (see Chemical Index), it seems inconceivable that juniper oil could be responsible for the reproductive toxicity noted above.

Nutmeg oil, which is apparently non-abortifacient, contains 10–25% *alpha*-pinene, 2-3% myrcene, 12% *beta*-pinene, 5–6% terpinen-4-ol, 15–35% sabinene and 3–4% limonene. These same components constitute some 75% of juniper oil. The administration of up to 0.56 g/kg of nutmeg oil to pregnant rodents for 10 consecutive days had no effect on nidation or on maternal or fetal survival; no teratogenic effect was observed [NTIS 1972, cited in 422]. Myrcene showed no reproductive toxicity when tested on pregnant rats at 250 mg/kg, which is equivalent to a human dose of 135 g of juniper oil [318].

Comments: There are two likely reasons why juniper oil acquired a 'tainted' reputation, which has since been quoted and re-quoted. Firstly, there has undoubtedly been some confusion between juniper (*Juniperus communis*) and savin (*Juniperus sabina*).

One research paper, published in 1928, has the sub-heading: 'Emmenagogue Oils (Pennyroyal, Tansy and Juniper)'. In the body of the text it later states: 'The popular idea has always been that pennyroyal, tansy, savin and other oils produce abortion.' [175]. It is very obvious that savin and juniper oils have been confused. Any such confusion would readily explain why juniper oil might have been thought of as being dangerous in pregnancy, since savin certainly is so.

Secondly, if juniper berries are abortifacient, and if the component responsible for this is unknown, then suspicion could naturally fall on the essential oil. Gin has a reputation as an abortifacient, but again juniper oil is very unlikely to be responsible for any such effects, since the average maximum concentration of juniper oil in alcoholic beverages is only 0.006% [156].

The authors believe, therefore, that there is no reason to regard juniper oil as being hazardous in any way.

Khella

Botanical name: *Ammi visnaga*

Family: Apiaceae (= Umbelliferae)

Oil from: Seeds

Notable constituents:
Borneol
Khellin   1%
Furanocoumarins

Acute oral LD$_{50}$: No data could be found

Hazards:
Toxic?
Phototoxic?
Dermal sensitiser?

Cautions (all routes): Use with caution generally, sensitisation possible

Toxicity data & recommendations: Khella oil contains furanocoumarins but of unknown quantity [34]. It does not appear to have been tested for phototoxicity, but there is a clear risk. The oil also contains lactones, and is suspected of sensitisation potential [34].

In Australia, khella oil is classed as a poison, due to its khellin content which, however, is very low. The oral LD$_{50}$ in rats of khellin is 0.08 g/kg.

Comments: It it unlikely that the very low khellin content of khella oil would be sufficient to render the oil itself toxic. Khella oil is commercially available, and is used occasionally in aromatherapy. It should be used with caution until further data become available.

Compare: Massoia, opopanax, rue

Lanyana

Synonym: African wormwood

Botanical name: *Artemisia afra*

Family: Asteraceae (= Compositae)

Oil from: Flowering herb

Notable constituents:
Alpha-thujone   52%
Beta-thujone   13%
Camphor   7%
Artemisia ketone   2%

Acute oral LD$_{50}$: No data could be found

Hazards:
Toxic* (severe)
Neurotoxic*

*Assumed from thujone content

Contraindications: Should not be used in therapy, either internally or externally

Toxicity data & recommendations: Thujone acute oral LD$_{50}$ is 0.21 g/kg; chronic administration leads to fatty degeneration of the liver [354]. Intraperitoneal thujone was found to be convulsant and lethal to rats above 0.2 ml/kg [234, 325]. Both the UK and EC 'standard permitted proportion' of *alpha*- and/or *beta*-thujone in food flavourings is 0.0005 g/kg of food [455, 456]. No toxicity data on lanyana oil could be found. However, considering its content of thujone, this essential oil is probably too toxic to be safely used in aromatherapy.

Comments: The compositional profile for lanyana is typical for the commercially available oil. Variations in composition are possible, but all seem equally toxic, with thujone, camphor or artemisia ketone present in quantity. 'Lanyana' should not be confused with 'lantana'.

Compare: Armoise, *Artemisia arborescens*, sage (Dalmatian), tansy, thuja, western red cedar, wormwood

Laurel

Synonym: Bay leaf

Botanical name: *Laurus nobilis*

Family: Lauraceae

Oil from: Leaves

Notable constituents:

1,8-cineole	30–36%
Linalool	6%
Phenols	10%
Lactones	

Hazards:
Dermal sensitiser (variable)
Mucous membrane irritant (moderate)

Cautions (dermal): Hypersensitive, diseased or damaged skin, all children under 2 years of age. Some laurel leaf oils may cause skin sensitisation

Contraindications (IRV): Do not use at more than 3% on mucous membranes

Toxicity data & recommendations: There are many reports of contact dermatitis or allergy from laurel leaf oil [83, 388, 460, 461]. The RIFM monograph on laurel leaf oil reports that three different samples produced no sensitisation reactions when tested on 25, 25 and 49 volunteers [422]. Some laurel leaf oils are thought to contain small amounts of costunolide [388].

Comments: It is not clear from the literature whether or not laurel leaf oil is likely to cause sensitisation problems. This may be because some laurel oils contain sensitising agents and others do not. Laurel leaf oil should not be confused with the oil from the berries of *Pimenta racemosa*, West Indian bay oil, used in 'bay rum', and also known as bay leaf oil.

Compare: Bay (W. Indian), cade

Lavandin

Botanical name: *Lavandula × intermedia*

Family: Lamiaceae (= Labiatae)

Oil from: Flowering tops

Notable constituents:

Linalool	20–60%
Linalyl acetate	15–50%
Camphor	5–15%
1,8-cineole	5–25%

Acute oral LD$_{50}$: 5.0 g/kg [422]

Hazards: Neurotoxic—assumed from camphor content (mild)

Contraindications (oral): Pregnancy

Cautions (oral): Epilepsy, fever

Toxicity data & recommendations: Camphor readily causes epileptiform convulsions if taken in sufficient quantity [269, 272]. Lavandin oil should therefore be used with caution in oral doses.

Comments: Should be used with caution, due to camphor content, which is significantly higher than that of true lavender oil (<1%).

Compare: Camphor, *Lavandula stoechas*, rosemary, spike lavender, yarrow (camphor CT)

Lavandula stoechas

Botanical name: *Lavandula stoechas*

Family: Lamiaceae (= Labiatae)

Oil from: Flowering tops

Notable constituents:
Camphor	15–30%
Fenchone	45–50%

Hazards: Neurotoxic—assumed from camphor content (mild)

Contraindications (oral): Epilepsy, fever, pregnancy

Cautions (dermal, IRV): Epilepsy, fever, pregnancy

Toxicity data & recommendations: Camphor readily causes epileptiform convulsions if taken in sufficient quantity [269, 272]. *Lavandula stoechas* oil should therefore be used with caution, especially in oral doses.

Comments: Camphor content is significantly higher than that of true lavender oil (< 1%).

Compare: Camphor, lavandin, rosemary, spike lavender, yarrow (camphor CT)

Lavender cotton

Synonyms: Santolina, cotton lavender

Botanical name: *Santolina chamaecyparissus*

Family: Asteraceae (= Compositae)

Oil from: Flowering tops

Notable constituents:
Artemisia ketone	10–45%
Longiverbenone	9–17%

Acute oral LD$_{50}$: No data could be found

Hazards:
Toxic?
Neurotoxic?

Contraindications (oral): Should not be taken in oral doses

Cautions: Use with caution generally, due to uncertain toxicity: use with caution in epilepsy and pregnancy

Toxicity data & recommendations: It is historically assumed, from anecdotal material, that essential oils rich in artemisia ketone are toxic. However, no toxicity data for this compound could be found. In the circumstances it would be prudent to only use lavender cotton oil with extreme caution.

Comments: The majority of artemisia oils are thujone-rich, and consequently present some degree of toxicity (see armoise, *Artemisia arborescens*, lanyana, wormwood). Lavender cotton oil is occasionally produced on a small scale.

Compare: Annual wormwood

Lemon

Botanical name: *Citrus limonum*

Family: Rutaceae

Oil from: Peel by expression

Notable constituents:

d-limonene	70%
Furanocoumarins	2%

Hazards:
Phototoxic (moderate)
Photocarcinogenic?

Contraindications (dermal): If applied to the skin at over maximum use level, skin must not be exposed to sunlight or sunbed rays for 12 hours

Maximum use level: 2%

Toxicity data & recommendations: Phototoxicity is due to bergapten, and up to ten other furanocoumarins [216]. Bergapten is photocarcinogenic [483] (see p. 87). IFRA recommends that, for application to areas of skin exposed to sunshine, expressed lemon oil be limited to a maximum of 2% in the final product except for bath preparations, soaps and other products which are washed off the skin [430]. See note on combinations of phototoxic oils (p. 114).

Comments: Lemon oil is also produced by distillation, notably for particular flavour applications, such as soluble essences for lemonades. Distilled lemon oil is non-phototoxic [420]. It has an inferior odour to that of expressed lemon oil.

Compare: Grapefruit, orange

Lemongrass

Botanical name: *Cymbopogon citratus* (West Indian lemongrass), *Cymbopogon flexuosus* (East Indian lemongrass)

Family: Poaceae (= Gramineae)

Oil from: Grass

Notable constituents:

Citral	75%
Limonene	5%

Acute oral LD$_{50}$: >5 g/kg [422]

Hazards: Hormone-like action (mild—assumed from citral content)

Contraindications (oral): Glaucoma

Cautions (all routes): Prostatic hyperplasia

Cautions (dermal): Hypersensitive, diseased or damaged skin, children under 2 years of age

Toxicity data & recommendations: Citral can cause a rise in ocular tension, which would be dangerous in cases of glaucoma. In tests a very low daily oral dose (2–5 µg) produced an increase in ocular pressure within 2 weeks in monkeys [348]. It would be prudent, therefore, to avoid oral use of lemongrass oil in cases of glaucoma.

In rats, 185 mg/kg/day for 3 months produced benign prostatic hyperplasia [507] which may be testosterone-dependent [508]. Other studies have suggested that the effect occurs via competition with steroid receptors and that citral has an oestrogenic effect in causing vaginal hyperplasia in rats [510]. A receptor-mediated mechanism may also be responsible for citral's reported ability to cause sebaceous gland proliferation [509]. In all the above studies citral was applied topically, in amounts equivalent to 10 ml in a human. This is at least 15 times the amount used topically in aromatherapy, in an essential oil with 75% citral, but is very close to an oral dosage level. It is not known whether citral-rich oils such as lemongrass will have a hormonal effect as used in aromatherapy nor, if they do, whether that effect will be oestrogenic or androgenic.

It has been reported that citral can impair reproductive performance in female rats by reducing the number of normal ovarian follicles [199]. The effect has been seen only after a series of intraperitoneal injections at a dose of 0.3 g/kg. This is equivalent to injecting around 25 ml of lemongrass oil into the abdomen every 4–5 days for 60 days. Clearly it has no bearing on the use of essential oils in aromatherapy.

Citral is able to induce sensitisation reactions, and this effect can be reduced by the co-presence of *d*-limonene [104]. *d*-limonene is present in lemongrass. Both East Indian and West Indian lemongrass oils were found to be non-irritant, non-sensitising and non-phototoxic [422].

Comments: Here we see one of the few instances in which the toxic effect of a major component (in this case sensitisation) does not carry over to the essential oil. Sensitisation reactions from lemongrass oil are still possible, however, in rare instances.

Compare: May chang, melissa

Lime

Botanical name: *Citrus aurantifolia* (= *Citrus latifolia*)

Family: Rutaceae

Oil from: Peel by expression

Notable constituents:

d-limonene	42–64%
Bergapten	0.1–0.3%

Hazards:
Phototoxic (strong)
Photocarcinogenic?

Contraindications (dermal): If applied to the skin at over maximum use level, skin must not be exposed to sunlight or sunbed rays for 12 hours

Maximum use level: 0.7%

Toxicity data & recommendations: Phototoxicity is due to bergapten, and up to ten other furanocoumarins [216]. Bergapten is photo-carcinogenic [483] (see p. 87). IFRA recommends that, for application to areas of skin exposed to sunshine, lime oil be limited to a maximum of 0.7% in the final product except for bath preparations, soaps and other products which are washed off the skin [430]. See note on combinations of phototoxic oils (p. 114).

Comments: Distilled lime oil is not phototoxic [420]. It has an inferior odour to that of expressed lime oil.

Compare: Angelica, bergamot, cumin, rue

Mace

Botanical name: *Myristica fragrans*

Family: Myristicaceae

Oil from: Aril, or pericarp of fruit

Notable constituents:

Alpha-pinene	15–27%
Myrcene	12–21%
Myristicin	1.5–3.8%
Elemicin	0.2–2%
Safrole	0.2–2%

Acute oral LD$_{50}$: 3.64 g/kg [425]

Hazards: Psychotropic?

Contraindications (oral): Should not be taken in oral doses

Toxicity data & recommendations: Psychotropic effects have been reported for nutmeg, and myristicin and elemicin are thought to be responsible. However, the data suggest that other synergistic elements may need to be present for a psychotropic effect to take place (see pp. 71, 72). There is some suspicion that elemicin might be carcinogenic (see elemi profile). Myristicin shows high activity as an inducer of glutathione *S*-transferase (this correlates with tumour inhibition) [466] and it inhibits benzo[*a*]pyrene-induced tumours in mice [467].

Myristicin has been shown to be an inhibitor of monoamine oxidase (MAO) in rodents [367]. Nutmeg oil was less potent an inhibitor of MAO than myristicin [367]. MAO inhibitors should not be given in conjunction with pethidine. 'Very severe reactions, including coma, severe respiratory depression, cyanosis and hypotension have occurred in patients receiving monoamine oxidase inhibitors and given pethidine' [5]. Oral dosages of mace oil, in conjunction with pethidine, would be inadvisable. The safety of non-oral dosages is uncertain.

Comments: It would be reasonable to assume that findings for nutmeg oil would also apply to mace oil, since the two are so close, both chemically and botanically. It is feasible that the myristicin in mace oil might counter the potentially carcinogenic effect of safrole (and, possibly, elemicin). The level of safrole in mace oil is not considered to present a hazard in aromatherapy, except in oral dosages. A maximum use level of 5% would be appropriate but is not necessary, since this is the assumed maximum concentration for external use.

Compare: Elemi, nutmeg, parsley leaf, parsleyseed

Massoia

Botanical name: *Cryptocaria massoia* (= *Cryptocarya massoy* = *Massoia aromatica*)

Family: Lauraceae

Oil from: Bark

Notable constituents: Massoia lactone 68%

Hazards:
Dermal irritant* (strong)
Mucous membrane irritant (strong)

*Assumed from massoia lactone content

Contraindications (dermal): Hypersensitive, diseased or damaged skin, children under 2 years of age

Contraindications (IRV): Do not use on mucous membrane

Cautions: Might be best avoided altogether until further data become available, due to probable skin irritancy, and possible sensitisation

Toxicity data & recommendations: There is no RIFM monograph, nor an IFRA guideline for massoia, which has been suspected of irritation, sensitisation and phototoxicity. In a private communication from RIFM, the authors were informed that, in private tests, 'Massoia lactone was found to be so irritating that the standard tests for sensitisation and phototoxicity could not be conducted.' One company which sells a massoia CO_2 extract recommends a maximum use level of 0.01%.

Comments: Although no published data can be referenced, the authors are confident that the information given here is sound. If massoia lactone, the major component, is such a strong skin irritant, the essential oil will almost certainly cause similar problems. Massoia oil may be a sensitiser, similar to other lactone-rich oils such as costus and elecampane.

Compare: Cade, garlic

May chang

Synonym: Litsea cubeba

Botanical name: *Litsea cubeba*

Family: Lauraceae

Oil from: Fruit

Notable constituents:
Citral	75%
Limonene	5%

Acute oral LD$_{50}$: >5 g/kg [426]

Hazards: Hormone-like action (mild—assumed from citral content)

Contraindications (oral): Glaucoma

Cautions (all routes): Prostatic hyperplasia

Cautions (dermal): Hypersensitive, diseased or damaged skin, children under 2 years of age

Toxicity data & recommendations: Citral can cause a rise in ocular tension, which would not be good in cases of glaucoma. In tests a very low daily oral dose (2–5 µg) produced an increase in ocular pressure within 2 weeks in monkeys [348]. It would be prudent, therefore, to avoid oral use of may chang oil in cases of glaucoma.

In rats, 185 mg/kg/day for 3 months produced benign prostatic hyperplasia [507] which may be testosterone-dependent [508]. Other studies have suggested that the effect occurs via competition with steroid receptors and that citral has an oestrogenic effect in causing vaginal hyperplasia in rats [510]. A receptor-mediated mechanism may also be responsible for citral's reported ability to cause sebaceous gland proliferation [509]. In all the above studies citral was applied topically, in amounts equivalent to 10 ml in a human. This is at least 15 times the amount used topically in aromatherapy, in an essential oil with 75% citral, but is very close to an oral dosage level. It is not known whether citral-rich oils such as may chang will have a hormonal effect as used in aromatherapy nor, if they do, whether that effect will be oestrogenic or androgenic.

It has been reported that citral can impair reproductive performance in female rats by reducing the number of normal ovarian follicles [199]. The effect has been seen only after a series of intraperitoneal injections at a dose of 0.3 g/kg. This is equivalent to injecting around 25 ml of may chang oil into the abdomen every 4–5 days for 60 days. Clearly it has no bearing on the use of essential oils in aromatherapy.

Citral is able to induce sensitisation reactions, and this effect can be reduced by the co-presence of *d*-limonene [104]. *d*-limonene is present in may chang oil. May chang oil was found to be non-irritant and non-phototoxic; skin sensitisation reactions to a 2% concentration of may chang oil were reported in 3 out of 200 people with dermatitis [426].

Comments: As with lemongrass, the majority of people do not react allergically to may chang oil, but occasional reactions are possible.

Compare: Lemongrass, melissa

Melaleuca bracteata

Botanical name: *Melaleuca bracteata*

Family: Myrtaceae

Oil from: Leaves

Notable constituents: Methyleugenol (major component)

Acute oral LD$_{50}$: No data could be found

Hazards: Carcinogenic*

*Assumed from methyleugenol content

Contraindications: Should not be used in therapy, either internally or externally

Toxicity data & recommendations: Methyleugenol is genotoxic [22, 289, 291, 479]; it comes into the same category as safrole and estragole, and is carcinogenic in rodents [309, 310]. Neither methyleugenol nor *Melaleuca bracteata* oil are restricted by any agencies, but methyleugenol is clearly carcinogenic in rodents, and so *Melaleuca bracteata* oil should be avoided, at least until there is good evidence that it is non-carcinogenic.

Comments: This oil should not be confused with that from *Melaleuca alternifolia*, or tea tree, which contains no methyleugenol.

Compare: Snakeroot

Melissa

Botanical name: *Melissa officinalis*

Family: Lamiaceae (= Labiatae)

Oil from: Leaves

Notable constituents:

Citral	35–55%
Citronellal	4–39%

Hazards: Hormone-like action (mild—assumed from citral content)

Contraindications (oral): Glaucoma

Cautions (all routes): Prostatic hyperplasia

Cautions (dermal): Hypersensitive, diseased or damaged skin, children under 2 years of age

Toxicity data & recommendations: Citral can cause a rise in ocular tension, which would be dangerous in cases of glaucoma. In tests a very low daily oral dose (2–5 µg) produced an increase in ocular pressure within 2 weeks in monkeys [348]. It would be prudent, therefore, to avoid oral use of melissa oil in cases of glaucoma.

In rats, 185 mg/kg/day for 3 months produced benign prostatic hyperplasia [507] which may be testosterone-dependent [508]. Other studies have suggested that the effect occurs via competition with steroid receptors and that citral has an oestrogenic effect in causing vaginal hyperplasia in rats [510]. A receptor-mediated mechanism may also be responsible for citral's reported ability to cause sebaceous gland proliferation [509]. In all the above studies citral was applied topically, in amounts equivalent to 10 ml in a human. This is at least 15 times the amount used topically in aromatherapy, in an essential oil with 75% citral, but is very close to an oral dosage level. It is not known whether citral-rich oils such as melissa will have a hormonal effect as used in aromatherapy nor, if they do, whether that effect will be oestrogenic or androgenic.

It has been reported that citral can impair reproductive performance in female rats by reducing the number of normal ovarian follicles [199]. The effect has been seen only after a series of intraperitoneal injections at a dose of 0.3 g/kg.

This very high dose has no bearing on the use of essential oils in aromatherapy.

Citral is able to induce sensitisation reactions, and this effect can be reduced by the co-presence of *d*-limonene [104]. However, *d*-limonene is present only in trace amounts in melissa oil. Citronellal has also been a cause of occasional sensitisation [421].

Comments: As with lemongrass and may chang, occasional skin sensitisation reactions are possible with melissa.

Compare: Lemongrass, may chang

Mustard

Botanical name: *Brassica nigra* (= *Brassica sinapioides* = *Sinapis nigra*)

Family: Cruciferae

Oil from: Black mustard seeds

Notable constituents: Allyl isothiocyanate 99%

Acute oral LD$_{50}$: No data could be found

Hazards:
   Toxic (severe)*
   Dermal irritant (severe)
   Mucous membrane irritant (severe)

   *Assumed from allyl isothiocyanate content

Contraindications: Should not be used in therapy, either internally or externally

Toxicity data & recommendations: Inhalation of mustard oil produces extremely unpleasant sensations in the head and greatly irritates the eyes and mucous membranes of the nose and respiratory system [414]. On the skin, mustard oil produces almost instant blistering [156].
   Allyl isothiocyanate is extremely toxic, and a violent irritant to mucous membranes and skin. Acute oral LD$_{50}$ of a 10% solution in corn oil was 0.34 g/kg in rats, producing jaundice [338]. IFRA recommends that it should not be used as a fragrance ingredient or in fragrance ingredients [430].

Comments: Mustard is one of the most hazardous essential oils.
   The essential oil is not present in the free state in the seed or powdered seed, but in the form of glycosides, so preparations made from these by mechanical means (i.e. pressing) do not contain allyl isothiocyanate. The essential oil is only formed after fermentation, then distillation. When the mustard comes into contact with water, this hydrolyses a glycoside contained within the seeds' cells under the influence of an enzyme which is also present. White mustard seeds do not produce any allyl isothiocyanate under these conditions [155].

Compare: Almond (bitter), horseradish

Nutmeg

Botanical name: *Myristica fragrans*

Family: Myristicaceae

Oil from: Kernel

Note: There are two basic types of nutmeg oil, with important chemotype differences.

Notable constituents:

	West Indian	East Indian
Sabinene	42–50.7%	15.4–36.3%
Alpha-pinene	10.6–13.2%	18–26.5%
Myristicin	0.5–0.9%	3.3–14%
Elemicin	1.2–1.4%	0.1–4.6%
Safrole	0.1–0.2%	0.6–3.3%
Methyleugenol	0.1–0.2%	tr.–1.2%

Acute oral LD$_{50}$: 2.6–6.0 g /kg [422]

Hazards:
   Psychotropic
   Potentially carcinogenic*

*Assumed from safrole and methylengenol content

Contraindications (oral, East Indian): Should not be taken in oral doses

Cautions (oral, West Indian): Dose levels over 1 ml per 24 hours in an adult not recommended

Maximum use level (East Indian): 2.25%

Toxicity data & recommendations: Psychotropic effects have been reported for whole nutmeg, and myristicin and elemicin are thought to be responsible. However, the data suggest that other synergistic elements may need to be present for a psychotropic effect to take place (see pp. 71,72). There is very little information concerning psychotropic effects for the essential oil. One of us (RT) has ingested 1 ml of nutmeg oil with no noticeable effect, and 1.5 ml with a definite, moderately strong psychotropic effect. Three other individuals each ingested 1.5 ml of nutmeg oil, and two experienced psychotropic effects (see p. 71). See page 70 for detail on cases of intoxication.
   Myristicin has been shown to be an inhibitor of monoamine oxidase (MAO) in rodents [367]. Nutmeg oil was less potent an inhibitor of MAO than myristicin [367]. MAO inhibitors should not be given in conjunction with pethidine. 'Very

severe reactions, including coma, severe respiratory depression, cyanosis and hypotension have occurred in patients receiving monoamine oxidase inhibitors and given pethidine' [5]. Oral dosages of nutmeg oil, in conjunction with pethidine, would be inadvisable. The safety of non-oral dosages is uncertain.

The administration of up to 0.56 g/kg of nutmeg oil to pregnant rodents for 10 consecutive days had no effect on nidation or on maternal or fetal survival; no teratogenic effect was observed [NTIS 1972, cited in 422].

A 30-year-old woman, at 30 weeks of gestation, ingested several cookies made with 7 g of ground nutmeg, instead of the recommended eighth of a teaspoon [474]. 4 hours later she experienced a sudden onset of palpitations, blurred vision, agitation and a sense of impending doom. The fetal heartbeat was 160–170 b.p.m., and returned to a baseline of 120–140 within 12 hours. The fetal response was attributed to the myristicin content of nutmeg oil, and its anticholinergic (i.e. stimulant) effect. It is assumed that myristicin readily crosses the placenta. This case, with its assumed conclusions, is not a sufficient basis to contraindicate nutmeg oil in pregnancy.

Both safrole and methyleugenol are potentially carcinogenic (see p. 93). Due to its content of these two compounds, a maximum use level for external use of East Indian nutmeg oil is recommended. West Indian nutmeg oil is to be preferred for aromatherapy, due to its lower content of safrole and methyleugenol. There is some suspicion that elemicin might be weakly carcinogenic (see elemi profile). Myristicin shows high activity as an inducer of glutathione S-transferase (this correlates with tumour inhibition) [466] and it inhibits benzo[a]pyrene-induced tumours in mice [467].

Comments: It is feasible that the myristicin in nutmeg oil might counter the potentially carcinogenic effect of safrole, methyleugenol (and, possibly, elemicin). East Indian nutmeg oil is considered innocuous when applied externally but should not be taken in oral doses because of its possible psychotropic and carcinogenic effects. If myristicin is important in any of the therapeutic actions of nutmeg oil, the East Indian oil might be more useful than the West Indian oil.

Compare: Elemi, mace, parsley leaf, parsleyseed

Oakmoss[A]

Botanical name: *Evernia prunastri*

Family: Usneaceae

Absolute from: Dried moss

Notable constituents:
Alpha-thujone } Major components
Beta-thujone

Acute oral LD$_{50}$: No data could be found

Hazards:
Dermal sensitiser (strong)
Toxic (moderate/severe?)

Contraindications (dermal): Hypersensitive, diseased or damaged skin, children under 2 years of age

Contraindications (IRV): Do not use on mucous membrane

Cautions (all routes): Use with caution generally, due to uncertain toxicity: use with caution in epilepsy and pregnancy

Maximum use level: 0.6%

Toxicity data & recommendations: IFRA recommends a maximum use level of 0.6% in the final product, in order to avoid allergic reactions; this recommendation is based on as yet unpublished RIFM test data on the sensitising potential of oakmoss absolute [430]. The major components of oakmoss absolute are reported as being *alpha*- and *beta*-thujone, with no quantitative data given. This is likely to mean that oakmoss absolute should be regarded as toxic in an aromatherapy context.

Comments: A RIFM monograph on oakmoss *concrete* concludes that it is non-phototoxic, non-sensitising at 10%, and non-irritant at 10%. Oral toxicity is reported as 2.9 g/kg [421]. Since the absolute is prepared directly from the concrete, it could be surmised that the concrete might contain a substance which quenches the sensitising action of the absolute. It could also be surmised that the absolute is likely to be more toxic than the concrete. Oakmoss resinoids are also available.

Compare: Cassia, cinnamon bark, perilla, thuja, treemoss[A]

Ocimum gratissimum

Botanical name: *Ocimum gratissimum*

Family: Lamiaceae (= Labiatae)

Oil from: Leaves

Notable constituents: Eugenol 60–95%

Acute oral LD$_{50}$: No data could be found

Hazards:
Mucous membrane irritant (strong)
Hepatotoxic*
Inhibits blood clotting*

*Assumed from eugenol content

Cautions (oral): Alcoholism, anticoagulants, haemophilia, kidney disease, liver disease, paracetamol, prostatic cancer, SLE

Cautions (IRV): Do not use at more than 3% concentration on mucous membrane

Toxicity data & recommendations: Eugenol is a powerful inhibitor of platelet activity [370, 434]. Platelet activity is essential for blood clotting. This being the case, it would be prudent to avoid oral administration of *Ocimum gratissimum* oil in people whose blood clots only slowly, whether or not this is caused by hereditary haemophilia. Caution is recommended in anyone taking anticoagulant drugs, such as aspirin, heparin or warfarin.

Eugenol has been shown to cause liver damage in mice whose livers have been experimentally depleted of glutathione [400]. As in paracetamol poisoning, pretreatment with N-acetylcysteine prevents glutathione loss and hence cell death [401]. Although the oral doses of eugenol used were high (5 ml/kg) the data suggest that oral doses of eugenol-rich oils such as *Ocimum gratissimum* might be best avoided in those with impaired liver function and in anyone taking paracetamol (= acetaminophen).

Comments: *Ocimum gratissimum* is commercially available in Russia, and occasionally outside its borders. Eugenol is also the major component of *Ocimum sanctum*, eugenol chemotype, which is not commercially available.

Compare: Clove

Onion

Botanical name: *Allium cepa*

Family: Alliaceae

Oil from: Seeds

Notable constituents:
Dipropyl disulphide (major component)
Propyl disulphide
Allyl disulphide
Allyl monosulphide
Methyl disulphide

Hazards:
Inhibits iodine metabolism
Inhibits blood clotting
Dermal irritant (strong?)
Dermal sensitiser?

Contraindications (oral): Thyroid disease

Cautions (oral): Anticoagulants, haemophilia, kidney disease, liver disease, prostatic cancer, SLE

Contraindications (IRV): Use with caution on mucous membrane

Cautions (dermal): Use only with very great caution on skin, especially if hypersensitive, diseased or damaged

Toxicity data & recommendations: Many of the volatile components of onion, including methyl disulphide, allyl disulphide and propyl disulphide, inhibit iodine metabolism in rats at low concentrations [358, 359]. The effect is probably only significant for oral administration in low-iodine areas and where there is no use of iodised salt (rare) and in cases of thyroid disease. Allyl disulphide is the most active compound in this group of anti-thyroid essential oil constituents. There is no indication that *Allium* consumption by humans decreases thyroid function [363, 364].

Onion oil demonstrates anti-platelet activity [363, 365, 366, 372]. Platelet activity is essential for blood clotting. It is up to ten times more potent than garlic oil [363]. Onion oil should be avoided in people whose blood clots only slowly, whether or not this is caused by hereditary haemophilia. Caution is recommended in anyone taking anticoagulant drugs, such as aspirin, heparin or warfarin.

Since onion and garlic oils have a great deal in common, there is a suspicion that onion oil might, like garlic oil, present some risk of sensitisation or irritation. Safe levels for dermal application of onion oil are not known and patch testing is recommended before application to the skin.

Comments: Sulphur-rich oils such as onion are not used in fragrances, hence the lack of RIFM data; they are more commonly administered internally than externally in aromatherapy.

Compare: Garlic, massoia

Opopanax

Botanical name: *Commiphora erythraea*

Family: Burseraceae

Oil from: Gum

Typical constituents:
　Cis-alpha-bisabolene
　Alpha-santalene

Hazards: Phototoxic (degree undetermined)

Contraindications (dermal): If used topically, the skin must not be exposed to sunlight or sunbed rays for 12 hours

Toxicity data & recommendations: A 1977 RIFM report on phototoxicity testing for 160 fragrance raw materials found opopanax (= opoponax) oil to be phototoxic [418].

Comments: RIFM have not published a monograph on opopanax oil, nor have IFRA issued a guideline on recommended use levels for the oil. After initial testing, it was apparently decided that opopanax oil was not of significant interest to the fragrance industry. Since very little work has been published on the composition of this oil, we have indicated 'typical constituents'.

Compare: Khella, massoia

Orange (bitter)

Synonym: Seville orange

Botanical name: *Citrus aurantium*

Family: Rutaceae

Oil from: Peel by expression

Notable constituents:

d-limonene	89–96%
Bergapten	0.07%

Hazards: Phototoxic (moderate)

Contraindications (dermal): If applied to the skin at over maximum use level skin must not be exposed to sunlight or sunbed rays for 12 hours

Maximum use level: 1.4%

Toxicity data & recommendations: Bitter orange oil contains up to seven furanocoumarins [216]. The furanocoumarin content makes the oil phototoxic [420]. IFRA recommends that, for application to areas of skin exposed to sunshine, bitter orange oil be limited to a maximum of 1.4% in the final product, except for bath preparations, soaps and other products which are washed off the skin [430]. See note on combinations of phototoxic oils (p. 114).

Comments: Expressed *sweet* orange oil is not phototoxic [420]. It might, therefore, be preferable for aromatherapy use in some instances.

Compare: Grapefruit, lemon

Oregano

Synonym: Origanum

Botanical name: *Origanum vulgare, Coridothymus capitatus (= Thymus capitatus)* and others (see below)

Family: Lamiaceae (= Labiatae)

Oil from: Dried flowering herb

Notable constituents:

Origanum vulgare:	
Thymol	1-2%
Carvacrol	0.5%
Thymus capitatus:	
Thymol	0–15%
Carvacrol	44–75%
Origanum smyrnaeum:	
Thymol	1%
Carvacrol	83%
Origanum gracile:	
Thymol	60%
Carvacrol	9%
Origanum maru:	
Thymol	31%
Carvacrol	44%
Origanum vulgare subsp. *hirtum*:	
Thymol	1–85%
Carvacrol	3–84%

Acute oral LD$_{50}$: 1.85 g/kg [420]

Hazards:
 Skin irritant (moderate)
 Mucous membrane irritant (strong)

Cautions (dermal): Hypersensitive, diseased or damaged skin, children under 2 years of age

Cautions (IRV): Do not use at more than 1% concentration on mucous membrane

Toxicity data & recommendations: Undiluted oregano oil was severely irritating when applied to mouse skin, and moderately irritating when applied to rabbit skin. Tested at 2% it produced no reaction after a patch test on human subjects [420].

Comments: True oregano is *Origanum vulgare*, the herb used in the kitchen. However, the essential oil most commonly used in Europe is distilled from *Thymus capitatus*, and most safety data refer to this oil. Oils from many origins, as detailed above, may be offered as 'oregano' or 'origanum' oils.

Compare: Ajowan, savory, thyme

Parsley leaf

Botanical name: *Petroselinum sativum* (= *Petroselinum crispum*)

Family: Apiaceae (Umbelliferae)

Oil from: Leaves

Notable constituents:

p-mentha-1,3,8-triene	6–60%
Myristicin	7–33%
Parsley apiol	tr.–18%
Elemicin	0.2–3%

Acute oral LD$_{50}$: 3.3 g/kg [427]

Hazards:
  Abortifacient*
  Hepatotoxic*

*Assumed from apiol content

Contraindications (all routes): Pregnancy

Contraindications (oral): Should not be taken in oral doses

Toxicity data & recommendations: Both apiol and various preparations of parsley have been used for many years to procure illegal abortion in Italy [500, 502, 503]. The lowest daily dose of apiol which induced abortion was 0.9 g, taken for 8 consecutive days. This is approximately equivalent to 6 ml of parsley leaf oil. The inevitable conclusion is that parsley leaf oil presents a very high risk of abortion if taken in oral doses, and that external use would also seem inadvisable in pregnancy.

There are several recorded cases of fatal poisoning from apiol ingestion [501, 502, 503]. The lowest total dose of apiol causing death is 6.3 g (2.1 g/day for 3 days) [530]; the lowest fatal daily dose is 0.77 g, which was taken for 14 days [533]; the lowest single fatal dose is 8 g [500]. At least 19 g has been survived [530], but if 1 or 2 g of apiol can be fatal to an adult, it would seem prudent to completely avoid oral dosages of parsley leaf oil, which may be more toxic in humans than the animal data would indicate.

Common signs of apiol poisoning are fever, severe abdominal pain, vaginal bleeding, vomiting and diarrhoea [501, 530]. Post-mortem examination invariably reveals considerable damage to both liver and kidney tissue, often with gastrointestinal inflammation and sometimes damage to heart tissue [501, 503, 533].

Myristicin has been shown to be an inhibitor of monoamine oxidase (MAO) in rodents [367]. MAO inhibitors should not be given in conjunction with pethidine. 'Very severe reactions, including coma, severe respiratory depression, cyanosis and hypotension have occurred in patients receiving monoamine oxidase inhibitors and given pethidine' [5]. Oral dosages of parsley leaf oil, in conjunction with pethidine, would be inadvisable. The safety of non-oral dosages is uncertain.

There is some suspicion that elemicin might be weakly carcinogenic (see elemi profile). Myristicin shows high activity as an inducer of glutathione S-transferase (this correlates with tumour inhibition) [466] and it inhibits benzo[a]pyrene-induced tumours in mice [467].

Psychotropic effects have been reported for whole nutmeg, and myristicin and elemicin are thought to be responsible. However, the data suggest that other synergistic elements may need to be present for a psychotropic effect to take place (see p. 71).

Undiluted parsley leaf oil was severely irritating to rabbit skin, but produced no irritation when tested on human subjects at 2%; it was non-phototoxic and non-sensitising [427].

Comments: In spite of its sometimes high myristicin content, it seems unlikely that even oral doses of parsley leaf oil are psychotropic. The level of the suspect carcinogen elemicin is not considered to represent a hazard, especially considering the relatively high level of antitumoral myristicin.

Compare: Elemi, mace, nutmeg, parsleyseed

Parsleyseed

Botanical name: *Petroselinum sativum*

Family: Apiaceae (= Umbelliferae)

Oil from: Seeds

Notable constituents:

Typical commercial oil:
Myristicin	28%
TMAB*	23%
Parsley apiol	21%

Apiol chemotype:
Parsley apiol	58–80%
Myristicin	9–30%
TMAB*	tr.–6%

Myristicin chemotype
Myristicin	49–77%
TMAB*	1–23%
Parsley apiol	0–3%

TMAB* chemotype:
TMAB*	52–57%
Myristicin	26–37%
Parsley apiol	0–tr.

*2,3,4,5-tetramethoxyallylbenzene

Acute oral LD$_{50}$: 1.52 g/kg (mouse); 3.96 g/kg (rat) [421]

Hazards:
  Abortifacient*
  Hepatotoxic*

*Assumed from apiol content

Contraindications (all routes): Pregnancy

Contraindications (oral): Should not be taken in oral doses

Toxicity data & recommendations: Both apiol and various preparations of parsley have been used for many years to procure illegal abortion in Italy [500, 502, 503]. The lowest daily dose of apiol which induced abortion was 0.9 g, taken for 8 consecutive days. This is approximately equivalent to between 1.5 ml and 6 ml of parsleyseed oil. The inevitable conclusion is that parsleyseed oil presents a very high risk of abortion if taken in oral doses, and that external use would also seem inadvisable in pregnancy.

There are several recorded cases of fatal poisoning from apiol ingestion [501, 502, 503]. The lowest total dose of apiol causing death is 6.3 g (2.1 g/day for 3 days) [530]; the lowest fatal daily dose is 0.77 g, which was taken for 14 days [533]; the lowest single fatal dose is 8 g [500]. At least 19 g has been survived [530], but if 1 or 2 g of apiol can be fatal to an adult, it would seem prudent to completely avoid oral dosages of parsleyseed oil, which may be more toxic in humans than the animal data would indicate.

Common signs of apiol poisoning are fever, severe abdominal pain, vaginal bleeding, vomiting and diarrhoea [501, 530]. Post-mortem examination invariably reveals considerable damage to both liver and kidney tissue, often with gastrointestinal inflammation and sometimes damage to heart tissue [501, 503, 533].

Myristicin has been shown to be an inhibitor of monoamine oxidase (MAO) in rodents [367]. MAO inhibitors should not be given in conjunction with pethidine. 'Very severe reactions, including coma, severe respiratory depression, cyanosis and hypotension have occurred in patients receiving monoamine oxidase inhibitors and given pethidine' [5]. Oral dosages of parsleyseed oil, in conjunction with pethidine, would be inadvisable. The safety of non-oral dosages is uncertain.

Psychotropic effects have been reported for whole nutmeg, and myristicin and elemicin are thought to be responsible. However, the data suggest that other synergistic elements may need to be present for a psychotropic effect to take place (see p. 71). One of us (RT) has ingested 1 ml of parsleyseed oil, with no apparent psychotropic effects. Parsleyseed oil is non-irritant, non-sensitising and non-phototoxic [421].

Comments: Three chemotypes for parsleyseed oil have been identified, with either apiol, TMAB or myristicin as the major component [216]. The majority of commercial oils contain significant amounts of apiol.

Compare: Parsley leaf, sage (Spanish), savin

Pennyroyal (European and N. American)

Botanical name:
 Mentha pulegium (European)
 Hedeoma pulegioides (N. American)

Family: Lamiaceae (= Labiatae)

Oil from: Fresh herb

Note: Due to their compositional similarity these two essential oils are treated here in one profile.

Notable constituents:
 d-pulegone   55–95% (*M. pulegium*)
              60–80% (*H. pulegioides*)

Acute oral LD$_{50}$: 0.4 g/kg [420]

Hazards: Toxic (severe)

Contraindications: Should not be used in therapy, either internally or externally

Toxicity data & recommendations: Toxicity is due to the content of pulegone, which is toxic to the liver because it is metabolised to epoxides. Pulegone acute oral LD$_{50}$ is 0.5 g/kg in rats [424]. Both the UK and EC 'standard permitted proportion' of pulegone in food flavourings is 0.025 g/kg of food [455, 456].

There have been several recorded cases of pennyroyal oil poisoning, three with a fatal outcome. In two of these cases about 30 ml was ingested by adults. For further detail on cases of human poisoning see page 53.

The plant has enjoyed folk status as an abortifacient from ancient times [171]. In two separate cases, 7.5 ml of pennyroyal oil, taken orally, failed to induce miscarriage [168]. In a third case, 10 ml, taken orally, also failed [170]. In a fourth case, it was not determined how much oil was taken, but abortion did take place, with death following massive urea leakage into the blood [169].

Pennyroyal oil, it seems, is not abortifacient, unless taken in such massive quantities that it causes acute hepatotoxicity in the mother. She miscarries only because she is so poisoned by pulegone that the pregnancy cannot be maintained [242].

Comments: Pennyroyal oil is not abortifacient, but the oil is too highly toxic to be used in aromatherapy, and most especially during pregnancy.

Compare: Buchu, plectranthus, savin

Peppermint

Botanical name: *Mentha piperita*

Family: Lamiaceae (= Labiatae)

Oil from: Leaves

Notable constituents:

l-menthol	40–50%
Menthone	19%
d-pulegone	0.1–2%

Acute oral LD$_{50}$: 4.4 g/kg [153]

Hazards:
Neurotoxic?
Mucous membrane irritant (moderate)

Contraindications (oral): G6PD deficiency

Contraindications (all routes): Cardiac fibrillation

Cautions (oral): Epilepsy, fever; dose levels over 1 ml per 24 h in an adult not recommended

Cautions (IRV): Do not use at more than 3% concentration on mucous membrane

Toxicity data & recommendations: Peppermint oil produces microscopic dose-related lesions in the brains of rats when fed to them in the diet at 40 mg/kg and 100 mg/kg; no effect was seen at 10 mg/kg [142, 143]. The doses causing lesions would be equivalent to the consumption of 3–6 g of peppermint oil by an adult. The lower dose level, producing no lesions, is equivalent to 0.65 ml in an adult. A proprietary menthol-containing oil has been reported as producing incoordination, confusion and delirium when 5 ml of the product (35.5% peppermint oil) was inhaled over a long time period [141].

Peppermint oil has been known to produce convulsions and ataxia, with paralysis, loss of reflexes and very slowed breathing in rats [337]. However, the doses were extremely high (0.5–2 ml/kg i.p.). This toxicity may be due to the pulegone and/or menthone content; pulegone causes histopathological changes in the white matter of the cerebellum above 80 mg/kg [329]. Menthone produces similar cerebellar changes, but the lowest dosage level used was 0.2 g/kg/day [327]. There is no indication that peppermint oil can produce these toxic effects when used externally in aromatherapy, but some caution is required regarding oral dosage.

There have been reports of apnoea and instant collapse in infants following the local application of menthol to their nostrils [5].

Menthol has been shown to provoke severe neonatal jaundice in babies with a deficiency of the enzyme glucose-6-phosphate dehydrogenase (G6PD). This is a fairly common inheritable enzyme deficiency, particularly in Chinese, West Africans, and in people of Mediterranean or Middle Eastern origin [109]. Usually, menthol is detoxified by a metabolic pathway involving G6PD. When babies deficient in this enzyme were given a menthol-containing dressing for their umbilical stumps, menthol was found to build up in their bodies [109]. Peppermint oil should be avoided orally by males with a deficiency of this enzyme. Such people can be recognised because they will characteristically have had abnormal blood reactions to at least one of the following drugs, or will have been advised to avoid them: antimalarials; sulphonamides (antimicrobial); chloramphenicol (antibiotic); streptomycin (antibiotic); aspirin.

Large doses of menthol (above 0.2 g/kg) can produce long-term change in the appearance of hepatocytes when given orally to rats [329]. Menthone caused some liver toxicity at high dosage levels (over 0.2 g/kg/day for 28 days) [327]. Other studies have concluded that menthone is not hepatotoxic [111, 117]. It is unlikely to cause problems at dosage levels used in aromatherapy.

Menthol has been shown to dilate capillaries in the nasal mucosa upon inhalation in some individuals; this effect correlates with its reported ability to dilate systemic blood vessels after intravenous administration [222]. Mentholated cigarettes and peppermint confectionery have been responsible for several instances of cardiac fibrillation in patients prone to the condition who are being maintained on quinidine, a stabiliser of heart rhythm [236]. Bradycardia has been reported in a person addicted to menthol cigarettes [388]. It is recommended that peppermint oil should be avoided altogether in cases of cardiac fibrillation.

Menthol is a mild irritant [422]. No skin irritancy data could be found for peppermint oil, but cornmint oil, which has a similar menthol content, is non-irritant [421]. Urticarial hypersensitivity has been reported for menthol [85]. Peppermint may occasionally produce contact dermatitis [38].

Comments: The levels of pulegone in peppermint oil are not high enough to cause concern. Any neurotoxicity in peppermint oil is likely to be caused by a combination of pulegone and menthone.

Compare: Cornmint

Perilla

Botanical name: *Perilla frutescens*

Family: Lamiaceae (= Labiatae)

Oil from: Leaves and flowering tops

Notable constituents:

Perillaldehyde	50%
Perilla ketone	15–38%

Acute oral LD$_{50}$: 2.77 g/kg (mouse); 5.0 g/kg (rat) [428]

Hazards:
Sensitiser (mild)
Pulmonary toxin?

Cautions (all routes): Use with caution generally, and especially in pregnancy due to uncertain toxicity

Cautions (dermal): Hypersensitive, diseased or damaged skin, children under 2 years of age

Toxicity data & recommendations: Perilla ketone is a potent lung toxin to laboratory animals, and it often poisons grazing cattle which eat perilla leaves [332, 333]. The perilla ketone content of perilla oil has been found to be a potent pulmonary toxin in mice, rats, heifers and sheep [428]. Some unusual chemotypes of perilla oil contain egomaketone or *iso*-egomaketone, both of which are similar pulmonary toxins in animals to perilla ketone [332, 333].

Perilla oil is used as a flavour ingredient in Japan, and there are no reports which suggest that it is hazardous to man in the amounts consumed. Since the pulmonary toxic effects have been seen in several mammals, and since the amounts used in aromatherapy are higher than those used in food flavourings, caution is recommended.

In a test for sensitisation, perilla oil produced one questionably positive reaction out of 26 people when tested at 4%; out of 152 perilla workers, dermal effects were observed in about 50%; perilla oil produced allergic reactions in all of 17 perilla workers with dermatitis [428].

IFRA recommends a maximum use level in fragrance compounds of 0.5% for perillaldehyde, due to its sensitisation potential [430]. However, this should not be taken to mean that perilla oil should be used at a maximum level of 1% in fragrance compounds (equivalent to 0.2% in the final product) since, at 4%, it produced only one questionable reaction. Other components of the essential oil presumably have a quenching effect on the perillaldehyde content.

Comments: *l*-perillaldehyde is about 2000 times sweeter than sucrose, and is used as a sweetening agent in Japan.

Compare: Oakmoss[A], treemoss[A]

Pimento

Botanical name: *Pimenta dioica* (= *Pimenta officinalis*)

Family: Myrtaceae

Oil from: Berries or leaves

Note: Because they are so similar in terms of chemical composition and toxicology, pimento berry and pimento leaf oils are treated here in one profile.

Notable constituents:

Eugenol	60–95% (leaf)
	67–83% (berry)

Acute oral LD$_{50}$:

Leaf	3.6 ml/kg [420]
Berry	No data could be found

Hazards:
Mucous membrane irritant (strong)
Hepatotoxic*
Inhibits blood clotting*

*Assumed from eugenol content

Cautions (oral): Alcoholism, anticoagulants, haemophilia, kidney disease, liver disease, paracetamol, prostatic cancer, SLE

Cautions (IRV, leaf): Do not use at more than 1% concentration on mucous membrane

Cautions (IRV, berry): Do not use at more than 3% concentration on mucous membrane

Toxicity data & recommendations: Eugenol is a powerful inhibitor of platelet activity [370, 434]. Platelet activity is essential for blood clotting. This being the case, it would be prudent to avoid oral administration of pimento leaf oil or pimento berry oil in people whose blood clots only slowly, whether or not this is caused by hereditary haemophilia. Caution is recommended in anyone on anticoagulant drugs, such as aspirin, heparin or warfarin.

Eugenol has been shown to cause liver damage in mice whose livers have been experimentally depleted of glutathione [400]. As in paracetamol poisoning, pretreatment with *N*-acetylcysteine prevents glutathione loss and hence cell death [401]. Although the oral doses of eugenol used were high (5 ml/kg) the data suggest that oral doses of eugenol-rich oils such as pimento oils might be best avoided in those with impaired liver function and in anyone taking paracetamol (= acetaminophen).

Tested undiluted on rabbit skin, pimento leaf oil was severely irritating, but tested at 12% on human subjects, it produced no irritation after a 48-hour closed-patch test; it is non-sensitising [420].

Comments: The berries of this plant are known as allspice, and yield pimento berry oil, which has a superior odour and taste to the leaf oil. This should not be confused with the oil from the berries of *Pimenta racemosa*, or West Indian bay oil (used in 'bay rum') which in turn is sometimes confused with *Laurus nobilis* ('laurel leaf', or 'bay leaf') oil.

Compare: Clove

Pine (dwarf)

Synonyms: Dwarf, or mountain pine

Botanical name: *Pinus pumilio* (= *Pinus mugo* var. *pumilio*)

Family: Pinaceae

Oil from: Needles and twigs

Notable constituents:
  d-limonene
  Alpha-pinene
  Delta-3-carene
  Phellandrene

Acute oral LD$_{50}$: >5 g/kg [422]

Hazards: Dermal irritant (if oxidised)

Cautions: Hypersensitive, damaged or diseased skin

Toxicity data & recommendations: Dwarf pine oil produced irritation in three out of 22 human subjects when tested at 12%; *delta*-3-carene (probably oxidised) is suspected of producing sensitisation in dwarf pine oil [422].

Comments: Dwarf pine oil is a frequent cause of contact dermatitis, although it is almost certainly only oxidised oils which cause problems, due to terpene oxidation. It is very likely that all pine oils are potential irritants when oxidised, but not when fresh.

Compare: Terebinth

Plectranthus

Botanical name: *Plectranthus fruticosus*

Family: Lamiaceae (= Labiatae)

Oil from: Leaves

Notable constituents: Sabinyl acetate >60%

Hazards:
  Abortifacient?
  Embryotoxic
  Fetotoxic

Contraindications (all routes): Pregnancy

Toxicity data & recommendations: Plectranthus oil has been shown to be both embryotoxic and fetotoxic in rodents. It dramatically increased the rate of resorption after oral administration at doses equivalent to about 1 g of oil on each of 2 days of a human pregnancy [197]. The oil caused fetal malformation in mice, mainly causing anophthalmia (lack of eyes) [189]. Sabinyl acetate is thought to be responsible for these effects [197].

Comments: Plectranthus oil has never been commercially available, although the leaves are used in traditional Romanian medicine. The profile is included because of its implications for other oils rich in sabinyl acetate.

Compare: Sage (Spanish), savin

Ravensara anisata

Botanical name: *Ravensara anisata*

Family: Lauraceae

Oil from: Bark

Notable constituents:

Estragole	88%
Trans-anethole	7%

Acute oral LD$_{50}$: No data could be found

Hazards:
Carcinogenic*
Hepatotoxic*

*Assumed from estragole content

Contraindications: Should not be used in therapy, either internally or externally

Toxicity data & recommendations: Estragole (= methyl chavicol) produces DNA abnormalities in the livers of rodents [122, 289]. It is carcinogenic in mice [309, 312] because it is metabolised, in vivo, to the carcinogenic compound 1'-hydroxyestragole [62, 123, 312, 478]. The same metabolic process is believed to take place in humans [30]. High doses are potentially carcinogenic, but very low doses are not, since they are readily detoxified [62, 306]. See page 98.

Estragole is not restricted by any regulatory agencies. However, the carcinogenic potential of estragole suggests that *Ravensara anisata* oil should be avoided altogether in aromatherapy, due to its high estragole content.

Comments: *Ravensara anisata* is commercially available, and should not be confused with *Ravensara aromatica*, often referred to as 'ravensara oil'.

Compare: Basil, fennel, tarragon

Rosemary

Botanical name: *Rosmarinus officinalis*

Family: Lamiaceae (= Labiatae)

Oil from: Leaves

Notable constituents:

	Tunisian	Spanish
Alpha-pinene	10–12%	19–27%
Camphor	10–13%	13–20%
1,8-cineole	40–44%	17–25%

Acute oral LD$_{50}$: 5.0 ml/kg [420]

Hazards: Neurotoxic—assumed from camphor content (mild)

Contraindications (oral): Pregnancy

Cautions (oral): Epilepsy, fever

Toxicity data & recommendations: Camphor readily causes epileptiform convulsions if taken in sufficient quantity [269, 272]. Rosemary oil should therefore be used with caution, in oral doses.

Comments: While rosemary oil is produced in many parts of the world the two principal areas are Spain and Tunisia. Due to its lower camphor content, the Tunisian oil might be preferable for general use in aromatherapy.

Compare: Camphor, lavandin, *Lavandula stoechas*, spike lavender, yarrow (camphor CT)

Rue

Botanical name: *Ruta graveolens*

Family: Rutaceae

Oil from: Fresh herb

Notable constituents:

2-undecanone*	31–49%
2-nonanone	18–25%
Furanocoumarins	

* = methylnonyl ketone

Acute oral LD$_{50}$: 2.5–5.0 g/kg [421]

Hazards:
Abortifacient?
Neurotoxic?
Phototoxic (strong)
Photocarcinogenic?

Contraindications (dermal): If applied to the skin at over maximum use level, skin must not be exposed to sunlight or sunbed rays for 12 hours

Cautions (oral): Epilepsy, fever

Cautions (all routes): Pregnancy

Maximum use level: 0.78%

Toxicity data & recommendations: According to the Merck Index 'Ingestion of large quantities [of rue oil] causes epigastric pain, nausea, vomiting, confusion, convulsions, death; may cause abortion' [158]. A single oral dose at 0.4 g/kg has been reported as causing death in guinea pigs due to liver, kidney and adrenal gland damage [156]. This is inconsistent with the (mouse) LD$_{50}$ range given above.

There are isolated reports that rue oil has a direct action on the uterus, but the details of the research on which they are based are obscure [176]. Used in massive amounts on pregnant rabbits (12 ml/kg) and guinea pigs (50 ml/kg) rue oil, not surprisingly, was equally toxic to mother and fetus, causing widespread tissue damage and some fatalities [177]. These dosage levels are equivalent to human ingestion of 800 ml and 3.2 litres. Two pregnant guinea pigs, each fed 12 drops of rue oil, aborted; rue oil was found in the fetal tissue, and was considered toxic to it [335]. This is still a very high dosage, around 3 ml/kg.

There are anecdotal reports that rue oil can cause abortion in humans. It is possible that any abortifacient activity that rue oil has is due to maternal toxicity.

The FDA consider that rue oil is safe for human consumption as it is currently used in food flavourings, 'but not necessarily under different conditions of use'; in particular, further teratological research is recommended [335]. Our recommendation is that rue oil should be used with caution during pregnancy until further data become available.

2-undecanone is non-toxic, non-irritant, non-sensitising and non-phototoxic [421]. The RIFM monograph on rue oil reports it as being non-toxic, non-irritant and non-sensitising [421]. There are anecdotal reports of rue oil being an irritant to both skin and mucous membrane [155]. Our own tests found it to be non-irritant to mucous membrane.

In phototoxicity tests, distinct positive results were obtained with concentrations of 100%, 50%, 25%, 12.5%, 6.25% and 3.125%. Borderline results were obtained with 1.56%, and negative results with 0.78% [421]. The three dominant furanocoumarins in rue oil are bergapten, xanthotoxin, and psoralen [217]. All three possess similar phototumorigenic properties [496] (see p. 87).

IFRA recommends that, for application to areas of skin exposed to sunshine, rue oil be limited to a maximum of 3.9% in fragrance compound (equivalent to 0.78% in the final product) except for bath preparations, soaps and other products which are washed off the skin [430]. See note on combinations of phototoxic oils (p. 114).

The toxicity of rue oil in humans is unclear, but there seems no reason to restrict its use, except during pregnancy and the normal cautions for phototoxicity. It should be used with caution in people with epilepsy.

Comments: Inconsistent reports regarding toxicity and irritation could be due to compositional differences between different rue oils. Data on the composition of rue oil are scarce.

Compare: Angelica, bergamot, cumin, hyssop, lime, sage (Dalmatian)

Sage (Dalmatian)

Botanical name: *Salvia officinalis*

Family: Lamiaceae (= Labiatae)

Oil from: Leaves

Notable constituents:

Alpha-thujone	29–44%
Beta-thujone	3–53%
Camphor	26%

Acute oral LD$_{50}$: 2.6 g/kg [420]

Hazards:
Toxic?
Neurotoxic

Contraindications: Should not be used in therapy, either internally or externally

Toxicity data & recommendations: Animal data indicate sage oil to be non-toxic [420]. The normal total thujone content of commercial sage oil is around 50%. This is marginally lower than that of tansy and thuja, and similar to that of wormwood. It would therefore be prudent to consider sage oil as being equally toxic to these other oils, because their toxic properties arise from their content of thujone.

Tests in rats found that convulsions appeared for sage oil at dose levels over 0.5 g/kg [234, 325]. Convulsions were induced in an adult who accidentally ingested one 'swallow' of sage oil [140]. Presumably these were caused by the camphor and/or the thujone content. Camphor readily causes epileptiform convulsions if taken in sufficient quantity [269, 272].

Thujone acute oral LD$_{50}$ is 0.21 g/kg; chronic administration leads to fatty degeneration of the liver [354]. Intraperitoneal thujone was found to be convulsant and lethal to rats above 0.2 ml/kg [234, 325]. Both the UK and EC 'standard permitted proportion' of *alpha*- and/or *beta*-thujone in food flavourings is 0.0005 g/kg of food [455, 456].

A 24-hour patch test using undiluted sage oil produced one irritation reaction in 20 human subjects [420].

Although the rodent LD$_{50}$ of 2.6$_8$/kg does not indicate great toxicity, the usual thujone content is sufficiently high to warrant its exclusion from aromatherapy [414, 486]. In addition, the camphor content presents a potential risk of neurotoxicity. The combination of thujone and camphor in this oil is especially worrying, and toxicity problems from prolonged use of Dalmatian sage oil seem likely.

Comments: Dalmatian sage oil should not be confused with clary sage oil, which is non-toxic.

Compare: Armoise, camphor, lanyana, sage (Spanish)

Sage (Spanish)

Botanical name: *Salvia lavandulifolia* (= *Salvia lavandulaefolia*)

Family: Lamiaceae (= Labiatae)

Oil from: Flowering tops

Notable constituents:

Camphor	11–36%
Cineole	18–29%
Sabinyl acetate	0.1–24%

Acute oral LD$_{50}$: > 5.0 g/kg [422]

Hazards:
Neurotoxic* (mild)
Abortifacient

*Assumed from camphor content

Contraindications (all routes): Pregnancy

Contraindications (oral): Epilepsy, fever

Cautions (dermal, IRV): Epilepsy

Toxicity data & recommendations: A fraction of Spanish sage oil, containing 50% sabinyl acetate, 41% 1, 8-cineole, and 5% camphor, caused dose-dependent maternal toxicity in rodents, and demonstrated a dose-dependent abortifacient effect, but was not fetotoxic [194]. Cineole is not teratogenic [193]. Camphor readily causes epileptiform convulsions if ingested in sufficient quantity [269, 272].

Comments: It would appear that oils rich in sabinyl acetate are among the most dangerous ones in pregnancy. The combination of camphor and sabinyl acetate in Spanish sage oil makes it especially hazardous in pregnancy.

Variation in composition of individual oils will give varying levels of toxicity. The wide variation of sabinyl acetate content might be due to the existence of chemotypes [379]. Certain subspecies may be much more potentially hazardous in pregnancy than others. Spanish sage oil should not be confused with Dalmatian sage, or clary sage.

Compare: Camphor, plectranthus, sage (Dalmatian), savin

Sassafras

Botanical name:
Sassafras albidum (sassafras)
Ocotea pretiosa (Brazilian sassafras)
Cinnamomum porrectum (Chinese sassafras)
Cinnamomum rigidissimum (Chinese sassafras)

Family: Lauraceae

Oil from:
Dried root bark (*S. albidum*)
Wood (*O. pretiosa, C. porrectam, C. rigidissimum*)

Notable constituents: Safrole 85–90%

Note: Due to their compositional similarity these essential oils are treated here in one profile.

Acute oral LD$_{50}$:
S. albidum: 1.9 g/kg [426]
O. pretiosa: 1.6 g/kg [424]

Hazards: Carcinogenic

Contraindications: Should not be used in therapy, either internally or externally

Toxicity data & recommendations: A few drops of sassafras oil have been estimated to be sufficient to kill a toddler [269] and the US Dispensatory of 1888 records the case of a male adult who died following the ingestion of a teaspoon of sassafras oil; symptoms of poisoning were those of CNS depression (stupor, ataxia), vomiting, nausea, possibly dilated pupils [269]. These fatal quantities are not consistent with the rodent LD$_{50}$ data; either the human data are incorrect, or sassafras oil is more toxic to humans than to rodents. See page 54 for more detail on cases of poisoning.

The minimum lethal oral dose of safrole in the rabbit is 1.0 g/kg [38]. Safrole is highly toxic on chronic dosing; 95% lethal to rats at 10 000 p.p.m. after 19 days' dosing [154]. Safrole has a hepatocarcinogenic effect when given subcutaneously or by stomach tube to male mice during infancy [153]. When administered orally, safrole is a low-level hepatic carcinogen in the rat [320]. Oesophageal tumours have also been reported in rats given safrole in the diet [319]. Sassafras oil and safrole, given in the diet, produced no tumours after 22 months, but a large percentage of the same animals showed initial tumour development at 24 months [343].

A metabolite of safrole, 1'-hydroxysafrole, is a much more potent hepatic carcinogen than safrole itself [32, 153, 286]. Other toxic metabolites may also be formed in vivo [59, 72, 287]. Safrole and 1'-hydroxysafrole can both cause hepatic cell enlargement in experimental animals [313]. There is some evidence that safrole may be able to activate a cancer-causing virus, the polyoma virus, at least in rats. It is not known whether safrole has a similar action in humans [344].

Natural root beer, made with sassafras, used to be popular in the USA and is still made illicitly. It has been banned for a number of years because of its safrole content. Safrole was itself banned as a food additive in the USA in 1961.

IFRA recommends that safrole should not be used as a fragrance ingredient. It recommends a maximum use level of 0.05% in fragrance compounds for safrole (equivalent to 0.01% in the final product), when safrole-containing essential oils are used [430]. Both the UK and EC 'standard permitted proportion' of safrole in food flavourings is 0.001 g/kg of food [455, 456].

Comments: Very little oil is produced from *Sassafras albidum* in the USA. Brazilian sassafras oil is sometimes incorrectly referred to as originating from *Ocotea cymbarum*. It actually comes from *Ocotea pretiosa*. Chinese sassafras oil is sometimes the safrole-rich fraction of camphor oil, 'brown camphor oil', although it is also obtained from *Cinnamomum porrectum* and *Cinnamomum rigidissimum* [512]. The Chinese oil has taken over from the Brazilian in the world market, because of over-logging in Brazil. China is now in the process of destroying its own sassafras trees.

Compare: Basil, calamus, camphor (brown and yellow), tarragon

Savin

Botanical name: *Juniperus sabina*

Family: Cupressaceae (= Coniferae)

Oil from: Twigs and leaves

Notable constituents:

Sabinyl acetate	20–53%
Sabinene	20–42%

Acute oral LD$_{50}$: No data could be found

Hazards:
Toxic?
Abortifacient?
Embryotoxic

Contraindications (all routes): Pregnancy

Cautions (all routes): Use with great caution due to uncertain toxicity

Toxicity data & recommendations: There are very worrying anecdotal reports of savin oil causing convulsions, fatal poisoning, abortion, and being able to cross the placenta [159, 435]. Savin oil has been reported to cause liver lesions in guinea pigs suggestive of a degenerative hepatitis [386].

Savin oil is embryotoxic, but not fetotoxic in rodents [190]. Sabinyl acetate is probably responsible for the embryotoxic and fetotoxic action of plectranthus oil [197] and the abortifacient action of Spanish sage oil [194].

IFRA recommends that savin oil should not be used at all as a fragrance ingredient [430].

Comments: Savin oil is periodically confused with either *Juniperus phoenicea*, or *Juniperus communis*. In the latter case this has contributed to the (in our opinion, mistaken) belief that common juniper oil is toxic and abortifacient. *Juniperus pfitzeriana* also contains sabinyl acetate (2–17%) but is not commercially available [497].

Compare: Buchu, juniper, pennyroyal, plectranthus, sage (Spanish)

Savory (summer and winter)

Botanical name:
Satureia hortensis (summer)
Satureia montana (winter)

Family: Lamiaceae (= Labiatae)

Oil from: Whole dried herb

Note: Due to their compositional similarity these two essential oils are treated here in one profile.

Notable constituents:

Carvacrol	3–67%
Thymol	1–49%
Para-cymene	7–26%

Acute oral LD$_{50}$: 1.37 g/kg [422]

Hazards:
Dermal irritant (moderate)
Mucous membrane irritant (strong)

Cautions (dermal): Hypersensitive, diseased or damaged skin, children under 2 years of age

Cautions (IRV): Do not use at more than 1% concentration on mucous membrane

Toxicity data & recommendations: Undiluted savory oil was strongly irritating to rabbit skin, and severely irritating to mouse skin; tested at 6%, savory oil did not irritate the skin of human subjects [422].

Comments: Winter and summer savory oils are very similar. Although toxicity data generally refer to summer savory, the winter variety can be considered identical for the purposes of aromatherapy.

Compare: Ajowan, oregano, thyme

Snakeroot

Synonyms: Canadian snakeroot, wild ginger

Botanical name: *Asarum canadense*

Family: Artistolochiaceae

Oil from: Dried roots and rhizomes

Notable constituents:

Methyleugenol	36-45%

Acute oral LD$_{50}$: 4.48 ml/kg [424]

Hazards: Carcinogenic*

*Assumed from methyleugenol content

Contraindications: Should not be used in therapy, either internally or externally

Toxicity data & recommendations: Methyleugenol is genotoxic [122, 289, 291, 479]; it comes into the same category as safrole and estragole and is carcinogenic in rodents [309, 310]. Neither methyleugenol nor snakeroot oil are restricted by any agencies, but methyleugenol is clearly carcinogenic in rodents, and so snakeroot oil should be avoided, at least until there is good evidence that it is non-carcinogenic. Snakeroot oil is non-irritant, non-sensitising, and non-phototoxic [424].

Compare: *Melaleuca bracteata*

Spike lavender

Botanical name: *Lavandula latifolia*

Family: Lamiaceae (= Labiatae)

Oil from: Flowering tops

Notable constituents:

Linalool	30–35%
1,8-cineole	25–32%
Camphor	10–20%

Acute oral LD$_{50}$: 3.8 g/kg [422]

Hazards: Neurotoxic—assumed from camphor content (mild)

Contraindications (oral): Pregnancy

Cautions (oral): Epilepsy, fever

Toxicity data & recommendations: Camphor readily causes epileptiform convulsions if taken in sufficient quantity [269, 272]. Spike lavender oil should therefore be used with caution, especially in oral doses.

Comments: Camphor content is significantly higher than that of true lavender oil (< 1%).

Compare: Camphor, lavandin, *Lavandula stoechas*, rosemary, yarrow (camphor CT)

Star anise

Botanical name: *Illicium verum*

Family: Apiaceae (= Umbelliferae)

Oil from: Fruit

Notable constituents:

Trans-anethole	75–90%
Safrole	< 1%

Acute oral LD$_{50}$: 2.57 g/kg [421]

Hazards: Oestrogen-like action (mild)

Cautions (oral): Alcoholism, liver disease, paracetamol use, breast-feeding, pregnancy, endometriosis, oestrogen-dependent cancers, prostatic hyperplasia

Cautions (dermal): Hypersensitive, diseased or damaged skin, children under 2 years of age

Toxicity data & recommendations: *Trans*-anethole and its derivatives are oestrogen-like in effect, as is anise oil [360, 361]. The effect is relatively weak, but it might be prudent to avoid oral dosing during pregnancy and breast-feeding in order to avoid influencing oestrogen levels. Endometriosis, prostatic hyperplasia and oestrogen-dependent cancers could be adversely affected by increased oestrogenic activity.

Trans-anethole shows a dose-dependent cytotoxicity to rat liver cells in culture, and causes glutathione depletion [452]. The mechanism of its action is unknown, but it is not due to a DNA-damaging effect. A primary metabolite of *trans*-anethole, anethole–1,2-epoxide, also depletes glutathione [504]. Because of this effect, anethole-rich oils such as star anise should be avoided orally in alcoholism, liver disease, or if also taking paracetamol (= acetaminophen).

Undiluted star anise oil was non-irritating to rabbit skin; tested at 4% it was non-irritant to human skin, and produced no sensitisation reactions in a maximation test at 4% [420]. It is believed that *trans*-anethole can enhance dermatitis.

Comments: The level of safrole present in star anise oil is not considered to represent a hazard except in oral dosages. *Cis*-anethole, which exists in synthetic anethole, is considerably more toxic than *trans*-anethole.

Compare: Anise, fennel

Taget

Synonym: Tagetes

Botanical name: *Tagetes patula, Tagetes erecta, Tagetes minuta* (= *Tagetes glandulifera*)

Family: Asteraceae (= Compositae)

Oil or absolute from: Flowering tops

Notable constituents (oil):

Tagetone	50–60%
Furanocoumarins	

Hazards: Phototoxic (severe)

Contraindications (dermal): If applied to the skin at over maximum use level, skin must not be exposed to sunlight or sunbed rays for 12 hours

Maximum use level: 0.05%

Toxicity data & recommendations: IFRA recommends that, for application to areas of skin exposed to sunshine, taget oil be limited to a maximum of 0.25% in fragrance compounds (equivalent to 0.05% in the final product) except for bath preparations, soaps and other products which are washed off the skin [430].

The IFRA guideline is based on as yet unpublished research by RIFM, using both oils and absolutes from *Tagetes minuta* and *Tagetes patula*. See note on combinations of phototoxic oils (p. 114). The RIFM monograph on taget oil, published in November 1982, concludes that it is non-phototoxic [426]. The IFRA guideline is presumably based on subsequent testing. Taget oil is non-toxic, non-irritant, and non-sensitising [426]. Taget oil is hypotensive in rats at a dose of 0.05 g/kg [352].

Comments: Great caution needs to be exercised with taget oil, since there is a very strong risk of phototoxicity if the oil is applied to the skin in the amounts normally used in aromatherapy. *Tagetes patula* is also known as 'French marigold' which frequently leads to confusion with *Calendula officinalis*, the common marigold. Undiluted taget oil is an orange colour, and may stain clothing.

Compare: Fig leaf[A], verbena

Tansy

Botanical name: *Tanacetum vulgare* (= *Chrysanthemum vulgare*)

Family: Asteraceae (= Compositae)

Oil from: Whole herb

Notable constituents:

Thujones	66–81%
Camphor	5%

Acute oral LD$_{50}$: 0.30 g/kg (dog); 1.15 g/kg (rat) [422]

Hazards:
Toxic (severe)
Neurotoxic

Contraindications: Should not be used in therapy, either internally or externally

Toxicity data & recommendations: Toxic signs produced by tansy oil poisoning include: convulsions; irregular heartbeat; vomiting; rigid pupils; gastroenteritis; uterine bleeding; flushing; hepatitis; cramps; loss of consciousness; and rapid breathing [159]. Fatal cases of poisonings from infusions of the herb have been reported [12].

Thujone acute oral LD$_{50}$ is 0.21 g/kg; chronic administration leads to fatty degeneration of the liver [354]. Intraperitoneal thujone was found to be convulsant and lethal to rats above 0.2 ml/kg [234,325]. Both the UK and EC 'standard permitted proportion' of *alpha*- and/or *beta*-thujone in food flavourings is 0.0005 g/kg of food [455, 456].

Comments: In the light of its high thujone content, tansy oil should be avoided altogether in aromatherapy. Averaging the dog and rat LD$_{50}$ values gives 0.73 g/kg.

Compare: Armoise, *Artemisia arborescens*, lanyana, sage (Dalmatian), thuja, western red cedar, wormwood

Tarragon

Synonym: Estragon

Botanical name: *Artemisia dracunculus*

Family: Asteraceae (= Compositae)

Oil from: Leaves

Notable constituents (French):
Estragole	70-87%
Methyleugenol	0.1-1.5%

Notable constituents (Russian):
Estragole	0.1-17%
Methyleugenol	5-29%

Acute oral LD$_{50}$: 1.9 ml/kg [420]

Hazards:
Carcinogenic*
Hepatotoxic*
*Assumed from estragole and methyleugenol content

Contraindications: Should not be used in therapy, either internally or externally

Toxicity data & recommendations: Estragole (= methyl chavicol) is genotoxic [122, 289]. Tarragon oil is mutagenic, with activity residing in the estragole fraction of the oil [281]. Estragole is carcinogenic in mice [309,312] because it is metabolised, in vivo, to the carcinogenic compound 1'-hydroxyestragole [62, 123, 312, 478]. The same metabolic process is believed to take place in humans [30]. High doses are potentially carcinogenic, but low doses are not, since they are readily detoxified [62, 306]. Methyleugenol is genotoxic [122, 289, 291, 479]; it comes into the same category as safrole and estragole, and is carcinogenic in rodents [309,310].

Neither estragole nor methyleugenol are restricted by any regulatory agencies. However, the carcinogenic potential of these compounds suggests that both the French and Russian types of tarragon oil should be avoided altogether in aromatherapy

Comment: Commercially available tarragon oils are mostly estragole-rich, and all possess at least 10% of estragole or methyleugenol or both. Those oils low in one component tend to be high in the other. It might be possible to find a tarragon oil which is very low in both compounds and therefore safe to use in aromatherapy, perhaps with a restriction on use level.

Compare: Basil, fennel, *Ravensara anisata*

Tea[A]

Botanical name: *Camellia sinensis* (= *Thea sinensis*)

Family: Theaceae (= Camelliaceae)

Absolute from: Leaves

Notable constituents: No data could be found

Acute oral LD$_{50}$: No data could be found

Hazards: Dermal sensitiser (severe)

Contraindications: Should not be used in therapy, either internally or externally

Toxicity data & recommendations: Tea absolute has been found to cause sensitisation in dilutions as low as 0.001% in guinea pigs; it has not been tested on humans (private communication from RIFM). IFRA recommends that tea absolute is not used as a fragrance ingredient or in fragrance ingredients. Best avoided in aromatherapy.

Comments: It is very unlikely that tea absolute has ever been used in aromatherapy, and it is of very little interest today to the fragrance industry.

Compare: Costus, elecampane

Tejpat leaf

Botanical name: *Cinnamomum tamala*

Family: Lauraceae

Oil from: Leaves

Notable constituents: Eugenol 75–80%

Hazards:
  Dermal irritant* (moderate)
  Mucous membrane irritant*
  Hepatotoxic*
  Inhibits blood clotting*

*Assumed from eugenol content

Cautions (oral): Alcoholism, anticoagulants, haemophilia, kidney disease, liver disease, paracetamol, prostatic cancer, SLE

Cautions (IRV): Do not use at more than 3% concentration on mucous membrane

Toxicity data & recommendations: Eugenol is a powerful inhibitor of platelet activity [370, 434]. Platelet activity is essential for blood clotting. This being the case, it would be prudent to avoid oral administration of tejpat oil in people whose blood clots only slowly, whether or not this is caused by hereditary haemophilia. Caution is recommended in anyone taking anticoagulant drugs, such as aspirin, heparin or warfarin.

Eugenol has been shown to cause liver damage in mice whose livers have been experimentally depleted of glutathione [400]. As in paracetamol poisoning, pretreatment with *N*-acetylcysteine prevents glutathione loss and hence cell death [401]. Although the oral doses of eugenol used were high (5 ml/kg) the data suggest that oral doses of eugenol-rich oils such as tejpat oil might be best avoided in those with impaired liver function and in anyone taking paracetamol (= acetaminophen).

Comments: Tejpat leaf oil is commercially available in India, and occasionally further afield.

Compare: Clove

Terebinth

Synonym: Yarmor

Botanical name: *Pinus palustris*

Family: Pinaceae

Oil from: The oleoresin

Notable constituents:

Alpha-pinene	50%
Beta-pinene	25–35%
Delta-3-carene	20–60%

Hazards: Dermal irritant (if oxidised)

Cautions: Hypersensitive, damaged or diseased skin

Toxicity data & recommendations: Terebinth oil is a fairly frequent cause of dermatitis, although probably only from oxidised oils; oxidised *delta*-3-carene is thought to be responsible [74, 84]. Tested at 12% terebinth oil was both non-irritant and non-sensitising [427].

Comments: It is important to use relatively fresh oil for aromatherapy. It is almost certainly only oxidised oils which cause problems, due to terpene oxidation. It is very likely that all pine oils are potential irritants when oxidised, but not when fresh.

Compare: Pine (dwarf)

Thuja

Synonyms: White cedar, eastern arborvitae

Botanical name: *Thuja occidentalis*

Family: Cupressaceae

Oil from: Fresh leaves and twigs

Notable constituents:

Alpha-thujone	31–65%
Beta-thujone	8–15%
Fenchone	7–15%
Camphor	2–3%

Acute oral LD$_{50}$: 0.83 g/kg [420]

Hazards:
Toxic (severe)
Neurotoxic

Contraindications: Should not be used in therapy, either internally or externally

Toxicity data & recommendations: Toxic signs from thuja oil ingestion include convulsions, gastroenteritis, flatulence and hypotension. In severe cases, coma is followed by death [156]. Thuja oil caused epileptiform convulsions in a 50-year-old woman who took 20 drops twice a day for 5 days [234].

Thujone acute oral LD$_{50}$ is 0.21 g/kg; chronic administration leads to fatty degeneration of the liver [354]. Intraperitoneal thujone was found to be convulsant and lethal to rats above 0.2 ml/kg [234, 325]. Both the UK and EC 'standard permitted proportion' of *alpha*- and/or *beta*-thujone in food flavourings is 0.0005 g/kg of food [455, 456].

Comments: Poisoning has only been reported after oral ingestion. However, the high thujone content suggests that the oil should be avoided altogether. Thuja oil is also known as cedarleaf oil.

Compare: Armoise, *Artemisia arborescens*, lanyana, sage (Dalmatian), tansy, western red cedar, wormwood

Thyme

Botanical name: *Thymus zygis, Thymus vulgaris, Thymus serpyllum* (= wild thyme)

Family: Lamiaceae (= Labiatae)

Oil from: Herb

Notable constituents:

Thymol chemotype:
Thymol	32–63%
Carvacrol	1–5%

Carvacrol chemotype:
Thymol	1–13%
Carvacrol	23–44%

Thymol/carvacrol chemotype:
Thymol	26%
Carvacrol	26%

Wild thyme:
Thymol	1–16%
Carvacrol	21–37%

Acute oral LD$_{50}$: 4.7 g/kg [420]

Hazards:
Dermal irritant (moderate)
Mucous membrane irritant (strong)

Cautions (dermal): Hypersensitive, diseased or damaged skin, children under 2 years of age

Cautions (IRV): Do not use at more than 1% concentration on mucous membrane

Toxicity data & recommendations: Animal tests make thyme oil appear less toxic than would be expected from its content of thymol and carvacrol [420]. (Compare with toxicity levels of oregano and savory oils.) Undiluted thyme oil was severely irritating to both mouse and rabbit skin; tested at 12% it produced no irritation on human subjects [420].

Comments: The source of the Spanish thyme oil of commerce is *Thymus zygis*. Oils from *Thymus vulgaris* are available from France and other countries. Spanish *Thymus vulgaris* does exist; its relatively rare essential oil contains very little thymol, and is rich in limonene and cineole.

Several chemotypes have been identified for thyme. There are three important ones, from a safety standpoint: the thymol, carvacrol, and thymol/carvacrol chemotypes, shown above. Other chemotypes include: linalool, *alpha*-terpinyl acetate, and geranyl acetate. Moroccan thyme oil is distilled from *Thymus satureoides*, and is rich in borneol.

Compare: Ajowan, oregano, savory

Treemoss[A]

Botanical name: *Evernia furfuracea, Usnea barbata*

Family: Usneaceae

Absolute from: Dried moss

Notable constituents: Thujone?

Hazards:
  Dermal sensitiser (strong)
  Toxic?

Contraindications (dermal): Hypersensitive, diseased or damaged skin, all children under 2 years of age

Contraindications (IRV): Do not use on mucous membrane

Cautions (all routes): Use with caution generally, due to uncertain toxicity; use with caution in epilepsy and pregnancy

Maximum use level: 0.6%

Toxicity data & recommendations: IFRA recommends a maximum use level of 0.6% in the final product, in order to avoid allergic reactions; this recommendation is based on as yet unpublished RIFM test data on the sensitising potential of treemoss [430].

A RIFM monograph on treemoss *concrete* concludes that it is non-phototoxic, non-sensitising at 10%, and non-irritant at 10%. Oral toxicity was reported as 4.33 ml/kg [421]. Since the absolute is prepared directly from the concrete, we could surmise that the concrete might contain a substance which quenches the sensitising action of the absolute. We could also surmise that the absolute is likely to be more toxic than the concrete.

Comments: No information could be found concerning the composition of treemoss absolute. Considering that it is so similar to oakmoss absolute, it may also contain thujone as a major component. Resinoids are also available.

Compare: Cassia, cinnamon bark, oakmoss[A], perilla, thuja

Verbena

Synonym: Lemon verbena

Botanical name: *Lippia citriodora (=Aloysia triphylla)*

Family: Verbenaceae

Oil from: Leaves

Note: There is a separate monograph for verbena absolute.

Notable constituents:

Geranial	20%
Neral	13%
Limonene	10%
Photocitrals	1.5%

Hazards:
  Dermal irritant (moderate?)
  Dermal sensitiser (strong)
  Phototoxic (moderate?)

Contraindications: Should not be used in therapy, either internally or externally

Toxicity data & recommendations: Six different samples of verbena oil, tested at 12% on 159 volunteers, produced two irritation reactions. Six different samples of verbena oil, all tested at 12%, produced differing results in maximation tests. The six tests produced 0/30, 2/28, 4/26, 13/25, 15/25 and 18/25 sensitisation reactions [429].

Verbena oil is not very powerfully phototoxic. Out of six samples tested, three were phototoxic when applied undiluted; one of these was not phototoxic at 12.5%, and another was not phototoxic at 50% [429]. There are three photocitrals in verbena oil, which are presumably responsible for its phototoxicity.

IFRA recommends that verbena oil should not be used as a fragrance ingredient, because of its sensitising and phototoxic potential.

Comments: Different verbena oils seem to present different levels of sensitisation risk. For practical reasons, the highest level of risk must be assumed. This level would be unacceptable in an aromatherapy context. IFRA's recommendation that verbena oil is not used in fragrances is presumably more due to its sensitisation potential than its fairly modest phototoxicity.

Compare: Cassia, cinnamon bark, costus, elecampane, verbena[A]

Verbena[A]

Synonym: Lemon verbena

Botanical name: *Lippia citriodora (=Aloysia triphylla)*

Family: Verbenaceae

Absolute from: Leaves

Note: There is a separate monograph for verbena essential oil.

Notable constituents: No data could be found

Hazards: Dermal sensitiser (strong)

Contraindications: Should not be used in therapy, either internally or externally

Toxicity data & recommendations: Verbena absolute, tested at 12% on three groups of volunteers, resulted in 0/27, 0/26 and 2/26 sensitisation reactions. Verbena absolute is non-irritant and non-phototoxic [429].

IFRA recommends that verbena absolute should not be used at over 1% in fragrance compounds (equivalent to 0.2% in the final product) [430].

Compare: Cassia, cinnamon bark, oakmoss[A], treemoss[A], verbena

Western red cedar

Synonyms: Pacific thuja, western arborvitae

Botanical name: *Thuja plicata*

Family: Cupressaceae

Oil from: Leaves

Notable constituents:

Alpha-thujone	5–10%
Beta-thujone	70–80%

Acute oral LD$_{50}$: No data could be found

Hazards:
Toxic* (severe)
Neurotoxic*

*Assumed from thujone content

Contraindications: Should not be used in therapy, either internally or externally

Toxicity data & recommendations: Thujone acute oral LD$_{50}$ is 0.21 g/kg; chronic administration leads to fatty degeneration of the liver [354]. Intraperitoneal thujone was found to be convulsant and lethal to rats above 0.2 ml/kg [234, 325]. Both the UK and EC 'standard permitted proportion' of *alpha-* and/or *beta*-thujone in food flavourings is 0.0005 g/kg of food [455, 456].

Comments: Western red cedar oil is not widely available, but exists. There are periodical attempts to introduce it as a substitute for *Thuja occidentalis*. Even rarer is an oil from the *wood* of *Thuja plicata*. According to Arctander, this oil 'is poisonous due to the presence of a ketone, *gamma*-thujaplicin.' [155].

Compare: Armoise, *Artemisia arborescens*, lanyana, sage (Dalmatian), tansy, thuja, wormwood

Wintergreen

Synonym: Gaultheria

Botanical name: *Gaultheria procumbens*

Family: Ericaceae

Oil from: Leaves

Notable constituents: Methyl salicylate 98%

Acute oral LD$_{50}$: 1.2 g/kg (methyl salicylate) [424]

Hazards: Toxic (moderate/severe)

Contraindications: Should not be used in therapy, either internally or externally; do not use if taking anticoagulants

Toxicity data & recommendations: Numerous cases of methyl salicylate poisoning have been reported, with a 50–60% mortality rate; 4–8 ml is considered a lethal dose for a child [424]. Methyl salicylate could be three to five times more toxic in humans than in rodents (see p. 50). In the years 1926, 1928 and 1939–1943, 427 deaths occurred in the USA from methyl salicylate poisoning [237].

Common signs of methyl salicylate poisoning are: CNS excitation; rapid breathing; fever; high blood pressure; convulsions; coma. Death results from respiratory failure after a period of unconsciousness [424]; 0.5 ml of methyl salicylate is approximately equivalent to a dose of 21 aspirins. Methyl salicylate can be absorbed transdermally in sufficient quantities to cause poisoning in humans [100].

Topically applied methyl salicylate can potentiate the anticoagulant effect of warfarin, causing side-effects such as internal haemorrhage [521]. A similar interaction is possible, but by no means certain, with other anticoagulants such as aspirin and heparin. Many linaments contain methyl salicylate or wintergreen oil.

Comments: Virtually all commercial 'wintergreen oil' is in fact synthetic methyl salicylate. There have been sufficient cases of poisoning by methyl salicylate or by oils containing it that it would be prudent to avoid all use of this oil.

Compare: Birch (sweet)

Wormseed

Synonym: Chenopodium

Botanical name: *Chenopodium ambrosioides*

Family: Chenopodiacae

Oil from: Whole plant, including fruit

Notable constituents: Ascaridole 60–80%

Acute oral LD$_{50}$: 0.25–0.38 g/kg [422]

Hazards:
   Toxic (severe)
   Neurotoxic

Contraindications: Should not be used in therapy, either internally or externally

Toxicity data & recommendations: Wormseed oil is toxic even at low doses because of its depressant action on the heart. Cases of fatal poisoning in children have occurred after ingestion of as little as 6 drops of wormseed oil [271, 415]. Inhalation of the oil for even a short period is likely to induce headache [12]. Of the recorded cases of poisoning, 70% have resulted in death [153]. Damage to the CNS is a common feature of poisoning. See page 55 for more detail on cases of poisoning.

Toxic signs include: skin and mucous membrane irritation; headache; vertigo; tinnitus; double vision; nausea; vomiting; constipation; deafness; blindness; damage to the kidneys, liver and heart [422]. IFRA recommends that wormseed oil should not be used as a fragrance ingredient or in fragrance ingredients.

Wormseed oil can explode when heated [5].

Comments: One of the most toxic essential oils. In the past, wormseed oil was commonly used as a remedy for intestinal parasites in children. If the amount given was a mere two or three times the recommended dose of 1 drop per year of age, a fatal outcome was possible. The toxicity of both boldo and wormseed oils is undoubtedly due to their content of ascaridole.

Compare: Boldo

Wormwood

Synonyms: Absinthe, artemisia

Botanical name: *Artemisia absinthium*

Family: Asteraceae (= Compositae)

Oil from: Leaves and flowering tops

Notable constituents: Thujone 34–71%

Acute oral LD$_{50}$: 0.96 g/kg [421]

Hazards:
Toxic (severe)
Neurotoxic

Contraindications: Should not be used in therapy, either internally or externally

Toxicity data & recommendations: Thujone acute oral LD$_{50}$ is 0.21 g/kg; chronic administration leads to fatty degeneration of the liver [354]. Intraperitoneal thujone was found to be convulsant and lethal to rats above 0.2 ml/kg [234, 325]. Both the UK and EC 'standard permitted proportion' of *alpha*- and/or *beta*-thujone in food flavourings is 0.0005 g/kg of food [455, 456]. Toxic signs include: vomiting; convulsions; vertigo; insomnia; nightmares.

In 1915, France banned the production of *absinthe* containing wormwood oil. It was claimed, with some justification, that the oil acted as a narcotic in higher doses, and was habit-forming. It was, and is still believed that the thujone in wormwood oil was largely or solely responsible for these effects [439]. It has been suggested that thujone and *delta*-9-tetrahydrocannabinol, the most active ingredient in cannabis, interact with a common receptor in the CNS and so have similar pharmacological effects [201]. Absinthe survives in the form of *Pernod*, but without the wormwood.

Comments: Today, wormwood oil is widely used as a food flavour at a concentration of about 60 p.p.m., and as an ingredient in alcoholic drinks at around 250 p.p.m. [12]. Wormwood oil is often confused with armoise oil.

Compare: Armoise, *Artemisia arborescens*, lanyana, sage (Dalmatian), tansy, thuja, western red cedar

Yarrow (camphor chemotype)

Synonym: Milfoil

Botanical name: *Achillea millefolium*

Family: Asteraceae (= Compositae)

Oil from: Flowering herb

Notable constituents:
Camphor	10–20%
1,8-cineole	14%
Borneol	9%

Hazards: Neurotoxic—assumed from camphor content (mild)

Contraindications (oral): Pregnancy

Cautions (oral): Epilepsy, fever

Toxicity data & recommendations: Camphor readily causes epileptiform convulsions if taken in sufficient quantity [269, 272]. The camphor chemotype of yarrow oil should therefore be used with caution in oral doses.

Comments: Yarrow oil exists in many different chemotypic forms, and the composition of commercial oils can vary according to country of origin. Other chemotypes are rich in sabinene, 1,8-cineole or chamazulene. None of these is likely to present a hazard in aromatherapy. Yarrow oil has been used as a substitute for German chamomile oil.

Compare: Camphor, lavandin, *Lavandula stoechas*, rosemary, spike lavender

11

Chemical index

The purpose of this index is to provide a brief outline of the important essential oil constituent chemicals, most of which are referred to in other sections of the book. The index also provides additional reference material to complement, in particular, the essential oil profiles. At the same time it may provide a useful rough guide to the safety of any essential oil whose composition is known, but which is not included in this book.

Virtually all of the listed chemicals are available as either isolates or synthetics (or both) as well as occurring naturally in essential oils. Isolates are chemicals which are (in this case) taken from essential oils as opposed to being synthesised.

The index lists those chemicals which are either known to be hazardous, occur frequently in essential oils, or are present in large amounts in a few essential oils. This, however, is not a comprehensive catalogue of essential oil components—such a list would include several hundred additional materials.

Other chemicals, and classes of chemicals, which are of relevance to essential oil safety are also listed.

Especially hazardous materials are marked with the symbol:

For these components, the essential oils they are most commonly found in, and the percentage of component in each oil are listed where this is known. This listing is also given in a few other particularly useful cases, such as citral and eugenol, which are frequently referred to in the book.

Note that some essential oil constituents are dangerous when pure, but may not present a hazard in the amounts in which they occur in essential oils. In such cases we have stated: 'Its presence in an essential oil should not be cause for concern'. Examples include benzyl acetate and guaiacol.

Recommendations from the International Fragrance Association (IFRA) Code of Practice are included where applicable. Note that where maximum use levels are given these apply to the final product. In some instances IFRA gives maximum use levels for fragrance compounds, i.e. for the fragrance in a product. This is always based on the assumption that fragrance compounds are used at 20% in the final product.

For the sake of simplicity, LD_{50} levels are not given in every instance. Where the acute oral LD_{50} is over 2.0 g/kg, we have stated 'non-toxic'.

⚠ Alantolactone

A sesquiterpenoid lactone which is a major component (52%) of elecampane oil [518]. Powerful skin sensitiser [422].

Iso-alantolactone

Similar to alantolactone, constituting 33% of elecampane oil [518].

Alcohols

Common class of essential oil terpenoid with no particular toxicity associated. Alcohols possess an -OH functional group.

Aldehydes

Common class of essential oil terpenoid. Implicated in some dermal sensitivity reactions. Contain -CHO functional group.

Alkenylbenzene

Class of chemical compound including safrole and estragole some of which possess carcinogenic potential. Also known as alkylbenzene and alkenebenzene.

Allicin (= diallyl thiosulphinate)

Sulphur-containing principle of garlic with true garlic odour, and a minor component of garlic oil. Converted to diallyl disulphide and related sulphide materials when exposed to air or on decomposing on distillation (see Diallyl disulphide). Acute s.c. LD_{50} for allicin 0.12 g/kg; acute i.v. LD_{50} 0.06 g/kg [356]. Allicin may be the cause of sensitisation in people allergic to garlic [362].

⚠ Allyl isothiocyanate

Found in:

Horseradish	50%
Mustard	99%

A pale yellow, pungent, irritating liquid with an acrid taste found in oils of mustard and horseradish. (Also found in raw cabbage and used in food flavourings, but in both cases at very low concentrations.) Extremely toxic, and a violent irritant to mucous membranes and skin. Acute oral LD_{50} of a 10% solution in corn oil was 0.34 g/kg in rats, producing jaundice [338]. IFRA recommends that it should not be used as a fragrance ingredient or in fragrance ingredients [430].

Trans-Anethole (= anethole)

Found in:

Anise	80–90%
Fennel	52–86%
Star anise	75–90%

Terpenoid which is commonly used in food flavourings. The rare *cis*-isomer (i.p. rat LD_{50} 0.093 g/kg) is many times more toxic than the common *trans*-isomer. *Trans*-anethole is non-toxic, non-irritant and non-sensitising [419]. Weak oestrogen-like activity. Anethole-rich oils should be avoided altogether in people with oestrogen-dependent cancers, and should not be taken in oral doses by women who are pregnant, breast-feeding, or have endometriosis. *Trans*-anethole has a zero- to low-level genotoxic activity [122, 289, 291] and is not significantly carcinogenic [304, 309]. A metabolite, 3-hydroxyanethole, is not carcinogenic [309]. Another metabolite, anethole 1,2-epoxide, is non-genotoxic [504].

Angelates

Esters (e.g. isoamyl angelate, isobutyl angelate) found chiefly in Roman chamomile oil. No evidence of toxicity.

p-Anisaldehyde (= anisic aldehyde)

Non-terpenoid aldehyde found as a minor constituent of anise, star anise, fennel and other oils. LD_{50} 1.5 g/kg [338]. Mildly irritant, non-sensitising [420].

Anisic aldehyde

See *p*-anisaldehyde.

Apiol (dill apiol)

Found in:

Dill herb	< 1%
Indian dill seed	20–30%
Dill root (NCA)	11%

Dill apiol and parsley apiol are isomers with very similar chemical structures, and presumably similar pharmacology. They are both sensitive to decomposition on storage. Dill apiol has a very low level of genotoxicity [122, 289] and is non-carcinogenic [309]. For toxicity of dill apiol, see parsley apiol (below).

Apiol (parsley apiol)

Found in:

Parsley leaf	< 18%
Parsleyseed	21–80%
Dill root (NCA)	20%

Both apiol and various preparations of parsley

have been used for many years to procure illegal abortion in Italy [500, 502, 503]. The lowest daily dose of apiol which induced abortion was 0.9 g, taken for 8 consecutive days. This is approximately equivalent to 6 ml of parsley leaf oil, between 1.5 ml and 6 ml of parsleyseed oil, or 5 ml of Indian dill oil. The inevitable conclusion is that all of these apiol-rich essential oils present a very high risk of abortion if taken in oral doses, and that external use would also seem inadvisable in pregnancy.

There are several recorded cases of fatal poisoning from apiol ingestion [501, 502, 503]. The lowest total dose of apiol causing death is 6.3 g (2.1 g/day for 3 days) [530]; the lowest fatal daily dose is 0.77 g, which was taken for 14 days [533]; the lowest single fatal dose is 8 g [500]. At least 19 g has been survived [530] but if 1 or 2 g can be fatal to an adult, it would seem prudent to completely avoid oral dosages of apiol-rich oils.

Common signs of apiol poisoning are fever, severe abdominal pain, vaginal bleeding, vomiting and diarrhoea [501, 530]. Post-mortem examination invariably reveals considerable damage to both liver and kidney tissue, often with gastrointestinal inflammation and sometimes damage to heart tissue [501, 503, 533].

Artemisia ketone (= santolinenone)

Found in:

Annual wormwood	35–63%
Lanyana	1–24%
Lavender cotton	10–45%

The major component of lavender cotton (santolina) oil. No toxicity data could be found for artemisia ketone, but some toxicity is suspected.

Asarone

Found in:

Acorus calamus var. *calamus*	8–19%
Acorus calamus var. *angustatus*	45–80%

There are *alpha* and *beta*-isomers, also known as *trans*- and *cis*-asarone, respectively. *Beta*-asarone is genotoxic [445] and is presumed to be the component of *Acorus calamus* which makes it carcinogenic in rodents (see p. 95).

Beta-asarone is banned in the USA as a pharmaceutical ingredient, while the *alpha*-isomer is permitted. Both the UK and EC 'standard permitted proportion' of *beta*-asarone in food flavourings is 0.0001 g/kg [455, 456]. IFRA recommends that *alpha*- and *beta*-asarone should not be used as fragrance ingredients, and that the level of asarone in consumer products containing calamus oil should not exceed 0.01% [430].

Ascaridole

Found in:

Boldo	16%
Wormseed	60–80%

Anthelmintic terpene peroxide which constitutes 60–80% of wormseed (chenopodium) oil. It is prone to explode when heated or mixed with organic acids [5]. Extremely toxic, as are many de-worming agents [12].

Azulene

See Chamazulene.

Benzaldehyde

The major component of bitter almond oil. A yellowish liquid with an almond smell and burning aromatic taste. Fairly water-soluble. Acute oral LD_{50} varies from 1.0 to 2.85 g/kg; some report it as non-irritant, others as strongly irritant; can produce allergic reactions in sensitive individuals; narcotic in high doses; oral fatal dose in humans about 50-60 ml [422]. Its presence in an essential oil should not be cause for concern.

Benzanol

See Benzyl alcohol.

Benzene

Clear, colourless, highly flammable liquid, present in automobile fuel. Acute oral LD_{50} 3.8 ml/kg. Carcinogenic; ingestion or inhalation in humans

can cause irritation of mucous membranes, restlessness, convulsions, excitement, depression; harmful amounts may be absorbed through the skin [158].

Used as a solvent for the extraction of concretes. Absolutes, which are processed from concretes, are sometimes used in aromatherapy, and may contain traces of benzene. IFRA recommends that the level of benzene should be kept as low as practicable, and should never exceed 10 p.p.m. in fragrance compounds [430].

Benzoic acid

Odourless material found in gum benzoin. Used as a preservative in foodstuffs and soft drinks. Excreted by most vertebrates except fowl. Found as a minor component in oils of ylang-ylang, neroli and vetiver. Mildly irritating to skin and mucous membranes; acute oral LD_{50} in rats 2.0 g/kg, risk of cumulative effects [158, 425]. Its presence in an essential oil should not be cause for concern.

⚠ Benzo[*a*]pyrene

Probably the best-known example of the carcinogenic polynuclear hydrocarbons typically found in smoke and tar. Many of them can be readily metabolised by the liver's P_{450} enzyme system to highly reactive and genotoxic compounds, often epoxides, which are known to be potent initiators and promoters of cancer.

Benzo[*a*]pyrene is one of the carcinogenic components of cigarette smoke. It is also found in unrectified cade oil at 8000 p.p.b. and in rectified cade oil at < 20 p.p.b. [462]. Rectified cade oil, in which most of the polynuclear hydrocarbons are removed by fractional distillation, is regarded by IFRA as safe for use in fragrances [430]. It is possible that birch tar oil also contains carcinogenic polynuclear hydrocarbons such as benzo[*a*]pyrene (see note on p. 203).

Benzyl acetate

Major constituent of jasmine absolute (10–25%) and a minor component of many essential oils such as ylang-ylang. Mucous membrane irritant which can cause vomiting and diarrhoea if ingested [158]. Non-toxic, non-sensitising, but the vapours are irritating to the eyes and respiratory passages [419].

Benzyl acetate has been shown to weakly promote pancreatic tumours in rats, but only when the animals had been previously exposed to a carcinogen. It is very unlikely that this represents a hazard in an aromatherapy context [303]. Its presence in an essential oil should not be cause for concern.

Benzyl alcohol (= benzanol)

Minor component of ylang-ylang and jasmine. Non-toxic [158, 338].

Benzyl benzoate

Ester widely used for the treatment of scabies. Found in ylang-ylang (5–15%) and many other essential oils. Non-toxic, non-sensitising, can cause irritation [419].

⚠ Bergapten (= 5-methoxypsoralen = 5-MOP)

Found in:

Bergamot	0.3–0.4%
Grapefruit (expressed)	0.012–0.013%
Lemon (expressed)	0.15–0.25%
Lime (expressed)	0.1–0.3%
Orange (bitter, expressed)	0.069–0.073
Rue	no reliable figures
Skimmia laureola	
(controversial)	no reliable figures
Taget	no reliable figures

Furanocoumarin found in bergamot oil and in some other cold-pressed citrus oils. Phototoxic [485] may increase damaging effects of ultraviolet light, although can also protect against them [441, 442], potentially photocarcinogenic [483]. IFRA recommends that the total bergapten content in consumer products should not exceed 15 p.p.m. (0.0015%). See Xanthotoxin.

Bisabolene

Sesquiterpene found in *Origanum vulgare* and opopanax oils. No evidence of toxicity.

Alpha-Bisabolol

Major sesquiterpenoid component of German chamomile and stenophylla oils. Toxicity is very low (toxicity threshold 1–2 ml/kg) [331].

Borneol (Borneo camphor)

There are two optical isomers, of which the d-form predominates. Found in Borneo camphor, (see p. 213) rosemary, citronella, valerian and other oils. Related chemically to camphor but an alcohol (unlike camphor, which is a ketone) and apparently less toxic. May cause nausea, confusion, dizziness and convulsions [158]. Non-toxic, non-irritant, non-sensitising [424].

Cadinene

Occurs in several isomeric forms, notably *alpha* and *beta*, which together occur in over 150 essential oils. *Delta*-cadinene is found in juniper oil (3%). Non-toxic, non-irritant, non-sensitising [419].

Camphene

Terpene found in turpentine, citronella, rosemary, yarrow, valerian and many other essential oils. High vapour pressure, so volatilises easily. Non-toxic, non-irritant, non-sensitising [421].

⚠ Camphor

Found in:

Armoise	30%
Balsamite (camphor CT)	72–91%
Camphor (white)	30–50%
Ho leaf (camphor/safrole CT)	42%
Lanyana	4–10%
Lavandin	5–15%
Lavandula stoechas	15–30%
Rosemary	10–20%
Sage (Dalmatian)	1–26%

Sage (Spanish)	11–35%
Spike lavender	10–20%
Yarrow (camphor CT)	10–20%

This ketone is found at highest levels in the essential oils of balsamite, camphor CT (72–91%), white camphor (35%) and armoise (30%). Ingestion may cause nausea, vomiting, vertigo, confusion, delirium, convulsions, coma, respiratory failure and death [158]. There have been reports of instant collapse in infants following the local application of camphor to their nostrils [5]. Camphor is related to borneol but more toxic. Very readily absorbed through skin and mucous membranes, and crosses the placenta freely; classified as very toxic, with a probable human lethal dose of 0.05–0.55 g/kg [424]. Camphor readily causes epileptiform convulsions if taken in sufficient quantity [269, 272]. In Great Britain, the occupational exposure standards for camphor are 12 mg/m³ (long term) and 18 mg/m³ (short term) [5].

Delta-3-carene

Terpene found in turpentines from various *Pinus* species and in many other essential oils, including angelica root (9%). Non-toxic, mildly irritant, non-sensitising; however, its auto-oxidation products (made very slowly on exposure to air) are powerfully eczematous [419]. This means that there is a risk of skin sensitisation from old or poorly stored batches of *delta*-3-carene-rich essential oils.

Carotol

Bicyclic sesquiterpenoid alcohol, which is the major component of carrot seed oil. No toxicology data could be found, and not suspected of any toxicity.

⚠ Carvacrol

Found in:

Oregano	0.5–84%
Savory	3–67%
Thyme	1–44%

Phenolic terpene which is irritating and corrosive because of its phenol-like nature, and can produce sensitisation. Carvacrol is toxic: acute oral LD_{50} 0.1–0.81 g/kg [425]. It lends a moderate irritancy and a moderate toxicity to those oils containing it in quantity.

Carvone

Terpenoid ketone with both optical isomers found in nature. The *d*-isomer is found in caraway and dill seed oils and has the odour of caraway. The *l*-isomer is the main component of spearmint oil. Acute oral LD_{50} 1.64 g/kg in rats [336]. Non-irritant, non-sensitising [419, 424]. Carvone is not hepatotoxic [111].

Caryophyllene

See *Beta*-caryophyllene.

Alpha-Caryophyllene

See Humulene.

Beta-Caryophyllene

Sesquiterpene occurring in many essential oils, especially clove and black pepper. Has an odour midway between clove and turpentine. Non-toxic, mildly irritating, non-sensitising [419].

Alpha-Cedrene

The major component of Virginian cedarwood oil, and a minor component of several other oils. Non-toxic, non-irritant, non-sensitising [424].

Cedrol

Occurs in cypress, Atlas cedarwood, and other coniferous essential oils. Non-toxic, non-irritant, non-sensitising [421].

Chamazulene

Intensely blue compound found in oils of German chamomile, yarrow and wormwood, with a very unusual molecular structure (seven-membered ring fused to a five-membered ring). Non-toxic. Crystalline when pure [157].

Chavicol

A relative of estragole, perhaps suspect by association, but no evidence of carcinogenicity. Found in West Indian bay oil (*Pimenta racemosa*) at 11–21%.

1,8-cineole (= cineole)

Also known as eucalyptol or cajeputol. Terpenoid which is the main constituent of most eucalyptus oils and is also found in rosemary and many other oils. Camphor-like odour, cooling taste. Non-toxic, non-irritant, non-sensitising [421]. Low oral toxicity according to some sources [488] but 1 ml has caused transient coma. Recovery has occurred after ingestion of 30 ml. Poisoning produces severe gastrointestinal and CNS effects. 1,4-cineole is found very rarely in essential oils.

⚠ Cinnamaldehyde (= cinnamic aldehyde)

Found in:

Cassia	75–90%
Cinnamon bark	60–75%

Oral LD$_{50}$ 2.2 g/kg in rats [158]. Powerful skin sensitiser [422]. IFRA recommends the use of cinnamaldehyde in conjunction with substances preventing sensitisation, for example an equal amount of *d*-limonene, which is present in large amounts in lemon, orange, grapefruit, etc. [430]. Cinnamaldehyde has been found to depress rat liver glutathione levels [475]. Cinnamaldehyde-rich oils would, therefore, be best avoided orally if also taking paracetamol (= acetaminophen).

Cinnamic acid

A minor, or trace component of cinnamon bark oil, and a minor component of tolu balsam oil. Produced in vivo by oxidation of cinnamaldehyde [38]. Cinnamic acid does not deplete hepatic glutathione [475]. Non-toxic, non-irritating, non-

sensitising on healthy skin, sensitising in some people with eczema [424].

⚠ Cinnamic aldehyde

See Cinnamaldehyde.

Citral

Found in:

Backhousia citriodora	90%
Cymbopogon jawarancusa subsp.	
olivieri (NCA)	< 60%
Eucalyptus staigeriana	16–40%
Lemongrass	75%
May chang	75%
Melissa	35–55%
Verbena	33%

The common name for two isomers of a widely distributed terpene aldehyde, geranial (citral a) and neral (citral b). Citral from natural sources is a mixture of the two isomers. Can cause dermal sensitisation but this effect is markedly reduced by the presence of *d*-limonene or *alpha*-pinene [422]. Non-toxic [338]. Citral can cause a rise in ocular tension, which would not be good in cases of glaucoma. In tests a very low daily oral dose (2–5 µg) produced an increase in ocular pressure within 2 weeks in monkeys [348].

In rats, the administration of 185 mg/kg/day of citral for 3 months produced benign prostatic hyperplasia [507] which may be tetosterone-dependent [508]. Other studies have suggested that the effect occurs via competition with steroid receptors and that citral has an oestrogenic effect in causing vaginal hyperplasia in rats [510]. A receptor-mediated mechanism may also be responsible for citral's reported ability to cause sebaceous gland proliferation [509]. In all the above studies, citral was applied topically, in amounts equivalent to around 10 ml in a human.

It has been reported that citral can impair reproductive performance in female rats by reducing the number of normal ovarian follicles [199]. The effect has been seen only after a series of i.p. injections at a dose of 0.3 g/kg.

Citronellal

Found in:

Citronella	30–48%
Combava leaf	58–82%
Eucalyptus citriodora	70–75%
Melissa	4–39%

Terpene aldehyde. Non-toxic, mildly irritant; can cause occasional sensitisation; believed to be cause of eczematous irritation from citronella oil [421].

Citronellic acid

Carboxylic acid terpenoid, occurring in three isomeric forms, *d, l* and *dl*. The *d* form is a minor constituent of Java citronella, geranium, petitgrain and lemongrass oils. Acute oral LD$_{50}$ 2.61 g/kg; mildly irritant [426].

Citronellol

Terpene alcohol sometimes called *beta*-citronellol or cephrol. The *l*-isomer is a constituent of rose and geranium oils; the *d*-isomer occurs in Java and Ceylon citronella oils. Non-toxic, mildly irritant, non-sensitising [421].

Citronellyl acetate

Minor component of Ceylon citronella, and about 20 other essential oils. Non-toxic, moderately irritating, non-sensitising [419].

Citronellyl formate

Found in geranium oil at around 7%. Non-toxic, moderately irritating, non-sensitising [419].

Copaene

There are both *alpha*- and *beta*-isomers of this sesquiterpene. Found in cubeb, lovage, copaiba and several other oils. Probably non-toxic [157].

⚠ Costunolide

Sesquiterpene lactone found in costus. No reliable figures on percentage. Dermal sensitiser [157].

⚠ Costuslactone

Sesquiterpene lactone found in costus. No reliable figures on percentage. Powerful dermal sensitiser [157].

Coumarin

Found in:

Deertongue[A]
Flouve
Hay[A]
Sweet clover[A]
Tonka[A]
Woodruff[A]

The above materials, most only available as concretes or absolutes, all contain high levels of coumarin, but there are no reliable published figures for percentage levels.

Coumarin is a lactone found in quantity in several absolutes, and in very small amounts in a few essential oils. Chronic oral administration of coumarin produced severe liver damage at a level of 2500 p.p.m. in the diet of rats and 100 mg/kg per day for dogs [347]. This hepatotoxicity was confirmed by subsequent research [336, 469] and one of these studies reported finding bile duct carcinomas in coumarin-fed rats [469]. Coumarin shows a high toxicity in several animals; its acute oral LD$_{50}$ is 0.2 g/kg in both guinea pigs and mice [420] and 0.68 g/kg in rats [338]. Based on all the animal data, the FDA prohibited the use of coumarin in food in 1954; both the UK and EC 'standard permitted proportion' of coumarin in food flavourings is 0.002 g/kg [455, 456].

It was noticed by several workers that there was a significant interspecies difference in the way coumarin was metabolised [108, 525, 526, 527]. For example, the percentage of orally administered coumarin excreted in the urine as 7-hydroxycoumarin is 1% in the guinea pig, 3% in the dog, 19% in the cat, and 60% in the baboon [527]. In the 1970s the relevance of the animal data to humans was being seriously questioned. One study, giving coumarin to baboons at 50 p.p.m. and 100 p.p.m. (3 weeks at each dosage level) did find histological damage in the liver [527]. In another study, coumarin was administered to baboons for between 16 and 24 months at 2.5, 7.5, 22.5 or 67.5 mg/kg/day [528]. At the highest dose level only, an increase in liver weight was noted, and ultrastructural examination of the liver revealed dilatation of the endoplasmic reticulum.

In clinical trials using coumarin as an anticancer agent only 0.37% of patients developed (reversible) abnormal liver function. The majority of 2173 patients received 100 mg/day of coumarin for 1 month, followed by 50 mg/day for 2 years. (The other patients received between 25 mg/day and 2000 mg/day.) Of the total, only eight patients developed elevated liver enzyme levels, which returned to normal on stopping the coumarin [520]. On this evidence, coumarin cannot be regarded as hepatotoxic in humans.

The second baboon study indicates a no-effect level of 22.5 mg/kg/day, which equates to around 1.5 g/day for an average (65 kg) adult human. This dosage level is 30 times higher than the 50 mg/day received by many of the cancer patients for 2 years.

Although coumarin derivatives are used as anticoagulant drugs (warfarin, dicoumarin), coumarin itself is not anticoagulant [529]. Coumarin is now used in the treatment of lymphoedema, has been tried in the treatment of various cancers, and is reported to be an immunostimulant [5]. Coumarin is non-sensitising [420].

Para-Cresol

A phenol found in cade oil. Percentage level uncertain. Non-toxic, non-sensitising, mildly irritant [157, 420].

Cuminaldehyde

Terpene aldehyde constituent of at least 50 essential oils, including cumin, cassia, cinnamon, eucalyptus and myrrh. Relatively non-toxic (acute oral LD_{50} in rats 1.39 g/kg) non-irritant, non-sensitising [420].

⚠ Cyanide

See Hydrocyanic acid.

Para-Cymene

Terpene hydrocarbon which occurs in many essential oils, such as frankincense, marjoram and tea tree. Non-toxic, non-sensitising, but a mild or moderate irritant [420].

2-decenal (= 2-decen-1-al)

Major component of coriander leaf oil (13–23%). Non-toxic, non-irritant, non-sensitising [425].

Diallyl disulphide

The major component of garlic oil (60%). Acute oral LD_{50} 0.26 g/kg, acute dermal LD_{50} 3.6 g/kg; some irritancy, non-sensitising [428]. May cause sensitisation in people allergic to garlic [362]. Diallyl disulphide inhibits thyroid function. Possible contraindication for oral garlic oil in patients with low thyroid function. See Allicin.

Diallyl thiosulphinate

See Allicin.

Diisopropyl disulphide

Sulphur-containing compound found as a major component in onion oil. Also present in garlic oil. May be irritating to skin and mucous membranes.

Diosphenol

Terpenoid ketone and phenol found in buchu oil. Likely to possess the irritancy generally associated with phenols.

Dipentene

Mixture of *d*- and *l*-limonene. Terpene hydrocarbon minor constituent of many oils including frankincense and palmarosa oils. Non-toxic, mildly irritant, non-sensitising [420].

Egomaketone

Egomaketone and *iso*-egomaketone are found as major components of unusual chemotypes of perilla oil. As with perilla ketone, they are both pulmonary toxins to cattle and laboratory animals, but there is no evidence that they have a similar effect on humans [332, 333]. Perilla oil is used as a food flavouring in Japan.

Elemicin

Found in:

Cinnamomum cecidodaphne level unknown	
Elemi	3.0–12%
Mace	0.2–2%
Nutmeg	1.5–5.5%
Parsley leaf	0.2–3%

Alkenylbenzene with possible, but unproven hepatotoxic and psychotropic properties. Similar chemical structure to the carcinogenic alkenylbenzenes estragole and safrole. Two studies revealed only a low to moderate level of genotoxicity [122, 289]. Recent research found it to be significantly genotoxic [445]. One study found both elemicin and its metabolite, 1´-hydroxyelemicin to be non-carcinogenic [309]. Other work found that 1´-hydroxyelemicin possessed a weak, but statistically significant carcinogenic activity [453]. Currently there is insufficient data to regard elemicin as carcinogenic.

Iso-Elemicin

Can be metabolised to chemically reactive epoxide, so may be very weak hepatic carcinogen of the same general type as safrole [61]. Likely to present the same genotoxicity as elemicin.

Epoxides

Organic compounds which are products of animal or plant metabolic oxidations. They are sometimes formed by liver metabolism of essential oil constituents, in which case they are frequently toxic to the liver.

Esters

Class of essential oil ingredient found in many oils. Formed by the reaction between an organic acid and an alcohol.

⚠ Estragole (= methyl chavicol)

Found in:

Basil	5–87%
Chervil (NCA)	70–80%
Fennel	2–7%
Ravensara anisata	88%
Tarragon (French)	70–87%
Tarragon (Russian)	0.1–17%

Terpenoid ether. Genotoxic in rat hepatocytes [122, 289, 291]. Estragole is carcinogenic in mice [309, 312] because it is metabolised, in vivo, to the carcinogenic compound 1´-hydroxyestragole [62, 123, 312, 478, 479]. The same metabolic process is believed to take place in humans [30]. High doses are potentially carcinogenic, but very low doses are not, since they are readily detoxified [62, 306]. See page 98. Estragole is not restricted by any regulatory agencies, even though its carcinogenic potential is similar to that of safrole, which is restricted. Acute oral LD_{50} in mice is 1.25 g/kg. Mildly irritating, non-sensitising [422].

Eucalyptol

See 1,8-cineole.

Eugenol

Found in:

Bay (W. Indian)	38–75%
Betel leaf	28–90%
Cinnamon leaf	70–90%
Clove bud	70–95%
Clove stem	70–95%
Clove leaf	70–95%
Ocimum gratissimum	12–20%
Pimento berry	67–83%
Pimento leaf	60–95%
Tejpat leaf	75–80%

Phenol with analgesic properties used in dentistry. Strong mucous membrane irritant, mild dermal irritant, non-sensitising. Eugenol has caused peripheral vascular collapse and osteoporosis in large oral doses given to animals [154]. Acute oral LD_{50} ranges from 1.93 to 3.0 g/kg [421]. The anti-platelet activity of eugenol means that eugenol-rich oils should not be taken in oral doses by those whose blood clots slowly [370, 434].

Eugenol has been shown to cause liver damage in mice whose livers have been experimentally depleted of glutathione [400]. Although the oral doses of eugenol used were high (5 ml/kg) the data suggest that eugenol-rich oils should be used with caution in those with impaired liver function and that they should be avoided alongside the use of paracetamol (= acetaminophen). As in paracetamol poisoning, pretreatment with *N*-acetylcysteine prevents glutathione loss and hence cell death [401]. Eugenol is not genotoxic [122, 289, 291] or carcinogenic [309].

 ## *Iso*-Eugenol

Found in:

Clove bud	0.14–0.23%
Clove stem	0.14–0.20%

An isomer of eugenol found in small quantities in clove and other essential oils in two isomeric forms, *cis* and *trans*. Acute oral LD_{50} values vary from 0.54 to 1.56 g/kg. *Iso*-eugenol is irritating to human skin; in a closed-patch test on normal human subjects at 5% dilution, primary irritation (erythema) was produced in 3 out of 35 volunteers; non-sensitising when tested on 25 volunteers at 8% [421]. Further research by RIFM showed *iso*-eugenol to be sensitising at 8% and 10% [430]. IFRA recommends a maximum use level for *iso*-eugenol of 0.2% in products which will come into contact with the skin; for those which will not, such as air fresheners, a maximum use level of 0.5% is recommended [430].

Eugenyl acetate

Minor component of clove oils. Acute oral LD_{50} 1.7 g/kg [338]. Non-irritating; non-sensitising [420].

Farnesol

A sesquiterpene alcohol found as a minor component in many essential oils, including citronella, linden blossom, neroli, lemongrass and rose. The *cis-trans* isomer is found in petitgrain oil. No evidence of toxicity.

Fenchone

Terpenoid ketone. Occurs in bitter and sweet fennel oils, and in *Lavandula stoechas*. Non-toxic, non-irritant, non-sensitising [422].

Functional groups

In essential oils, usually oxygen-containing combinations of atoms within a terpenoid structure which contribute to the oils' chemical and aromatic characteristics. For example: alcohol, ester, ketone, aldehyde.

Furanocoumarins

Also known as furocoumarins. A cyclic structure containing oxygen. Associated with phototoxicity on exposure to UV light. See Bergapten.

Furans

Furans are oxygenated compounds where the oxygen atom is part of a ring. They are only found in a few essential oils.

Geranial

Terpenoid aldehyde. An isomer of citral found in many essential oils, but particularly in melissa, lemongrass, verbena, *Backhousia citriodora* and *Litsea cubeba*. Apparently non-toxic (see Citral and Neral).

Geraniol

Terpene alcohol found in many essential oils, notably rose, palmarosa and geranium. An isomer of linalool. Non-toxic; hypersensitivity has been noted in a few individuals [420].

Geranyl acetate

Terpenoid ester found as a minor component of citronella, palmarosa and other oils. Non-toxic, non-sensitising [474].

⚠ Guaiacol

Found in:

Cade	no reliable figures
Guaiacwood	60–85%

Also found as a minor component of Cupressaceae and pine oils and the distillation water of orange leaves. Moderately to severely irritating; non-sensitising; fairly toxic—acute oral LD_{50} in rats 0.725 g/kg [426]. With the exception of unrectified cade oil, its presence in an essential oil should not be cause for concern.

Humulene

Sesquiterpene isomer of *beta*-caryophyllene also known as *alpha*-caryophyllene. See *Beta*-caryophyllene. Occurs in many essential oils, notably hop. No evidence of toxicity.

⚠ Hydrocyanic acid

Also known as prussic acid, hydrogen cyanide, or 'cyanide'. A gas at room temperature in its pure form. Released from amygdalin during distillation of bitter almond oil and present in the unrectified oil. Very weak acid with characteristic almond odour (which cannot therefore be used to detect contaminated samples of bitter almond oil). Poisoning may occur from inhalation, ingestion or dermal absorption [5]. Poisoning produces rapid and laboured breathing (causing increased intake), paralysis, unconsciousness, convulsions and death. The fatal dose for man is considered to be about 50 mg [5]. Removed during the production of bitter almond oil FFPA (free from prussic acid). Both the UK and EC 'standard permitted proportion' of hydrocyanic acid in food flavourings is 0.001 g/kg [455, 456]. In Great Britain, the maximum exposure limit of hydrocyanic acid is 10 mg/m³ (short term) and 5 mg/m³ (long term) [5].

Ionone

Occurs in *alpha*- and *beta*-isomers, usually found together in absolutes of boronia, cassie and osmanthus. Non-toxic, non-irritant, non-sensitising [421].

Alpha-Irone

Principal fragrant ingredient in violet flowers, and a major component of orris oil, where the *beta*- and *gamma*-isomers also occur. Non-toxic, non-irritant, non-sensitising [421].

Isomers

There are many different types of isomer. In the chemistry of essential oils, the types of isomer usually encountered are:

Optical isomers: The isomers are non-identical mirror images of one another, rather like a pair of gloves. The isomers can rotate the plane of plane-polarised light; one optical isomer will rotate it clockwise (the *d*-isomer), the other anti-clockwise (the *l*-isomer). Despite their chemical similarities, optical isomers can smell very different (e.g. *d*- and *l*-carvone).

Geometric isomers: The atoms in a molecule are joined up in the same order, but take up different positions in space (e.g. geraniol and nerol). See pages 13 and 14 for diagrams of isomers.

Cis-Jasmone

A terpenoid ketone found as a minor component of jasmine absolute. Non-toxic, mildly irritant, non-sensitising [425].

Ketones

Common terpenoid constituents of many essential oils. They have a reputation for toxicity, especially to the CNS, but this is almost certainly simplistic. Some are indeed toxic (thujone, pulegone, camphor) but others are non-toxic (fenchone, carvone).

⚠ Khellin

Highly toxic compound with rat oral LD_{50} of 0.08 g/kg. Found in Khella oil at around 1%.

Lavandulol

Alcohol constituent of lavender oil. No evidence of toxicity.

Lavandulyl acetate

Ester constituent of lavender oil. Non-toxic, non-irritant, non-sensitising [424].

Limonene

A terpene hydrocarbon widely spread amongst essential oils. The *d*-isomer is much more common, being found as a major component of all citrus oils. There is no indication that *d*-limonene is toxic to humans or indeed most experimental animals [48, 124]. Toxicity has only ever been demonstrated in one strain of male rat and it seems certain that this represents a species-specific reaction.

Limonene quenches (i.e. reduces) the ability of citral to cause sensitisation when present at 20% in a mixture with 80% citral [422]. Citral-rich oils generally contain sufficient limonene and other quenching agents to prevent sensitisation. It also quenches the sensitisation potential of cinnamaldehyde, when present in equal amounts. Not found in significant amounts in cinnamaldehyde-rich oils. The *d*-isomer of limonene is non-toxic and mildly irritant; it has been reported as a sensitiser, but it seems clear that the reported cases were due to substances other than limonene; a RIFM maximation test produced no reactions in 23 subjects [421].

The *l*-isomer is a minor component of a few pine, mint and other oils. It is non-toxic, mildly irritant, and non-sensitising [424].

Linalool

Terpenoid alcohol, and an isomer of geraniol. The chief constituent of rosewood, ho leaf and coriander oils; also found in some 200 other essential oils. Narcotic effects seen at about half the lethal dose in mammals and demonstrated in many experimental animals; non-toxic, non-irritating and non-sensitising [421].

Linalyl acetate

Terpenoid ester which is very widely distributed. Especially important in giving aroma to lavender, clary sage, petitgrain and bergamot oils. No evidence of toxicity. Oral LD_{50} 14.5 g/kg [338].

⚠ Menthofuran

Found in:

Peppermint	0.1–7.5%

A toxic metabolite of pulegone, menthofuran is produced by the liver's enzyme systems [112, 396]. Menthofuran destroys the enzyme cytochrome P_{450} [119]; it is both hepatotoxic and lung toxic in mice [110]. The amounts of menthofuran present in peppermint oil do not give cause for concern.

Menthol

A terpenoid alcohol. The *l*-isomer is obtained from peppermint and other mint oils. May be a CNS stimulant implicated in some cases of overdose and hyperexcitation. There have been reports of apnoea and instant collapse in infants following the local application of menthol to their nostrils [5, 250]. Menthol has been shown to dilate capillaries in the nasal mucosa upon inhalation in some individuals [222]. Menthol cannot be metabolised by people with G6PD deficiency [109, 523] and menthol-rich oils should be avoided orally by G6PD-deficient people. Relatively non-toxic (oral LD_{50} in rats 3.0 g/kg); non-sensitising, but a possible mild irritant [422].

Menthone

Found in:

Cornmint	15–30%
Pennyroyal	1–16%
Peppermint	19%

A terpenoid ketone also found in very small amounts in spearmint and geranium oils. The *l*-isomer is more common. Caused some CNS and liver toxicity at high dosage levels (over 0.2 g/

kg/day for 28 days) [327]. Other studies have concluded that menthone is not hepatotoxic [111, 117]. Unlikely to cause problems at dosage levels used in aromatherapy.

Iso-Menthone

An isomer of menthone found primarily in mint oils. Probably similar in activity to menthone.

Menthyl acetate

Terpenoid ester of menthol present in mint oils. Non-toxic, non-irritating, non-sensitising [422].

Methyl allyl trisulphide (= methyl-2-propenyl trisulphide)

Component of garlic oil which inhibits platelet aggregation [363, 373].

Methyl anthranilate

Terpenoid ester present in mandarin, ylang-ylang, neroli, bergamot, orange and other essential oils, as well as in jasmine absolute. Non-toxic, non-irritating, non-sensitising [420].

Methyl chavicol

See Estragole.

Methyleugenol

Found in:

Asarum	
europaeum (NCA)	major component
Cinnamomum	
oliveri (NCA)	major component
Doryphora sassafras bark (NCA)	47%
Doryphora sassafras leaf (NCA)	43%
Huon pine (NCA)	major component
Melaleuca bracteata	major component
Nutmeg (East Indian)	tr.-1.2%
Snakeroot	36-45%
Tarragon (French)	0.1-1.5%
Tarragon (Russian)	5-29%

Methyleugenol chemotypes also exist for *Cinnamomum longepaniculatum, Cinnamomum parthenoxylon, Ocimum sanctum* (42-62%), *Ocotea pretiosa* and West Indian bay (43%), none of which are commercially available. Methyleugenol sometimes occurs as a trace constituent of oils of basil, betel, calamus, cinnamon bark, citronella, clove, elemi, ho leaf, hyacinth, laurel, nutmeg (West Indian), pimento, rose, tejpat, ylang-ylang and others. The highest level in any of these oils that we could find reported was 0.5% in the common form of tejpat leaf oil [216]. Methyleugenol is genotoxic [122, 289, 291, 479]; it comes into the same category as safrole and estragole, and is carcinogenic in rodents [309,310]. The acute oral LD_{50} has been reported as 0.81 and 1.56 g/kg in rats; non-irritant and non-sensitising [421].

Methyl heptenone

Found as a minor component of lemongrass, citronella, verbena and geranium. Non-toxic, mildly irritant, non-sensitising [421].

Methylnonyl ketone

See 2-undecanone.

Methyl salicylate

Found in:

Cassie[A]	11–19%
Sweet birch	98%
Wintergreen	98%

Terpenoid ester which may be colourless, yellow or reddish. Doses as small as 4 ml have caused death in infants [5]. Serious systemic toxicity has resulted from applying oils containing methyl salicylate to damaged skin [100]. Causes convulsions and gastrointestinal irritation on chronic administration [336]. Numerous cases of methyl salicylate poisoning have been reported, with a 50–60% mortality rate; signs of poisoning include nausea, vomiting, acidosis, pulmonary oedema, pneumonia and convulsions; LD_{50} from 0.7 to 2.8 g/kg; severely irritating to guinea pig skin, but tested at 8% in humans, produced no reactions; non-sensitising [424]. Could be three to five times more toxic in humans than in rodents (see p. 50). Methyl salicylate has been shown to produce nephrotoxicity in rats during gestation retarding kidney development, but not usually causing permanent abnormality [129]. High doses (4 ml/kg) of methyl salicylate, applied to the skin of rabbits, 5 days a week for up to 96 days, caused early deaths and kidney damage; lower dosage levels (0.5, 1.0, or 2.0 ml/kg) caused a higher than normal incidence of 'spontaneous' nephritis [415]. Topically applied methyl salicylate can potentiate the anticoagulant effects of warfarin, causing side-effects such as internal haemorrhage [521]. A similar interaction is possible, but by no means certain, with other anticoagulants such as aspirin and heparin. Many liniments contain methyl salicylate or wintergreen oil.

Myrcene

Terpene hydrocarbon found in many essential oils, notably juniper. Only the *beta*-isomer is found in nature. Non-toxic, mildly irritant, non-sensitising [422]. No effect on peri- and post-natal development has been found at doses of up to 250 mg/kg given orally to pregnant rats [318]. Above 500 mg/kg, some adverse effect on birth weight, perinatal mortality and postnatal development was seen [318]. This does not represent a level of fetotoxicity which would present any problems in essential oils as used in aromatherapy.

 Myristicin

Found in:

Cinnamomum cecidodaphne	level unknown
Mace	1.5–3.8%
Nutmeg	2–10%
Parsley leaf	7–33%
Parsleyseed	28%
Parsnip	17–40%

Aromatic ether which apparently causes psychotropic effects in doses equivalent to oral

aromatherapy use. However, this only occurs in synergy with another unknown compound also present in nutmeg (see pp. 71, 72). Myristicin-rich oils other than nutmeg and mace do not appear to be psychotropic at these dosage levels. Probably partly responsible for several cases of nutmeg poisoning from ingestion of large amounts of ground nutmeg (see p. 70). Myristicin cat oral LD$_{100}$ 0.57 g/kg [38]. Myristicin is believed to be especially toxic to cats; nutmeg repels cats.

Myristicin inhibits monoamine oxidase (MAO) in rodents [367]. MAO inhibitors should not be given in conjunction with pethidine [5].

Myristicin is a non-carcinogenic alkenylbenzene. It shows at most a low to moderate level of genotoxicity [122, 289, 445] and is not carcinogenic [309]. It shows high activity as an inducer of glutathione-*S*-transferase (this correlates with tumour inhibition) [466] and it inhibits benzo[*a*]pyrene-induced tumours in mice [467].

Neral

Terpenoid aldehyde. Isomer of citral present in melissa, verbena, lemongrass, *Backhousia citriodora* and may chang. No evidence of toxicity. See Citral and Geranial.

Nerol

Terpenoid alcohol present in many essential oils, including rose, *Helichrysum*, ylang-ylang and neroli. Non-toxic, mildly irritating, non-sensitising [422].

Nerolidol

Sesquiterpenoid alcohol which is the main constituent of cabreuva oil and is present in neroli, Peru balsam, cubeb seed and other oils. Non-toxic, non-irritant, non-sensitising [421].

Neryl acetate

Ester terpenoid present in neroli, *Helichrysum* and many other essential oils. Non-toxic, non-irritant, non-sensitising [422].

Ocimene

Terpene hydrocarbon. Several isomers exist. Found in basil, tarragon and marjoram oils. Non-toxic, non-irritant, non-sensitising [424].

Optical isomer

See Isomers.

Patchouli alcohol

Tricyclic sesquiterpenoid alcohol found chiefly in patchouli oil. Also known as patchoulol. Non-toxic [157].

Perilla ketone

Found in:

Perilla	15–38%

Terpenoid ketone frequently found as the principal component of perilla oil (some chemotypes of perilla oil have other principal components). Perilla ketone is a highly potent lung toxin to laboratory animals, and it often poisons grazing cattle which eat perilla leaves [332, 333]. Intraperitoneal LD_{50} (always lower than oral) is 2.5–10 mg/kg [333]. It is not clear whether perilla ketone (and hence perilla oil) represents a hazard to humans. Perilla oil is used as a flavour ingredient in Japan, and no reports linking its use with pulmonary conditions in humans have as yet been noted [332].

Perillaldehyde (= perilla aldehyde)

Terpenoid aldehyde related chemically to perilla ketone but without its toxicity (acute oral LD_{50} 1.72 g/kg). The *l*-isomer constitutes some 50% of perilla oil and has caused weak skin sensitisation when used at a 4% dilution, but none at 1%; mildly irritating [426]. IFRA recommends a maximum use level of 0.5% in fragrance compounds due to sensitisation potential [430]. *l*-perillaldehyde is about 2000 times sweeter than sucrose, and is used as a sweetening agent in Japan.

Alpha-Phellandrene

Terpene hydrocarbon primarily found in dill herb oil (49–58%), *Eucalyptus dives* (one chemotype—30%) and angelica root (8–20%). Non-toxic; can be irritating to the skin and is well absorbed through it; ingestion can cause vomiting and diarrhoea; may increase the risk of cancer in those exposed to carcinogens, but not carcinogenic in its own right [424].

Beta-Phellandrene

Isomer of *alpha*-phellandrene and major component of angelica root oil (16–24%). The *d*-form is found in water-fennel oil, the *l*-form in Canada balsam oil [157]. No evidence of toxicity as used in aromatherapy.

Phenols

Compounds derived from phenol (carbolic acid) in which a hydroxyl group is directly attached to a benzene ring. Simple phenols (like those often found in essential oils) are generally irritant, toxic and even corrosive. However, many important plant substances are derived from phenols.

Phenylacetic acid

Minor constituent of rose, neroli and a few other oils. Non-toxic, non-irritant, non-sensitising [421].

Phenylethyl alcohol (= phenylethanol)

Minor constituent of several essential oils, including rose. Major component of rose absolute (60%). Non-irritant, non-sensitising; acute oral LD_{50} varies from 0.4 to 1.79 g/kg [421]. Shows some chronic oral toxicity (stomach irritation) [336]. Its presence in an essential oil should not be cause for concern.

Phenylethyl isothiocyanate

Found in:

Horseradish	45%

Major ingredient of horseradish oil, and related to

allyl isothiocyanate. Probably very toxic. Only produced when the root is crushed.

Phytol

A decomposition product of chlorophyll which constitutes 8–20% of jasmine absolute. Non-toxic, mildly irritating, non-sensitising [426].

Alpha-Pinene

Main constituent of turpentine (terebinth), cypress, juniper, frankincense and rosemary oils. Also occurs in over 400 other essential oils. The American form is usually *d-alpha*-pinene, whereas the European form is usually *l-alpha*-pinene. Low toxicity but extremely large quantities produce CNS depression, bronchitis and kidney damage. Can cause the appearance of benign tumours; non-irritating; when oxidised it can cause skin sensitisation [424]. Its presence in an essential oil should not be cause for concern, so long as the oil is not oxidised.

Beta-Pinene

A minor component of many essential oils. Usually occurs alongside *alpha*-pinene. No apparent toxicity other than occasional idiosyncratic dermatitis [424].

Pinocamphone

Found in:

Hyssop   40%

A ketone which is largely responsible for the neurotoxicity of hyssop oil [432]. Intraperitoneal pinocamphone was found to be convulsant and lethal in rats above 0.05 ml/kg [234, 325]. Almost certainly dangerous for people prone to epilepsy.

Iso-Pinocamphone

Found in:

Hyssop   30%

Isomer of pinocamphone with similar properties.

Piperitone

Terpenoid ketone found in *Mentha sylvestris*, *Mentha arvensis*, *Eucalyptus dives*, and other oils. Non-toxic, non-irritant, non-sensitising, CNS excitatory [424].

Prussic acid

See Hydrocyanic acid.

 *d*-Pulegone

Found in:

Buchu (*B. crenulata*)	50%
Cornmint	0.2–5%
Lesser calamint (NCA)	< 70%
Pennyroyal (Eur.)	55–95%
Pennyroyal (N. Am.)	60–80%
Peppermint	0.1–2%
Spearmint	0–1%

Terpenoid ketone which is toxic to the liver because it is metabolised to epoxides. Acute oral LD_{50} 0.5 g/kg in rats [424]. Both the UK and EC 'standard permitted proportion' of pulegone in food flavourings is 0.025 g/kg [455, 456]. Non-irritant, non-sensitising [424]. Essential oils rich in *d*-pulegone should not be used in aromatherapy.

Sabinene

Terpene constituent of nutmeg and juniper oils. No evidence of toxicity.

Sabinyl acetate

Found in:

Juniperus pfitzeriana (NCA)	2–17%
Plectranthus (NCA)	60%
Sage (Spanish)	0.1–24%
Savin	20–53%

Terpene ester which is responsible for the reproductive toxicity of plectranthus, savin and some Spanish sage oils. Apparently embryotoxic, fetotoxic, teratogenic and abortifacient [189, 190, 194, 197]. One of the very few toxic essential oil esters.

⚠ Safrole

Found in:

Camphor (brown)	< 80%
Camphor (yellow)	10–20%
Cangerana	level unknown
Cinnamomum barmannii f.	
heyneanum (NCA)	97–99%
Cinnamomum cecidodaphne	
	level unknown
Cinnamomum paucifolium (NCA)	68–96%
Cinnamomum petrophilum (NCA)	90%
Cinnamomum porrectum	
	major component
Cinnamomum rigidissimum	
	major component
Cinnamomum subavenium (NCA)	70%
Cinnamon bark	0–0.04%
Cinnamon leaf	0.7–1%
Illium difengpi bark (NCA)	22%
Mace	0.2–1.9%
Mango ginger (NCA)	9%
Nutmeg (E. Indian)	0.6–3.3%
Nutmeg (W. Indian)	0.1–0.2%
Sassafras (all types)	85–90%
Star anise	0–1%
Ylang-ylang	0–0.3%

Terpene ether which is genotoxic in rat hepatocytes [122, 289, 291] and carcinogenic in rodents [153, 319, 320]. Safrole rabbit oral minimum lethal dose 1.0 g/kg [38]. Highly toxic, especially on chronic dosing; 95% lethal to rats at 10 000 p.p.m. after 19 days' dosing [154]. Safrole produces kidney epithelial and liver tumours in the offspring of pregnant mice given it in the diet [161].

A metabolite of safrole, 1´-hydroxysafrole, is almost certainly the culprit [153, 286] and is a more potent hepatic carcinogen than safrole itself [32]. Other toxic metabolites may also be formed in vivo [59,72,287]. Safrole and 1´-hydroxysafrole can both cause hepatic cell enlargement in experimental animals [313]. There is some evidence that safrole may be able to activate a cancer-causing virus, the polyoma virus, at least in rats. It is not known whether safrole has a similar action in humans [344].

IFRA recommends that safrole as such should not be used as a fragrance ingredient. It recommends a maximum use level of 0.05% in fragrance compounds (equivalent to 0.01% in final product) for safrole, when safrole-containing essential oils are used [430]. Both the UK and EC 'standard permitted proportion' of safrole in food flavourings is 0.001 g/kg [455, 456].

Iso-Safrole

Isomer of safrole sometimes found in tiny amounts in ylang-ylang oil. Weakly genotoxic [289] or non-genotoxic [291]. Although often quoted as being weakly carcinogenic to the liver in mice there is no evidence for this (private communication with Dr. J. Caldwell). Rat acute oral LD_{50} 1.34 g/kg [338].

IFRA makes no recommendation on iso-safrole itself, but recommends that the total concentration of safrole, iso-safrole and dihydrosafrole should not exceed 0.05% in fragrance compounds (equivalent to 0.01% in final product) [430]. Both the UK and EC 'standard permitted proportion' of iso-safrole in food flavourings is 0.001 g/kg [455, 456].

Santalols (*alpha, beta, epi-beta* and *trans-beta*)

Sesquiterpene alcohols constituting approximately 90% of East Indian sandalwood oil. No evidence of toxicity.

Santolinenone

See Artemisia ketone.

Sclareol

Terpenoid alcohol found in clary sage oil (0.1–3%). Non-toxic, non-irritant, some risk of sensitisation, probably due to impurities in the sclareol tested [428].

Sesquiterpenes

Plant hydrocarbon molecules produced in

secondary metabolism which have a $C_{15}H_{24}$ molecular formula (one-and-a-half times the monoterpene formula).

Sulphur compounds

Pungent compounds found in only a few essential oils, with a tendency to be pharmacologically active. Diallyl disulphide (<> garlic) is a typical essential oil sulphur compound.

Shyobunones

Constituents of an unusual type of calamus oil having unknown toxicity and being present to a variable degree in the oil (13–45%). See Calamus profile.

Terpenes

Hydrocarbons produced by the plant's so-called secondary (non-essential) metabolism, all with the molecular formula $C_{10}H_{16}$, but with a huge variety in the way in which these 26 atoms are connected. Very common essential oil components; generally non-toxic. Allergic effect of terpene-rich oils possibly due to hydroperoxide formation on storage [11].

Terpenoid

Secondary metabolites (see Terpenes) based on the terpene $C_{10}H_{16}$ molecular formula in which one or more hydrogen atoms are replaced by oxygen-containing functional groups. Many different types; very common essential oil ingredients.

Alpha-Terpinene

Found in tea tree oil at about 8%, and over 20 other oils. Acute oral LD_{50} 1.68 g/kg [422]. Hepato-toxicity is referred to in one paper, but this appears to be unsubstantiated [469].

Gamma-Terpinene

Terpene hydrocarbon found in tea tree, lemon and bergamot oils, among many others. Non-toxic, non-irritant, non-sensitising [422].

Terpinen-4-ol (= terpinenol-4 = 4-terpinenol)

An alcohol which occurs in several essential oils, notably tea tree (27–58%). Acute oral LD_{50} 1.3 g/kg; mildly irritating, non-sensitising [426]. Terpinen-4-ol from juniperberry oil showed an acute oral LD_{50} of 1.85 ml/kg; after chronic administration in the mouse it caused no pathological changes [376].

Terpineol

A terpenoid alcohol which exists in three isomeric forms, *alpha, beta* and *gamma*. Found in a great many essential oils. Non-toxic, mildly irritant, non-sensitising [420].

Terpinyl acetate

Terpenoid ester found in cardamon, bay, cypress, cajeput, niaouli, and pine among others. Occurs in three isomeric forms, *alpha, beta* and *gamma*. The *alpha*-isomer is the major component of cardamon oil. Non-toxic, non-irritant, non-sensitising [336, 420].

⚠ Thujone

Found in:

Armoise	35%
Artemisia arborescens	30–45%
Artemisia austriaca (NCA)	31%
Artemisia brevifolia (NCA)	20%
Artemisia klotzchiana (NCA)	34%
Artemisia kurramensis (NCA)	55%
Artemisia maritima (NCA)	63%
Lanyana	65%
Mugwort (NCA)	major component
Sage (Dalmatian)	50%
Salvia pomifera (NCA)	47–83%
Southernwood (NCA)	major component

Tansy	66–81%
Thuja	39–80%
Western red cedar	85%
Wormwood	34–71%

Terpene ketone existing in two isomeric forms, *alpha* and *beta*, which are usually found together. A major constituent of wormwood, mugwort, sage, lanyana, tansy and thuja oils. The *alpha*-isomer is more toxic than the *beta*; *alpha*-thujone acute s.c. LD_{50} 0.087 g/kg, *beta*-thujone acute s.c. LD_{50} 0.44 g/kg [357]. Intraperitoneal thujone was found to be convulsant and lethal in rats above 0.2 ml/kg [234, 325]. Mouse thujone (*alpha* + *beta*) acute oral LD_{50} 0.23 g/kg, rat oral LD_{50} 0.19 g/kg [354], giving an average of 0.21 g/kg.

Convulsions occur, and fatty degeneration of the liver is found in some animals on chronic administration [354]. Both the UK and EC 'standard permitted proportion' of *alpha*- and/or *beta*-thujone in food flavourings is 0.0005 g/kg [455, 456].

⚠ Thymol

Found in:

Ajowan	40–50%
Oregano	1–85%
Savory	1–49%
Thyme	1–63%

A phenol found as a major component in thyme, wild thyme, savory and ajowan oils. An irritant, as would be expected of a phenol [153, 158]. Mouse acute oral LD_{50} 1.8 g/kg [38]. Acute oral LD_{50} reported as 0.98 g/kg in rat, and 0.88 in guinea pig [338].

2-undecanone (= methylnonyl ketone)

Major constituent of the essential oil of rue. Non-toxic, non-irritating and non-sensitising [421]. Might be implicated in the possible abortifacient action of rue oil.

Vanillin

Occurs chiefly in vanilla products, but also as a minor component of a few essential oils and absolutes. Widely used in food flavourings. Non-toxic, non-irritant and non-sensitising [423].

Verbenone

Ketone found in oils of frankincense (6.5%) and rosemary, verbenone CT (15–37%). Mouse i.p. LD_{50} 0.25 g/kg [6]. There is a possibility that verbenone might present some toxicity, which would also manifest in rosemary, verbenone CT.

Vetiverol

A complex mixture of isomeric alcohols found in vetiver oil. Non-toxic, non-sensitising; mildly irritating [425].

⚠ Xanthotoxin (= 8-methoxypsoralen = 8-MOP)

Compound with a furanocoumarin structure found in rue oil. Phototoxic [485] and potentially photocarcinogenic [496]. See Bergapten.

Zingiberenes (*alpha* and *beta*)

Major components of ginger oil. No apparent toxicity.

12

Safety index

This index is in two sections, and its purpose is to provide an easy reference to the safety of a large number of essential oils. In both sections the oils are listed in alphabetical order by common name. Where there is no well-known common name, the botanical name only is given.

Part One consists of information on 177 essential oils and absolutes which have had a monograph published on their safety by RIFM [419–429]. This section, which is presented in the form of a table (Table 12.1), summarises the RIFM data on acute oral toxicity, dermal irritation, dermal sensitisation and phototoxicity. Most of the data in the last column (mucous membrane irritation, use fresh, cancer, etc.) are not taken from the RIFM monographs. Source references are not given for information in the last column, as this is a summary of material which is referenced elsewhere in the book.

Part Two includes a further 133 materials for which monographs have not been published by RIFM. In order to give some kind of guideline, major components are given where known, and this is used as a basis for estimating the safety or otherwise of these materials. This is not an ideal approach, and it is possible that some of the materials indicated as probably safe may later be shown to be hazardous in some sense. However, we consider that the quality of the information given is likely to be high, and that it is therefore preferable to stating that nothing is known about the safety of these materials.

In assembling this Safety Index no judgement has been made as to whether a material is currently used, or likely to be used in aromatherapy. The rationale behind this is that there is no clearly defined list of the materials which are used in aromatherapy, and that any such judgements are largely arbitrary. A judgement has been made, where possible, as to the availability of materials. There are several hundred essential oils which have been experimentally distilled and analysed, but which are not produced commercially, even on a small scale. The majority of these are not included in this index.

PART ONE

All categories

(P) after a material indicates that it is also to be found in the Profiles chapter (the most hazardous oils will all be found detailed there). As in the rest of the book, materials commonly available in the form of absolutes, rather than essential oils, are marked with a superscript A. NT = not tested.

Oral toxicity

Rodent LD_{50} values are categorised as follows:

A < 1.0 g/kg Toxic (best avoided altogether)
B 1–2 g/kg Mildly toxic (some are safe to use)
C 2–5 g/kg Non-toxic ⎫ (safe to use unless there are other
D > 5 g/kg Non-toxic ⎭ reasons not to)

Skin irritation

All the materials were tested at concentrations ranging between 1% and 30%. This range gives a good indication of any problems which may arise from aromatherapy use.

A Severely irritant
B Strongly irritant
C Moderately irritant
D Very mildly irritant
E Non-irritant

Sensitisation

All the materials were tested at concentrations ranging between 1% and 30%. This range gives a very good indication of any problems which may arise from aromatherapy use. Oils are categorised as either causing sensitisation (Yes) or not causing sensitisation (No). The 'yes's' are all profiled.

Phototoxicity

Oils are categorised as either causing phototoxicity (Yes) or not causing phototoxicity (No). The yes's are all profiled.

Other

Any other important warnings or comments are indicated here; fuller information will generally be found in the Profiles chapter. 'Use fresh' means use within 6 months of purchase, or within 1 year if kept refrigerated. Other oils will keep twice as long. MM irritant = mucous membrane irritant; PC = potentially carcinogenic.

Table 12.1 Summary of information on essential oils and absolutes that have been the subject of RIFM monographs

	Oral toxicity	Skin irritation	Sensitisation	Photo-toxicity	Other
Abies alba (cones) *Abies alba*	D	C	No	No	Use fresh
Abies alba (needles) *Abies alba*	D	D	No	No	Use fresh
Ale Pine species	D	D	No	No	Use fresh
Almond (bitter, unrectified) (P) *Prunus amygdalus* var. *amara*	A	D	No	No	Neurotoxic, do not use
Almond (bitter) FFPA *Prunus amygdalus* var. *amara*	B	D	No	No	Use fresh
Ambrette seed *Hibiscus abelmoschus*	D	E	No	No	Use fresh
Angelica root (P) *Angelica archangelica*	C/D	D	No	Yes	Phototoxic
Angelica seed *Angelica archangelica*	D	D	No	No	
Anise (P) *Pimpinella anisum*	C	D	?	No	Some cautions
Armoise (P) *Artemisia herba-alba*	A	D	No	No	Do not use
Basil (P) *Ocimum basilicum*	B	D	No	No	PC, do not use unless very low estragole
Bay (W. Indian) *Pimenta racemosa*	B	D	No	NT	MM irritant, some cautions
Benzoin (resinoid) *Styrax benzoin*	D	NT	No	NT	
Bergamot (P) *Citrus bergamia*	NT	NT	No	Yes	Use fresh, phototoxic
Bergamot (FCF) *Citrus bergamia*	D	D	No	No	Use fresh
Bergamot mint *Mentha citrata*	D	D	No	No	
Birch (sweet) (P) *Betula lenta*	B	D	No	No	Do not use
Birch tar *Betula pendula*	D	D	No	No	See note*
Boldo leaf (P) *Peumus boldus*	A	D	No	No	Neurotoxic, do not use

*Birch tar oil is produced by steam distillation of the tar, which is obtained by dry (destructive) distillation of the bark and wood. Destructive distillation produces carcinogenic, polycyclic hydrocarbons in cade oil (see profile). No information on the presence or otherwise of potentially carcinogenic compounds in birch tar oil could be found.

Table 12.1 Continued

	Oral toxicity	Skin irritation	Sensitisation	Photo-toxicity	Other
Broom[A] (= genet) *Spartium junceum*	D	E	No	No	
Cabreuva *Myrocarpus fastigiatus*	D	D	No	No	
Cade (rectified) (P) *Juniperus oxycedrus*	D	C	No	No	Avoid unrectified oil (PC)
Cajeput *Melaleuca leucadendron*	C	E	No	No	
Calamus (P) *Acorus calamus*	A	E	No	No	PC, do not use
Camphor (white) (P) *Cinnamomum camphora*	D	D	No	NT	Some contraindications
Camphor (brown) (P) *Cinnamomum camphora*	C	E	No	No	PC, do not use
Camphor (yellow) (P) *Cinnamomum camphora*	C	D	No	No	PC, do not use
Cananga *Cananga odorata*	D	D	No	NT	
Caraway *Carum carvi*	C	D	No	No	MM irritant (moderate)
Cardamon *Elettaria cardamomum*	D	E	No	No	
Carrot seed *Daucus carota*	D	D	No	No	
Cascarilla *Croton eluteria*	D	E	No	No	
Cassia (P) *Cinnamomum cassia*	C	C	Yes	No	Sensitiser, do not use
Cedarwood (Atlas) *Cedrus atlantica*	D	D	No	No	
Cedarwood (Texas) *Juniperus mexicana*	D	D	No	No	
Cedarwood (Virginian) *Juniperus virginiana*	D	D	No	No	
Celery seed *Apium graveolens*	D	E	No	No	
Chamomile (German) *Chamomilla recutita*	D	D	No	No	
Chamomile (Roman) *Anthemis nobilis*	D	D	No	No	

Table 12.1 Continued

	Oral toxicity	Skin irritation	Sensitisation	Photo-toxicity	Other
Cinnamon bark (P) *Cinnamomum zeylanicum*	C	C	Yes	No	Sensitiser, MM irritant, do not use
Cinnamon leaf (P) *Cinnamomum zeylanicum*	C	C	No	No	MM irritant, some cautions
Citronella (P) *Cymbopogon nardus*	D	D	No?	NT	Occasional sensitisation
Clary sage (French) *Salvia sclarea*	D	D	No	NT	
Clary sage (Russian) *Salvia sclarea*	NT	E	No	No	
Clove bud (P) *Syzygium aromaticum*	C	C	No?	No	MM irritant, some cautions
Clove leaf (P) *Syzygium aromaticum*	B	C	No	No	MM irritant, some cautions
Clove stem (P) *Syzygium aromaticum*	C	C	No	No	MM irritant, some cautions
Copaiba *Copaifera officinalis*	D	D	No	NT	
Coriander seed *Coriandrum sativum*	C	D	No	NT	
Cornmint (P) *Mentha arvensis*	B	E	No	No	Some cautions and contraindications
Costus (P) *Saussurea costus*	C	D	Yes	No	Sensitiser, do not use
Cubeb *Piper cubeba*	D	D	No	No	
Cumin *Cuminum cyminum*	C	D	No	Yes	Phototoxic
Cypress *Cupressus sempervirens*	D	D	No	No	
Davana *Artemisia pallens*	D	D	No	No	
Deertongue[A] *Liatris odoratissima*	D	E	No	No	
Dill seed *Anethum graveolens*	C	D	No	No	
Dill weed *Anethum graveolens*	C	E	No	No	
Eau de brouts[A] *Citrus aurantium*	D	D	No	No	

Table 12.1 Continued

	Oral toxicity	Skin irritation	Sensitisation	Photo-toxicity	Other
Elecampane (P) *Inula helenium*	NT	D	Yes	NT	Sensitiser, do not use
Elemi (P) *Canarium luzonicum*	C	D	No	No	PC? use with caution
Eucalyptus *Eucalyptus globulus*	C	D	No	No	
Eucalyptus citriodora *Eucalyptus citriodora*	D	D	No	No	
Fennel bitter (P) *Foeniculum vulgare*	C	D	?	No	Use fresh. Skin sensitiser when oxidised
Fennel sweet (P) *Foeniculum vulgare*	C	C	No	No	Some contraindications
Fenugreek[A] *Trigonella foenum-graecum*	D	D	No	No	
Fig leaf[A] (P) *Ficus carica*	NT	C	Yes	Yes	Do not use
Fir needle (Canada) *Abies balsamea*	D	E	No	NT	Use fresh
Fir needle (Siberia) *Abies sibirica*	D	B/C	No	NT	Use fresh
Flouve (sweet vernalgrass) *Anthoxanthum odoratum*	C	D	No	No	
Frankincense[A] *Boswellia carterii*	NT	D	No	No	
Galbanum *Ferula galbaniflua*	D	D	No	NT	
Geranium (Algerian) *Pelargonium graveolens*	D	D	No	No	
Geranium (Reunion) *Pelargonium graveolens*	D	D	No	No	
Geranium (Moroccan) *Pelargonium graveolens*	NT	D	No	No	
Ginger *Zingiber officinale*	D	D	No	No	
Grapefruit (P) *Citrus paradisi*	D	D	No	Yes	Use fresh, phototoxic
Guaiacwood *Bulnesia sarmienti*	D	D	No	No	
Gurjun *Dipterocarpus tuberculatus*	D	E	No	No	

Table 12.1 Continued

	Oral toxicity	Skin irritation	Sensitisation	Photo-toxicity	Other
Hay[A †] (= foin)	D	D	No	No	
Hibawood *Thujopsis dolobrata*	D	D	No	No	
Ho leaf (P) *Cinnamomum camphora*	C	D	No	NT	One chemotype is PC
Honeysuckle[A] *Lonicera caprifolium*	NT	D	No	No	
Hyacinth[A] *Hyacinthus orientalis*	C	C	No	No	
Hyssop (P) *Hyssopus officinalis*	B	E	No	No	Neurotoxic, some contraindications
Immortelle *Helichrysum angustifolium*	D	D	No	No	
Immortelle[A] *Helichrysum angustifolium*	C	D	No	No	
Jasmine[A] *Jasminum officinale*	D	E	No	No	
Jonquil[A] *Narcissus jonquilla*	NT	D	No	No	
Juniper (= juniperberry) (P) *Juniperus communis*	D	D	No	No	Use fresh, no contraindications
Karo karoundé[A] *Leptactina senegambica*	B	D	No	No	
Labdanum *Cistus ladaniferus*	D	D	No	NT	
Labdanum[A] *Cistus ladaniferus*	D	E	No	No	
Laurel (P) *Laurus nobilis*	C	C	?	No	Some risk of sensitisation
Lavandin *Lavandula x intermedia*	D	D	No	No	Some cautions
Lavandin [A] *Lavandula x intermedia*	D	D	No	No	
Lavender *Lavandula angustifolia*	D	D	No	No	
Lavender[A] *Lavandula angustifolia*	C	D	No?	NT	
Lemon (expressed) (P) *Citrus limonum*	D	D	No	Yes	Use fresh, phototoxic

[†]Various grasses, including *Lolium perenne, Lolium italicum, Pheleum pratense, Poa pratense, Cynosurus cristatus* and *Anthoxanthum odoratum*

Table 12.1 Continued

	Oral toxicity	Skin irritation	Sensitisation	Photo-toxicity	Other
Lemon (distilled) *Citrus limonum*	D	D	No	No	Use fresh
Lemon leaf (= lemon petitgrain) *Citrus limonum*	D	D	No	No	
Lemongrass (W. Indian) (P) *Cymbopogon citratus*	D	D	No	No	Some cautions
Lemongrass (E. Indian) (P) *Cymbopogon flexuosus*	D	D	No	No	Some cautions
Lime (expressed) (P) *Citrus aurantifolia*	NT	NT	NT	Yes	Use fresh, phototoxic
Lime (distilled) *Citrus aurantifolia*	D	D	No	No	Use fresh
Linaloe wood *Bursera delpechiana*	D	D	No	No	
Lovage root *Levisticum officinale*	C	D	No	No	
Mace (P) *Myristica fragrans*	C	D	No	No	Avoid oral use
Mandarin (expressed) (P) *Citrus reticulata*	D	D	No	No	Use fresh
Marjoram (sweet) *Origanum marjorana*	C	E	No	NT	
Marjoram (Spanish) *Thymus masticina*	D	D	No	No	
Mastic[A] *Pistacia lentiscus*	D	D	No	No	Use fresh
May chang (P) *Litsea cubeba*	D	D	No?	No	Some cautions
Mimosa[A] *Acacia decurrens*	D	D	No	No	
Myrrh *Commiphora molmol*	B	E	No	No	
Myrrh[A] *Commiphora molmol*	NT	E	No	No	
Myrtle *Myrtus communis*	D	D	No	No	Use fresh
Narcissus[A] *Narcissus poeticus & N. tazetta*	D	D	No	No	
Neroli *Citrus aurantium*	C	E	No	No	Use fresh

Table 12.1 Continued

	Oral toxicity	Skin irritation	Sensitisation	Photo-toxicity	Other
Nutmeg (East Indian)(P) *Myristica fragrans*	C	D	No	NT	Avoid oral use
Orange flower[A] *Citrus aurantium*	D	D	No	NT	
Orange (bitter, expressed) (P) *Citrus aurantium*	D	D	No	Yes	Use fresh, phototoxic
Orange (sweet, expressed) *Citrus sinensis*	D	D	No	No	Use fresh
Oregano (Spanish) (P) *Thymus capitatus,* *Origanum vulgare*	B	C	No	No	MM irritant, some cautions
Orris[A] *Iris pallida*	D	E	No	No	
Palmarosa *Cymbopogon martinii* var. *motia*	D	D	No	No	
Parsley leaf (P) *Petroselinum sativum*	C	C	No	No	Abortifacient, avoid oral doses
Parsleyseed (P) *Petroselinum sativum*	B/C	C	No	No	Abortifacient, avoid oral doses
Patchouli *Pogostemon cablin*	D	D	No	No	
Pennyroyal (European) (P) *Mentha pulegium*	A	D	No	No	Toxic, do not use
Pepper (black) *Piper nigrum*	D	D	No	No	
Perilla (P) *Perilla frutescens*	C/D	D	?	No	Uncertain toxicity, some cautions
Peru balsam *Myroxylon pereirae*	D	D	No?	NT	
Peruvian pepper *Schinus molle*	D	D	No	No	
Petitgrain (Bigarade) *Citrus aurantium*	D	D	No	No	
Petitgrain (Paraguay) *Citrus aurantium*	D	D	No	No	
Phoenician juniper *Juniperus phoenicea*	D	C/D	No	No	
Pimento leaf (P) *Pimenta officinalis*	C	C	No	No	MM irritant, use fresh, some cautions

Table 12.1 Continued

	Oral toxicity	Skin irritation	Sensitisation	Photo-toxicity	Other
Pine (dwarf) (P) *Pinus pumilio*	D	B/C	No	No	Use fresh, irritant when oxidised
Pine (Scotch) *Pinus sylvestris*	D	E	No	No	Use fresh
Rose (Bulgarian) *Rosa damascena*	D	D	No	No	
Rose[A] (cabbage) *Rosa centifolia*	D	D	No	No	
Rose (Moroccan) *Rosa damascena*	D	D	No	No	
Rose (Turkish) *Rosa damascena*	D	D	No	No	
Rosemary (P) *Rosmarinus officinalis*	D	D	No	NT	Some cautions
Rosewood *Aniba rosaeodora*	C	E	No	No	
Rue (P) *Ruta graveolens*	C/D	C	No	Yes	Some cautions, phototoxic
Sage (Dalmatian) (P) *Salvia officinalis*	C	C	No	NT	Do not use
Sage (Spanish) (P) *Salvia lavandulaefolia*	D	E	No	No	Abortifacient, some contraindications
Sandalwood *Santalum album*	D	D	No	No	
Sassafras (P) *Sassafras albidum*	B	C	No	No	PC, do not use
Sassafras (Brazilian) (P) *Ocotea pretiosa*	B	C	No	No	PC, do not use
Savory (summer) (P) *Satureia hortensis*	B	C	No	No	MM irritant, some cautions
Snakeroot (P) *Asarum canadense*	C	E	No	No	PC, do not use
Spearmint *Mentha spicata*	D	D	No	No	MM irritant (moderate)
Spike lavender (P) *Lavandula latifolia*	C	D	No	NT	Some cautions
Spruce[‡] (= hemlock spruce) *Tsuga canadensis,* *Picea mariana* & *P. glauca*	D	D	No	No	Use fresh

[‡]Also known as hemlock oil. Not to be confused with hemlock, *Conium maculatum*, which contains toxic alkaloids, and is not a source of essential oil.

Table 12.1 Continued

	Oral toxicity	Skin irritation	Sensitisation	Photo-toxicity	Other
Star anise (P) *Illicium verum*	C	D	No	No	Some cautions
Taget (P) *Tagetes patula, T. minuta,* *T. erecta*	C	C	No	Yes	Use fresh, phototoxic
Tangelo *Citrus reticulata* x *Citrus paradisi*	D	D	No	No	Use fresh
Tangerine (expressed) *Citrus reticulata*	D	D	No	No	Use fresh
Tansy (P) *Tanacetum vulgare*	A	D	No	No	Do not use
Tarragon (P) *Artemisia dracunculus*	B	C	No	No	PC, do not use
Tea tree *Melaleuca alternifolia*	B	D	No	No	
Terebinth (= yarmor) (P) *Pinus palustris* etc.	C	B/C	No	No	Use fresh, irritant when oxidised
Thuja (P) *Thuja occidentalis*	A	D	No	No	Do not use
Thyme (P) *Thymus vulgaris, T. zygis*	C	C	No	No	MM irritant, some cautions
Tobacco leaf[A] *Nicotinia affinis*	D	D	No	No	
Tonka[A] *Dipteryx odorata*	B	D	No	No	
Turmeric *Curcuma longa*	D	D	No	No	
Verbena (P) *Lippia citriodora*	D	C	Yes	Yes	Do not use
Verbena[A] (P) *Lippia citriodora*	D	D	Yes	No	Do not use
Vetiver *Vetiveria zizanoides*	D	D	No	No	
Violet leaf[A] *Viola odorata*	NT	E	No	No	
Wormseed (P) *Chenopodium anthelminticum*	A	D	No	No	Do not use
Wormwood (P) *Artemisia absinthium*	A	D	No	No	Do not use
Ylang-ylang *Cananga odorata*	D	D	No?	No	

PART TWO

All the information on essential oil composition is taken from Brian Lawrence's *Essential Oils* [214, 215, 216, 217] unless otherwise referenced. Most of the percentages have been rounded up or down to the nearest whole number. (P) after a material indicates that it will also be found in the Essential Oil Profiles.

Agarwood (wood)

Aquilaria agallocha

This oil is rich in some very unusual furans (e.g. various isomers of agarofuran) and alcohols, from which information it would be difficult to make any useful deductions regarding safety.

Ajowan (seeds) (P)

Trachyspermum ammi (= Carum copticum)

Major components include thymol (45–48%), cymene (21–24%), *gamma*-terpinene (19–20%) and carvacrol (4–7%). Likely to be moderately irritating to skin and mucous membrane due to high content of thymol/carvacrol. See profile.

Ammoniac (gum)

Dorema ammoniacum (= Dorema aucheri = Diserneston gummiferum)

Major components include citronellyl acetate, doremone, and *alpha*-ferulene. Unlikely to present any hazard in aromatherapy. Occasionally available.

Amyris (wood)

Amyris balsamifera

Major components include valerianol (22%), *beta*-eudesmol (17%), 10-*epi-gamma*-eudesmol (11%) and elemol (10%). Acute oral LD_{50} 5.58 g/kg [338]. Unlikely to present any hazard in aromatherapy.

Annual wormwood (herb) (P)

Artemisia annua

Major component is artemisia ketone (35–63%). This essential oil may be toxic in an aromatherapy context. See profile.

Artemisia arborescens (herb) (P)

Major components are *beta*-thujone (30–45%) and camphor (12–18%). Due to thujone content should be regarded as toxic, and avoided in aromatherapy. See profile.

Artemisia caerulescens (herb)

Major components are camphor and *alpha*-thujone; quantities uncertain, but different chemotypes of the plant are known. The oil produces epileptiform convulsions in a variety of animals, notably cats [139]. It is not currently produced commercially. Best avoided altogether in aromatherapy.

Artemisia vestita (herb)

Major components include *alpha*-phellandrene, aromadendrene and citral, all at around 10–20%. The oil has been used to a limited extent in India. Toxicity uncertain, but probably safe to use in aromatherapy.

Asafoetida (gum resin)

Ferula asa-foetida

The major components are three sulphur compounds, 1-methylthiopropyl-(1-propenyl)-disulphide (36–84%), 1-methylpropyl-(1-propenyl)-disulphide (9–31%) and 1-methyl-propyl-(3-methylthio-2-propenyl)-disulphide (0–52%). This richness in sulphur compounds may imbue asafoetida oil with a degree of sensitisation or irritancy. See Garlic oil profile.

Atractylis (roots)

Atractylis lancea

Major components are *beta*-eudesmol, hinesol, and atractylodin. Toxicity uncertain, but unlikely to present any hazard in aromatherapy.

Ayou (wood)

Nectandra globosa

Major component is a cadinene (80%); annual production in 1990 was 1 ton. Unlikely to present any hazard in aromatherapy.

Backhousia (leaves)

Backhousia citriodora

Major components are geranial and neral, which together comprise some 90% of the oil. Based on this, there may be a slight risk of sensitisation. As with other citral-rich oils, backhousia oil should not be taken orally in cases of glaucoma.

Balsamite (herb) (P)

Crysanthemum balsamita

Both camphor (72–91%) and *l*-carvone (50–80%) have been found as major components, presumably in different chemotypes of this oil. Balsamite oils rich in camphor should be avoided in aromatherapy, due to their very high camphor content. Those rich in *l*-carvone are (like spearmint oil) probably safe. See profile.

Betel (leaf) (P)

Piper betle

Betel leaf oils are now being offered commercially. There are numerous cultivars, many of which contain eugenol as a major component (< 90%). Other major components include *alpha*-terpinyl acetate, chavibetol, *trans*-anethole (around 20%) and safrole (< 45%). Cultivars rich in eugenol are likely to be mucous membrane irritant (moderate–strong) and dermal irritant (moderate); oral caution in people with blood coagulation problems or liver disease. Those rich in safrole are likely to be carcinogenic and should be avoided in aromatherapy. See profile.

Blackcurrant bud (= cassis)

Ribes nigrum

Major components of the essential oil include sabinene, limonene, *beta*-phellandrene, *delta*-3-carene and terpinolene. Major components of the absolute include *delta*-3-carene, caryophyllene and myrcene. Based on this information, there is no reason to suspect either the oil or the absolute of being hazardous in aromatherapy, but both should be used fresh.

Boronia[A] (flowers)

Boronia megastigma

The major components of boronia absolute are dodecanol, dodecyl acetate, tetradecyl acetate and *beta*-ionone. Unlikely to present any hazard in aromatherapy.

Borneo camphor (wood)

Dryobalanops aromatica (=Dryobalanops camphora)

The crevices and fissures in the wood of about 1% of old trees contain virtually pure *d*-borneol (see Chemical Index) a crystalline substance, but not an essential oil [3]. The wood of all trees does contain an essential oil, which contains *alpha*-pinene, *beta*-pinene, camphene, dipentene, terpineol and *d*-borneol [7]. This oil is unlikely to be hazardous in aromatherapy, but has never been commercially available.

Buchu (leaves)(P)

Barosma betulina and *Barosma crenulata*

B. crenulata contains 50% of *d*-pulegone [391] and so is likely to demonstrate the same toxicity as pennyroyal oil, and so should not be used in aromatherapy. Neither type of buchu oil should be used in pregnancy. See profile.

Caboré (wood)

Micrastur ruficolis

Major component is nerolidol (22%). Produced in Brazil. Unlikely to present any hazard in aromatherapy.

Cangerana (wood) (P)

Cabralea cangerana (= Cabralea canjerana)

The major component is caryophyllene, also contains an unknown quantity of safrole. Since safrole is a potential carcinogen, cangerana oil should probably be avoided in aromatherapy until further data are available. Produced in Brazil. See profile.

Cannabis (leaves)

Cannabis sativa (= Cannabis indica)

Major component is caryophyllene (45%). Hallucinogenic cannabinoids (1–2%) have been reported in distilled cannabis oil [457, 458]. Cannabis oil is not commercially available, and its production, importation, possession and sale are unlawful in most countries. See page 72.

Carnation^A (flowers)

Dianthus caryophyllus

The major component of the absolute appears to be benzyl benzoate (14.6%) with 8% of pentacosene, and 3.6% of eugenol. From the data available, it would be difficult to make any useful judgment as to the safety of this highly complex product. Caution is recommended.

Carqueja (shrub)

Baccharis genistelloides

Major component is 69% carquejyl acetate. The plant is used in South America in some beers, and is widely used in traditional medicine. Unlikely to present any hazard in aromatherapy. Produced in Brazil.

Cassie^A (flowers)

Acacia farnesiana

The major component of the absolute is methyl salicylate (11–19%), with significant amounts of *p*-anisaldehyde and geraniol. Although methyl salicylate is somewhat toxic, it will not constitute a risk at this sort of level. Cassie absolute is unlikely to present any hazard in aromatherapy.

Cassis

See Blackcurrant bud.

Catnep (herb)

Nepeta cataria

Major component is nepetalactone (60–95%). Possible sensitiser by nature of being lactone-rich. Use with caution.

Cedarwood, Himalayan (wood)

Cedrus deodara

This oil is botanically, chemically and odoriferously similar to Atlas cedarwood (see Part One). Major components are *alpha*- and *gamma*-atlantones, and *alpha*- and *beta*-himachalenes. Unlikely to present any hazard in aromatherapy.

Cedrela (= cedro) (wood)

Cedrela odorata

Major components are *alpha*-copaene (10–15%) and *delta*-cadinene (11–19%). Unlikely to present any hazard in aromatherapy.

Celery (leaves)

Apium graveolens

Major components are myrcene (33%) and limonene (26%). Unlikely to present any hazard in aromatherapy.

Chamomile (Moroccan) (flowers)

Ormenis mixta (= Ormenis multicaulis)

Major components appear to be *alpha*-pinene and artemisia alcohol. Unlikely to present any hazard in aromatherapy.

Champaca[A] (= champa = champak) (flowers)

Michelia champaca

Components are reported to include phenylethyl alcohol, benzyl alcohol, *iso*-eugenol, benzaldehyde, *para*-cresol and methyl anthranilate [349]. However, without more definite data it is not possible to comment on safety. Commercially available.

Chervil (leaves)

Anthriscus cerefolium (= Chaerophyllum sativum)

Major component is estragole (70–80%) and so will demonstrate similar carcinogenic activity to oils of basil and tarragon. Best avoided in aromatherapy. No profile, because the oil is not commercially available.

Coffin wood

See Siam wood.

Cognac (= wine lees)

Distilled from the residue of cognac distillation.

Major components are ethyl esters, such as ethyl hexanoate, ethyl heptanoate, ethyl octanoate and ethyl decanoate. No quantitative data. Unlikely to present any hazard in aromatherapy. Not an essential oil according to normal definitions.

Combava (fruit)

Citrus hystrix

Major components of this expressed oil are *beta*-pinene (36–42%) and sabinene (24%); also contains an unknown quantity of furanocoumarins, and so may be phototoxic to some degree [34]. Unlikely to present any other problem in aromatherapy. Should probably be used fresh.

Combava (leaves)

Citrus hystrix

The major component is citronellal (58–82%) which might lead to very occasional allergic reactions. Apart from this the oil is unlikely to present any hazard in aromatherapy.

Coriander (leaves)

Coriandrum sativum

Major components include (E)-2-decenal (13–23%), decanal (10–17%) and linalool (23–38%). Unlikely to present any hazard in aromatherapy.

Cyperus (= nut grass) (roots)

Cyperus rotundus

Major components are *alpha*-cyperone (33–54%), cyperol (49%) and cyperene (32%) [349]. Not currently commercially available. Unlikely to present any hazard in aromatherapy.

Cyperus (roots)

Cyperus scariosus

Major components are cyperene (15%) and *iso*-patchoulenone (16%) [34]. Insufficient data to give any useful guidelines.

Damiana (leaves)

Turnera diffusa

Major components are 1,8-cineole (25%) and terpinen-4-ol (25%). Unlikely to present any hazard in aromatherapy.

Erigeron

See Fleabane.

Eriocephalus (flowers)

Eriocephalus punctulatus

Major components are esters similar to those found in Roman chamomile oil, such as 2-methylbutyl 2-methylpropionate (23%). Unlikely to present any hazard in aromatherapy. Commercially available on a small scale.

Eucalyptus camaldulensis (leaves)

Major component is 1,8-cineole (> 70%). Unlikely to present any hazard in aromatherapy.

Eucalyptus dives (leaves)

Major component is 1,8-cineole (70–75%). Unlikely to present any hazard in aromatherapy.

Eucalyptus leucoxylon (leaves)

Major component is 1,8-cineole (65–75%). Unlikely to present any hazard in aromatherapy.

Eucalyptus polybractea (leaves)

Major component is 1,8-cineole (80–90%). Unlikely to present any hazard in aromatherapy.

Eucalyptus macarthuri (leaves)

There are three chemotypes, with varying amounts of geranyl acetate (19–70%), 1,8-cineole, geraniol and eudesmol. None of the chemotypes is likely to present any hazard in aromatherapy.

Eucalyptus radiata (leaves)

Major component is 1,8-cineole (70–75%). Unlikely to present any hazard in aromatherapy.

Eucalyptus smithii (leaves)

Major component is 1,8-cineole (70–80%). Unlikely to present any hazard in aromatherapy.

Eucalyptus staigeriana (leaves)

Major components are citral (16–40%) and geranyl acetate. Possibly a slight risk of sensitisation due to citral content. Due to citral content, might be best avoided in prostatic hyperplasia.

Fenugreek (seeds)

Trigonella foenum-graecum

Major components of the essential oil are hexanol, dihydrobenzofuran, and dihydroactinodiolide (all 5–10%). The absolute is non-sensitising, non-phototoxic and non-toxic (see Part One) [424].

Fleabane (= erigeron) (flowering herb)

Coniza canadensis (= *Erigeron canadensis*)

Major components include methyl esters (20–30%). Unlikely to present any hazard in aromatherapy. The oil is a water-white, or pale yellow liquid, which becomes darker and more viscous on ageing [155]; probably best used fresh. Occasionally available.

Frankincense (gum resin)

Boswellia carterii

Major component is *alpha*-pinene (43%). Unlikely to present any hazard in aromatherapy, but probably important to use fresh. See Part One for details on frankincense absolute.

Galangal (roots)

Alpinia galanga and *Alpinia officinarum*

Major component of *Alpinia officinarum* (lesser galangal) is 1,8-cineole (50%) with the balance largely made up of terpenes. Major components of *Alpinia galanga* (greater galangal) include 1,8-cineole (5–47%), *trans-beta*-farnesene (0–18%) and *beta*-bisabolene (0–16%); methyl cinnamate has also been reported as a major component. Neither form of galangal is likely to present any hazard in aromatherapy.

Garlic (bulbs) (P)

Allium sativum

Rich in sulphur compounds, major component diallyl disulphide. Use with caution orally in those with blood coagulation problems and avoid orally in thyroid disease. A definite potential allergen, and some irritancy is suspected. Patch testing would be prudent before dermal application, which is not recommended. Should not be used on hypersensitive, diseased or damaged skin. Extreme caution also warranted regarding application to mucous membrane. See profile.

Gingergrass (grass)

Cymbopogon martinii var. *sofia* (grass)

Major component is geraniol (35–65%). Unlikely to present any hazard in aromatherapy.

Goldenrod (herb)

Solidago canadensis

Components include terpenes, such as *alpha*-pinene, *alpha*-phellandrene, myrcene, limonene; also bornyl acetate. Unlikely to present any problems in aromatherapy, but probably important to use fresh.

Hedychium (= sanna) (rhizomes)

Hedychium spicatum

Major component is the ethyl ester of *p*-methoxy cinnamic acid. Also ethyl cinnamate (10%) and 1,4-cineole (6%). Unlikely to present any hazard in aromatherapy. Occasionally available.

Hemp

See Cannabis.

Hiba (wood)

Thujopsis dolobrata

Used in Japan, its country of origin. No compositional or safety data could be found for this oil, so toxicity uncertain.

Hinoki (leaves)

Chamaecyparis obtusa

Major components from various different hinoki leaf oils include *alpha*-pinene, longifolene, sabinene, thujopsene and elemol. Unlikely to present any hazard in aromatherapy. Probably best used fresh.

Hinoki (roots)

Chamaecyparis obtusa

Major component is longi-*alpha*-nojigikualcohol (20%). Also contains 8% of longi-*beta*-camphenilan aldehyde. Unlikely to be toxic, but might present problems of skin sensitisation.

Ho (wood) (= shiu wood)

Cinnamomum camphora, Cinnamomum camphora var. *linaloolifera*

Major component is linalool (50–75%), also contains cineole, pinene, camphene and myrcene [512]. Unlikely to present any hazard in aromatherapy. Ho leaf is covered in Part One and is also profiled.

Hop (inflorescence)

Humulus lupulus

Major components include terpenes such as *alpha*-humulene, *beta*-caryophyllene and myrcene. Also includes a number of aliphatic esters, such as isobutyrates and propionates. Some sources believe that the oil has oestrogen-like properties [34]. However, the basis for this appears to be fragile, since investigations of oestrogenic properties in the herb have been inconclusive [121].

Horseradish (roots) (P)

Cochlearia armoracia

Major components allyl isothiocyanate (50%) and phenylethyl isothiocyanate (45%). One of the most toxic and irritant oils. Should not be used in aromatherapy. See profile.

Hyssopus officinalis var. decumbens (herb)

Major component is *trans*-linalool oxide (57%) [34]. Unlike common hyssop oil, this variety is unlikely to present any hazard in aromatherapy.

Indian dill (seed) (P)

Anethum sowa

Contains 20–30% of apiol, which is abortifacient, and toxic to both liver and kidneys. Should not be taken in oral doses. See profile.

Inula (herb)

Inula graveolens

Major components include bornyl acetate and camphene. Unlikely to present any hazard in aromatherapy. Occasionally available.

Jamrosa (grass)

Hybrid of *Cymbopogon nardus* var. *confertiflorus* and *Cymbopogon jawarancusa*

Major components are geraniol (54–83%) and geranyl acetate (17–25%). Commercially available in India. Unlikely to present any hazard in aromatherapy.

Kewda

See Pandanus.

Khella (seeds) (P)

Ammi visnaga

Contains furanocoumarins and so may be phototoxic. In Australia this oil is classed as a poison, due to its khellin (1%) content. (The oral LD$_{50}$ in rats of khellin is 0.08 g/kg). Also contains lactones, and is suspected of sensitisation potential. The oil should not be used in

pregnancy, and should always be used with caution. See profile.

Kuromoji (leaves)

Lindera umbellata

Major components are linalool, 1,8-cineole and geranyl acetate. Unlikely to present any hazard in aromatherapy. Commercially available in Japan.

Lanyana (flowering herb) (P)

Artemisia afra

Presumed toxic due to high content of thujone (< 65%). Should not be used in aromatherapy. See profile.

Lavandula stoechas (flowering tops) (P)

Contains 15–30% camphor, and so should be used with caution in epilepsy, and should not be taken orally during fever or pregnancy. See profile.

Lavender cotton (flowering tops) (P)

Santolina chamaecyparissus

Contains artemisia ketone (10–45%) and longiverbenone (9–17%). No toxicology data could be found, but suspected of some toxicity. Use with caution. See profile.

Leek (seeds)

Allium porrum

Possibility of skin irritation and/or sensitisation, due to major components being sulphur compounds. See Garlic oil profile.

Lesser calamint (herb)

Calamintha nepeta

There are various subspecies, such as *glandulosa*, which contains little or no *d*-pulegone, and *nepeta*, which contains 45–70% *d*-pulegone, and therefore should not be used in aromatherapy. Currently not commercially available. Not to be confused with catnep (*Nepeta cataria*).

Longoza[A] (flowers)

Hedychium flavum

No compositional or safety data could be found for longoza absolute, so toxicology uncertain. Commercially available.

Lovage (herb)

Levisticum officinale

The composition of the leaf oil is not dramatically different to that of the root oil, which is detailed in Part One. Lovage leaf oil contains myrcene, limonene, 1,8-cineole, *alpha*-terpineol, ligustilide and butylidene phlalide. Unlikely to present any hazard in aromatherapy.

Mandarin leaf (= tangerine leaf)

Citrus reticulata

Major components vary considerably between different cultivars, and also according to the time of year. Commonly found are sabinene (0–27%), *gamma*-terpinene (0–38%), linalool (0–78%) and methyl-*N*-methyl anthranilate (0–65%). Whatever the composition, this oil is unlikely to present any hazard in aromatherapy.

Mango ginger (roots)

Curcuma amada

Major components are ocimene (47%) and *alpha*-pinene (18%). Also contains 9% of safrole, which could render the oil potentially carcinogenic, depending on the amounts of oil used. Not commercially available.

Massoia (bark) (P)

Cryptocaria massoia

Major component is massoia lactone (68%) which is a powerful irritant. The oil may also be sensitising

and/or phototoxic. Use with great caution on skin. See profile.

Mastic (gum)

Pistacia lentiscus

Major components of the oil are *alpha*-pinene (7–78%) and myrcene (3–25%). Unlikely to present any hazard in aromatherapy. Probably best used fresh. See Part One for data on mastic absolute.

Melaleuca bracteata (leaves) (P)

Major component is methyleugenol, which is carcinogenic [309, 310]. Do not use in aromatherapy. See profile.

Melaleuca quinquenervia, linalool CT (leaves)

Major component is linalool. Unlikely to present any hazard in aromatherapy. Compare niaouli.

Melaleuca quinquenervia, nerolidol CT (leaves)

Major component is nerolidol. Unlikely to present any hazard in aromatherapy. Compare niaouli.

Melissa (herb) (P)

Melissa officinalis

Major components are geranial (15–48%), neral (11–36%) and citronellal (4–39%). Due to high citral content should not be taken orally by people with glaucoma, and should be used with caution in cases of benign prostatic hyperplasia. May occasionally give rise to sensitisation, due to the aldehyde content. See profile.

Mikan

See Satsuma.

Monarda (herb)

Monarda fistulosa

There are several essential oil-producing species of monarda, many of which are rich in thymol/carvacrol. Oils from these chemotypes will probably demonstrate the irritancy normally associated with thymol/carvacrol, although none of them is currently commercially available. A geraniol chemotype of *Monarda fistulosa*, which is commercially available (geraniol 90–95%) is unlikely to present any hazard in aromatherapy.

Muhuhu (wood)

Brachylaena hutchinsii (= Synchodendron hutchinsii)

Major components *alpha*-amorphene (16%), brachyl oxide (10%) and copaenol (7.5%). Unlikely to present any hazard in aromatherapy.

Mustard (seeds) (P)

Brassica nigra

Major component is allyl isothiocyanate (99%). One of the most toxic and irritant oils. Do not use in aromatherapy. See profile.

Niaouli (leaves)

Melaleuca quinquenervia

Major component is 1,8-cineole (38–58%) [34]. Unlikely to present any hazard in aromatherapy.

Nut grass

See Cyperus.

Oakmoss[A] (P)

Evernia prunastri

Major components are variable amounts of *alpha*- and *beta*-thujone, so some toxicity is very likely. A moderate dermal sensitiser, IFRA recommends a maximum use level of 0.6%. Probably best avoided in aromatherapy. See profile.

Ocimum gratissimum (leaves) (P)

Major component is eugenol (21–90%). The oil is a major source of eugenol in Russia, but is rarely available outside its borders. Likely to have a similar safety profile to oils with similar eugenol content, such as clove and cinnamon leaf, i.e. moderately irritant, and to be used with caution orally in anyone with coagulation problems and liver disease. There are also thymol and linalool chemotypes, which are not commercially available. See profile.

Onion (seeds) (P)

Allium cepa

May present risk of skin irritation and/or sensitisation similar to garlic oil. Use with caution orally in anyone with blood coagulation problems, and avoid in thyroid disease. See profile.

Opopanax (gum) (P)

Commiphora erythrea

Major components are *cis-alpha*-bisabolene and *alpha*-santalene. Phototoxic (to an unknown degree) but otherwise safe. See profile.

Osmanthus[A] (flowers)

Osmanthus fragrans

Major components are *cis*-linalool oxide (2–21%), *trans*-linalool oxide (4–15%) and *gamma*-decalactone (5–12%). Osmanthus absolute is unlikely to present any significant toxicity in aromatherapy, but its safety for dermal application is unknown. It should be used with caution.

Pandanus[A] (= kewda) (flowers)

Pandanus odoratissimus

Major components are reported to include methyl-*beta*-phenylethyl ether (67–75%) and linalool (19%). On this basis, the absolute (and the rarer distilled oil) are unlikely to present any hazard in aromatherapy.

Parsnip (roots)

Pastinaca sativa

Major component is myristicin (17–40%) which should not be taken in conjunction with pethidine (see p. 42). Otherwise, unlikely to present any hazard in aromatherapy.

Peimou

See Siam wood.

Pennyroyal (N. American) (herb) (P)

Hedeoma pulegioides

Very similar toxicity to European pennyroyal. Major component is *d*-pulegone (60–80%) which is both neurotoxic and hepatotoxic. Should not be used in aromatherapy. See profile.

Peppermint (herb) (P)

Mentha piperita

Major component is menthol (40–50%). Acute oral LD$_{50}$ 4.4 g/kg [153] so non-toxic. Some contraindications. See profile.

Phoebe (wood)

Dreodaphne porosa

Major component is phoebile acetate (85–90%). Unlikely to present any hazard in aromatherapy. Commercially available.

Pimento berry (= allspice) (P)

Pimenta officinalis

Major component is eugenol (67–83%). Likely to have similar safety profile to oils with similar eugenol content, such as clove and pimento leaf, i.e. mucous membrane irritant (moderate–strong) and dermal irritant (moderate); oral contraindication in people with haemophilia or liver disease. Pimento leaf is covered in Part One. See profile.

Plectranthus (leaves) (P)

Plectranthus fruticosus

Major component is sabinyl acetate (60%). Embryotoxic and fetotoxic in rodents and teratogenic in mice, so potentially very dangerous in pregnancy. However, not commercially available. See profile.

Pteronia (herb)

Pteronia incana

Major components are *beta*-pinene (33%), *alpha*-pinene (19%), *para*-cymene (11%) and myrcene (10%). Unlikely to present any hazard in aromatherapy, but probably best used fresh. Available on a small scale.

Ravensara aromatica (leaves)

The major components are 1,8-cineole and *beta*-pinene. Unlikely to present any hazard in aromatherapy.

Ravensara anisata (bark) (P)

The major component is estragole (88%) with 7% *trans*-anethole (Brian Lawrence, private communication). The high estragole content makes this oil potentially carcinogenic, and the oil should not be used in aromatherapy. See profile.

Rosemary, verbenone CT (herb)

Rosmarinus officinalis

The content of verbenone (15–37%) might imbue this oil with a moderate degree of toxicity; verbenone mouse i.p. LD$_{50}$ 0.25 g/kg [6]. However, probably safe to use.

Sanna

See Hedychium.

Satsuma (fruit)

Citrus unshiu

The satsuma, or mikan, is a type of mandarin and the oil is not generally available outside Japan. Major component is limonene (82–91%). No significant levels of psoralens have been reported. Unlikely to present any hazard in aromatherapy, but best used fresh.

Savin (leaves and twigs) (P)

Juniperus sabina

Traditionally regarded as highly toxic, although no published data could be found to support this. There is good evidence that savin oil is abortifacient. IFRA recommend it is not used in fragrances, and it would be best avoided in aromatherapy. See profile.

Savory (winter) (herb) (P)

Satureia montana

Very similar toxicity to summer savory, detailed in Part One. See profile.

Shiu wood

See Ho wood.

Siam wood (= coffin wood = peimou)

Fokiena hodginsii

Major components include nerolidol (24%) and fokienol (24%) [34]. Occasionally available. Unlikely to present any hazard in aromatherapy.

Skimmia laureola (leaves)

Major components include linalyl acetate (40–67%) and linalool (18–28%). The precence of bergapten in the oil is disputed (see pp. 85, 86) so there is a possible risk of phototoxicity. Use with caution. Produced in commercial quantities in India.

Southernwood (herb)

Artemisia abrotanum

The major component is thujone, and consequently this oil should be regarded as toxic, and avoided in aromatherapy. Not commercially available.

Spikenard (roots)

Nardostachys jatamansi

Major components are *beta*-gurjunene (30%) and *alpha*-patchoulene (30%). Unlikely to present any hazard in aromatherapy.

Stenophylla (herb)

Salvia stenophylla

Major components are (+)-*epi-alpha*-bisabolol (30%) and 3-carene (20%). Unlikely to present any hazard in aromatherapy. Available on a small scale.

St John's Wort (flowering herb)

Hypericum perforatum

Major components are 2-methyloctane (16–20%) and *alpha*-pinene (10–16%). Unlikely to present any hazard in aromatherapy.

Styrax (gum)

Liquidambar orientalis and *Liquidambar styraciflua*

Major components are cinnamyl cinnamate (38%) and 3-phenylpropyl cinnamate (32%). Unlikely to present any hazard in aromatherapy.

Sugandha kokila (wood)

Cinnamomum cecidodaphne and *Cinnamomum glaucescens*

Safrole, myristicin and elemicin have been identified as components in the wood oil. This oil is said to be rich in safrole, and to have an odour of sassafras (which itself is very rich in safrole) [349]. If safrole is present in quantity, which seems likely, this oil should be regarded as potentially carcinogenic. Not commercially available. There is also a fruit oil (see below).

Sugandha kokila (fruit)

Cinnamomum cecidodaphne, Cinnamomum glaucescens and *Cinnamomum polyandrum*

A fruit oil from *Cinnamomum glaucescens* was found to contain elemicin (0.2%) and myristicin (2.5%) but no safrole. Fruit oils are occasionally commercially available, and are also processed from *Cinnamomum polyandrum* which contains up to 16% of menthyl cinnamate. Probably safe to use in aromatherapy.

Sweet clover[A] (herb)

Melilotus officinalis

The absolute is rich in coumarin, which is hepatotoxic in animals [336, 347, 469] and was banned by the FDA for food use in 1954. However, recent evidence shows that coumarin is not hepatotoxic in man, its metabolism being very different from that in animals [520, 527]. See page 63.

Tangerine leaf

See Mandarin leaf.

Tea[A] (leaves) (P)

Thea sinensis

Tea absolute has been found to cause sensitisation in dilutions as low as 0.001% in guinea pigs; however, it has not been tested on humans (private communication from RIFM). IFRA recommend that tea absolute is not used as a fragrance ingredient or in fragrance ingredients. Best avoided in aromatherapy. See profile.

Tejpat (leaves) (P)

Cinnamomum tamala

Major component is eugenol (75–80%). Likely to have similar safety profile to oils with similar

eugenol content, such as clove and cinnamon leaf, i.e. mucous membrane irritant (moderate–strong) and dermal irritant (moderate); oral caution in anyone with blood coagulation problems or liver disease. See profile.

Thyme, cineole CT (herb)

Thymus vulgaris

Major component is cineole (> 70%). Unlikely to present any hazard in aromatherapy.

Thyme, geraniol CT (herb)

Thymus vulgaris

Major component is geraniol (> 41%). Unlikely to present any hazard in aromatherapy.

Thyme, linalool CT (herb)

Thymus vulgaris

Major component is linalool (60–75%). Unlikely to present any hazard in aromatherapy.

Thyme, *para*-cymene CT (herb)

Thymus vulgaris

Major component is *para*-cymene (>44%). Unlikely to present any hazard in aromatherapy.

Thyme, thujanol CT (herb)

Thymus vulgaris

Major component is thujanol. Unlikely to present any hazard in aromatherapy. Occasionally available.

Thyme (wild) (herb) (P)

Thymus serpyllum

Major component is carvacrol (21–37%). Some skin and mucous membrane irritancy. See profile.

Tolu balsam (balsam from wood)

Myroxylon balsamum

Essential oils, absolutes and resinoids of tolu balsam are all available. No useful compositional or safety data could be found, so toxicity is uncertain.

Tuberose[A] (flowers)

Polianthes tuberosa

No useful compositional or safety data could be found, so toxicity is uncertain.

Valerian (roots)

Valeriana officinalis (= Valeriana fauriei)

Major components are isobornyl acetate (15–31%), bornyl acetate (1–17%), valeranal (3–16%), valeranone (0–32%) and camphene (8–14%). Valerian oil is non-toxic [414] and is unlikely to present any hazard in aromatherapy.

Valerian (Indian) (roots)

Valeriana wallichi

No useful compositional or safety data could be found, but similar to *Valeriana officinalis* and spikenard, so unlikely to present any hazard in aromatherapy.

Valerian (Japanese, = kesso) (roots)

Valeriana officinalis (= Valeriana fauriei)

No useful compositional or safety data could be found, but similar to *Valeriana officinalis* and spikenard, so unlikely to present any hazard in aromatherapy.

Vassoura (leaves)

Baccharis dracunculifolia

Major components are nerolidol (13–30%) and *beta*-pinene (18–22%). Unlikely to present any hazard in aromatherapy. Commercially available.

Western red cedar (leaves) (P)

Thuja plicata

Major component is thujone (85%) so toxicity may be assumed. Should not be used in aromatherapy. See profile.

Wine lees

See Cognac.

Wintergreen (leaves) (P)

Gaultheria procumbens

Major component is methyl salicylate (98%). Most 'wintergreen oil' is synthetic methyl salicylate. Some toxicity, and may be more toxic in humans than in rodents. Best avoided in aromatherapy. See profile.

Woodruff[A] (flowering herb)

Galium odoratum

The absolute is rich in coumarin, which is hepatotoxic in animals [336, 347, 469] and was banned by the FDA for food use in 1954. However, recent evidence shows that coumarin is not hepatotoxic in man, its metabolism being very different from that in animals [520, 527]. See page 63.

Yarrow (flowering herb) (P)

Achillea millefolium

Different chemotypes are commercially available. Most are safe, but the camphor chemotype is best avoided orally in fever and pregnancy, and should be used with caution by people with epilepsy. See profile.

Yuzu (fruit)

Citrus junos

Major component is limonene (75–80%). The oil is occasionally available outside Japan. No significant levels of psoralens have been reported. Unlikely to present any hazard in aromatherapy, but use fresh.

Zdravetz (leaves)

Geranium macrorrhizum

Major component is germacrone (50%). Toxicology uncertain, although the plant has been widely used in Bulgarian folk medicine.

Zedoary (roots)

Curcuma zedoaria

Major components are sesquiterpene alcohols (unspecified) 50%, also 1,8-cineole (10%) and *d*-camphor (4%). Toxicology uncertain, but probably safe to use in aromatherapy.

13

Cautions and contraindications

The contraindications listed below are intended to apply to the methods of use indicated for the materials indicated. They may not apply, for instance, to unusual chemotypes, which do not contain the toxic compound(s) as listed in the appropriate profile.

The contraindications are not intended to, and therefore may not, apply to the use of these oils in a non-aromatherapy context, such as soap perfumery, room fresheners or other fragrance products. Ultimately, concentrations in the final product will be a determining factor. The contraindications are not intended to apply to the raw materials and plants from which the oils are derived, nor to other quite different preparations from the same plant origin.

In a few cases we have indicated 'caution' only. These are cases where the risk is strong enough to merit flagging, but either the risk or the evidence is not strong enough to warrant a contraindication. The materials which are flagged as being unsuitable for any use in aromatherapy do not reappear under other, specific contraindications.

Essential oils which are not commercially available are not listed in this chapter, even the ones that have been profiled.

DEFINITIONS

Oral. This includes the oral administration of essential oils as used in medicinal aromatherapy, and assumes a minimum dose in 24 hours of

0.5 ml. Oral doses which are significantly lower than this may not warrant the contraindications listed for oral dosage.

Dermal. This includes the application of essential oils to the skin in concentrations between 1% and 5%. Concentrations significantly lower than 1% may not warrant the contraindications listed for dermal administration.

Inhalational. Inhalation includes the specific, intentional inhalation of essential oils, as in steam inhalation; it does not include the use of essential oils in vaporisers, in baths, or the incidental inhalation of essential oils by any other means.

Rectal. Rectal administration includes the use of essential oils in enemas or suppositories.

Vaginal. Vaginal administration includes the use of essential oils in pessaries or douches.

ALL ROUTES OF ADMINISTRATION

This includes oral, dermal, inhalational, rectal, and vaginal, as defined above.

All situations (i.e. never use)

Almond, bitter (unrectified), Armoise, *Artemisia arborescens*, Basil (high estragole), Birch (sweet), Boldo, Buchu (*B. crenulata*), Cade (unrectified), Calamus, Camphor (brown), Camphor (yellow), Cassia, Cinnamon bark, Costus, Elecampane, Fig leaf[A], Horseradish, Lanyana, *Melaleuca bracteata*, Mustard, Pennyroyal, *Ravensara anisata*, Sage (Dalmatian), Sassafras, Snakeroot, Tansy, Tarragon, Tea[A], Thuja, Verbena, Verbena[A], Wintergreen, Wormseed, Wormwood.

Cardiac fibrillation

Cornmint, Peppermint.

Caution

Use only with great caution in aromatherapy, due to uncertain toxicity:

Annual wormwood, Buchu (*B. betulina*), Cangerana, Elemi, Khella, Lavender cotton, Oakmoss[A], Savin, Treemoss[A].

Epilepsy

Avoid these oils in anyone suspected of being vulnerable to seizures:

Annual wormwood, Balsamite (camphor CT), Camphor (white), Ho leaf (camphor/safrole CT), Hyssop, Lavender cotton.

Fever

Annual wormwood, Balsamite (camphor CT), Camphor (white), Ho leaf (camphor/safrole CT), Hyssop, Lavender cotton.

Kidney disease (caution only)

Indian dill, Parsley leaf, Parsleyseed.

Liver disease (caution only)

Indian dill, Parsley leaf, Parsleyseed.

Pregnancy

Add to this list all the oils which are toxic in a general sense (see 'All situations' above):

Balsamite (camphor CT), Camphor (white), Ho leaf (camphor/safrole CT), Hyssop, Indian Dill, Parsley leaf, Parsleyseed, Sage (Spanish), Savin.

Pregnancy (caution only)

Essential oils which should be used with caution throughout pregnancy:

Annual wormwood, Buchu (B. betulina), Cangerana, *Lavandula stoechas*, Lavender cotton, Oakmoss[A], Perilla, Rue, Treemoss[A].

Prostatic hyperplasia (caution only)

Backhousia citriodora, *Eucalyptus staigeriana*?, Lemongrass, May chang, Melissa.

DERMAL ADMINISTRATION

Cancers (to avoid)

Maximum use levels:

Basil (low estragole)	2.0%
Fennel	1.5%
Ho leaf (camphor/safrole CT)	2.0%
Nutmeg (East Indian)	2.25%

Epilepsy (caution only)

Use these oils with caution in anyone suspected of being vulnerable to seizures. Reactions to essential oils by people with epilepsy can be idiosyncratic.

Lavandula stoechas, Sage (Spanish).

Fever (caution only)

Lavandula stoechas, Sage (Spanish).

Hypersensitive, diseased or damaged skin

Due to their irritant and/or sensitising potential the following essential oils should be avoided altogether on hypersensitive, damaged or diseased skin, such as in (normal) infants, or in eczema, dermatitis or psoriasis. Some of these oils are also contraindicated for other reasons.

Garlic, Massoia, Oakmoss[A], Treemoss[A], Verbena[A].

Also any oxidised oils, in particular the following:

Abies alba (cones), Abies alba (needles), Fennel (bitter), Pine oils, Spruce oils, Terebinth.

Hypersensitive, diseased or damaged skin (caution only)

Due to their irritant and/or sensitising potential the following essential oils should be used with caution on hypersensitive, damaged or diseased skin, such as in (normal) infants, or in eczema, dermatitis or psoriasis:

Ajowan, Anise, *Backhousia citriodora*, Cade (rectified), Citronella, Clove bud, Clove leaf, Clove stem, *Eucalyptus citriodora*, *Eucalyptus staigeriana*?, Khella, Laurel, Lemongrass, May chang, Melissa, Onion, Oregano, Perilla, Savory, Star anise, Thyme.

Infants

Children up to the age of about 2 years: avoid, or use with caution as appropriate, the oils listed above under 'Hypersensitive, diseased or damaged skin'.

Sunlight/sunbeds

Do not expose skin to sunlight or sunbed rays for 12 hours, if these oils are used at over maximum recommended concentrations on the skin:

Angelica root (0.78%), Bergamot (0.4%), Cumin (0.4%), Grapefruit (expressed) (4.0%), Lemon (expressed) (2.0%), Lime (expressed) (0.7%), Opopanax (level unknown), Orange (bitter, expressed) (1.4%), Rue (0.78%), Taget (0.05%).

Sunlight/sunbeds (caution only)

There is reason to suspect these essential oils of phototoxicity, but it is not known whether or not they are phototoxic, and if so at what level they could be safely used:

Combava (fruit), Khella, Mikan, *Skimmia laureola*, Yuzu.

ORAL ADMINISTRATION

The contraindications given here do not necessarily apply to the use of essential oils in any other context, such as in foods, or external use in aromatherapy. Also see 'All routes'.

All uses

These oils should not be taken orally:

Annual wormwood, Balsamite (camphor CT), Basil, Fennel, Ho leaf (camphor/safrole CT), Hyssop, Indian dill, Lavender cotton, Mace, Nutmeg (E. Indian), Parsley leaf, Parsleyseed.

230 ESSENTIAL OIL SAFETY

Alcoholism

See 'Liver disease'.

Anticoagulants (caution only)

These include aspirin, heparin and warfarin.

Bay (W. Indian), Betel leaf, Cinnamon leaf, Clove bud, Clove stem, Clove leaf, Garlic, *Ocimum gratissimum*, Onion, Pimento berry, Pimento leaf, Tejpat leaf.

Aspirin

See 'Anticoagulants'.

Blood clotting deficiency (caution only)

Includes people with haemophilia, kidney disease, liver disease, prostate cancer and SLE.

Bay (W. Indian), Betel leaf, Cinnamon leaf, Clove bud, Clove stem, Clove leaf, Garlic, *Ocimum gratissimum*, Onion, Pimento berry, Pimento leaf, Tejpat leaf.

Breast-feeding (caution only)

It would be prudent to be very cautious about oral dosing during breastfeeding, especially with the following oils:

Anise, *Backhousia citriodora*, Fennel, Lemongrass, May chang, Melissa, Star anise.

Cancers (oestrogen-dependent) (caution only)

This only applies to a few types of cancer, such as some breast cancers.

Anise, Fennel, Star anise.

Cancer of prostate

See 'Blood clotting deficiency'.

Cirrhosis

See 'Liver disease'.

Doses over 1 ml per 24 hours in an adult not recommended

Cornmint, Nutmeg (W. Indian), Peppermint.

Endometriosis (caution only)

Anise, *Backhousia citriodora*, *Eucalyptus staigeriana*, Fennel, Lemongrass, May chang, Melissa, Star anise.

Epilepsy

Avoid these oils in anyone suspected of being vulnerable to seizures. Reactions to essential oils by people with epilepsy can be idiosyncratic.

Camphor (white), *Lavandula stoechas*, Sage (Spanish).

Epilepsy (caution only)

Cornmint, Lavandin, Peppermint, Rosemary, Rue, Spike lavender, Yarrow (camphor CT).

Fever

Lavandula stoechas, Sage (Spanish).

Fever (caution only)

Cornmint, Lavandin, Peppermint, Rosemary, Rue, Spike lavender, Yarrow (camphor CT).

G6PD deficiency

Avoid taking these oils orally, if also G6PD deficient:

Cornmint, Peppermint.

Glaucoma

Backhousia citriodora, Lemongrass, May Chang, Melissa.

Haemophilia

See 'Blood clotting deficiency'.

Heparin

See 'Anticoagulants'.

Hepatitis

See 'Liver disease'.

Kidney disease (caution only)

Bay (W. Indian), Betel leaf, Cinnamon leaf, Clove bud, Clove stem, Clove leaf, Garlic, Indian dill, *Ocimum gratissimum*, Onion, Parsley leaf, Parsleyseed, Pimento berry, Pimento leaf, Tejpat leaf.

Liver disease (cirrhosis, hepatitis), alcoholism, or concurrent paracetamol use (caution only)

Also see 'Blood clotting deficiency'.

Anise, Basil, Bay (W. Indian), Betel leaf, Cinnamon leaf, Clove bud, Clove stem, Clove leaf, Fennel, Indian dill, *Ocimum gratissimum*, Parsley leaf, Parsleyseed, Pimento berry, Pimento leaf, Star anise, Tejpat leaf.

Paracetamol

See 'Liver disease'.

Pethidine

Nutmeg, Parsley leaf, Parsleyseed, Parsnip.

Pregnancy

These oils are listed in particular because they are safe to use externally (but not orally, rectally or vaginally) in pregnancy. It is recommended that essential oils in general are administered orally with great caution in pregnancy.

Anise, Fennel, Lavandin, *Lavandula stoechas*, Rosemary, Spike lavender, Star anise, Yarrow (camphor CT).

Prostate cancer

See 'Blood clotting deficiency'.

Prostatic hyperplasia (caution only)

Anise, Fennel, Star anise.

Systemic lupus erythematosus (SLE)

See 'Blood clotting deficiency'.

Thyroid disease

Garlic, Onion.

Warfarin

See 'Anticoagulants'.

INHALATIONAL, RECTAL OR VAGINAL ADMINISTRATION

Undiluted essential oils should not be applied to mucous membrane (eyes, mouth, vagina, rectum).

Epilepsy (caution only)

Reactions to essential oils by people with epilepsy can be idiosyncratic.

Lavandula stoechas, Sage (Spanish).

Fever (caution only)

Lavandula stoechas, Sage (Spanish).

Do not use at all on mucous membrane

Garlic, Massoia, Oakmoss[A], Treemoss[A], Verbena[A].

Also any oxidised oils, in particular the following:

Abies alba (cones), Abies alba (needles), Fennel (bitter), Pine oils, Spruce oils, Terebinth.

Do not use at more than 1% concentration on mucous membrane

Ajowan, Cade (rectified), Clove leaf, Clove stem, Oregano, Pimento leaf, Savory, Thyme.

Do not use at more than 3% concentration on mucous membrane

Bay leaf (W. Indian), Betel leaf, Caraway, Cinnamon leaf, Clove bud, Cornmint, Laurel, *Ocimum gratissimum*, Peppermint, Pimento berry, Spearmint, Tejpat leaf.

Pregnancy (caution only)

It is recommended that essential oils are only used with great caution rectally or vaginally during pregnancy.

Use with caution on mucous membrane

Backhousia citriodora, Catnep, Citronella, Combava leaf, *Eucalyptus citriodora*, *Eucalyptus staigeriana*, Khella, Lemongrass, May chang, Melissa, Onion, Perilla, Star anise.

ADDITIONAL SAFETY SUGGESTIONS

Alcoholism

Certain essential oils should be avoided in oral doses by alcoholics. This is because the liver of an alcoholic is more susceptible to damage than the liver of a healthy person.

Allergies

People who have an allergy to some or all fragrances may be also allergic to some or all essential oils. Such people normally avoid aromatherapy. Very rarely, a person who is not aware of any allergy will react allergically to an essential oil. In this case that person should avoid that essential oil. Patch testing may help to identify the allergenic oil in case of doubt.

Baths

Whenever essential oils are 'mixed' with water, without a dispersing agent, there is a risk of irritation, since 'clumps' of essential oil can attach to the skin, sometimes in sensitive areas. A useful way to proceed is to use a small amount the first time, say 2 drops for an adult, 1 for a child, and increase by 1 drop per bath, up to a maximum of 5 or 10 drops.

Breast-feeding

It would be prudent to be very cautious about orally administered essential oils while breast-feeding, especially with a few particular oils (see p. 230). It would also be wise not to apply essential oils directly to the breasts, especially the nipples, unless absolutely necessary.

Cancer

Essential oils will not interfere with the action of drugs given for the treatment of cancer. There is an opinion that cancer patients may associate the smells of essential oils with vomiting after the administration of chemotherapy, if the two are given close together. This might, by association, cause vomiting later on, whenever similar essential oils are smelt. A solution might be to time aromatherapy treatments so that they do not closely coincide with the administration of chemotherapy.

Areas of skin exposed to radiation therapy should not be massaged with oils, since these areas of skin become very fragile, and massage can cause the skin to break down. Massage should not be given over or close to tumour sites, nor over areas of skin cancer. It is extremely unlikely that massage could cause any type of cancer to spread. Only very gentle massage should be given to a person with cancer.

Children

Because their skin is especially sensitive, certain oils are best avoided on children under 2 years of age (see p. 229). Essential oils should not be instilled into the noses of children under 5 years (see pp. 25, 26).

Cirrhosis

Certain essential oils should not be taken orally by people with cirrhosis. See page 230.

Compresses

As with baths and douches, there is a slight risk of irritation if essential oils are applied to the skin improperly dispersed.

Condoms

Latex condoms can be weakened by both vegetable oils and essential oils. Corn oil, for instance, has been shown to cause a loss of up to 77% of a condom's strength after only 15 minutes. Condoms are used either to avoid pregnancy, or to guard against spreading sexually transmissible disease. In either case it is important to avoid contact with essential or vegetable oils.

Douching

If douching with essential oils, it should be remembered that they will not adequately dissolve in water, and that mucous membrane tissue is more sensitive to irritation than skin. Even if only 1 drop is put in 1 pint of water, there is a risk of irritation, depending on the essential oils used.

Driving (after aromatherapy)

It has been said that clients should not drive following aromatherapy massage with certain essential oils, such as clary sage. Perhaps more sensible advice would be to recommend not driving if the client feels significantly disoriented, following aromatherapy massage with any essential oils. In the majority of cases disorientation following treatment is both mild and transitory. Only in rare circumstances is driving proficiency likely to be affected.

Drugs

Some essential oils should be avoided orally, or used with caution, if taking certain drugs. See page 43.

Endometriosis

The nature of endometriosis is such that it could be exacerbated by oestrogenic essential oils. Their effect is relatively weak, and may not be important for external application. Caution is recommended for oral dosing with these oils. See page 230.

Epilepsy

Some essential oils, if taken in oral doses, could cause convulsions in a vulnerable person. People with epilepsy who are taking suppressant medication are probably no more vulnerable than non-epileptics. Epileptics who are not on medication are vulnerable, as are people who do not realise they are epileptics. Externally used essential oils are no more likely to cause a fit than any other fragrance or strong odour.

Eyes

Undiluted essential oils should not be applied to or very near the eyes. Great care should be taken, even with diluted essential oils. See 'Glaucoma'.

Fever

Some essential oils, if taken in oral doses, could bring on convulsions in someone with a fever, because we are more susceptible to convulsions when feverish. See page 230.

Genitals (applying oils to)

Genitals, especially in women, are more sensitive to irritation than the rest of the body.

Glaucoma

Certain essential oils should not be taken orally by people with glaucoma. This is because they may exacerbate the condition by increasing pressure inside the eye. See page 230.

Goitre

Certain essential oils should not be taken orally by people with thyroid disease such as goitre. See page 231.

Haemophilia

Certain essential oils should not be taken orally by people with haemophilia, because, in oral doses, these oils hinder the processes which enable blood to clot. See page 230.

Heart disease

Cornmint and peppermint oils should not be taken orally by people prone to cardiac fibrillation.

Hepatitis

Certain essential oils should not be taken orally by people with hepatitis. See page 231.

Hormone replacement therapy (HRT)

It is unlikely that aromatherapy treatments could adversely affect hormone replacement therapy as any hormonal effect of the oil would almost certainly be considerably weaker than that of the HRT.

Hypertension

It is very unlikely that any essential oils could exacerbate hypertension. See page 64.

Hypotension

It is very unlikely that any essential oils could exacerbate hypotension. See page 64.

Inhalation

It is very unlikely that there is any risk involved in the normal use of a few drops of essential oil in burners, vaporisers or in steam inhalations. However, prolonged inhalation (several consecutive hours) of concentrated essential oil vapours (e.g. steam inhalation, or direct from bottle) can lead to headaches, vertigo, nausea and lethargy. In certain instances more serious symptoms might be experienced, such as incoherence and double vision. See page 31.

Migraine

In some cases of migraine strong odours, such as essential oils, may exacerbate an attack. Essential oils should be used with great caution in such cases during an attack, but they are not likely to precipitate one.

Oral administration

Great care is required when prescribing or taking essential oils orally, as there are more dangers to beware of. For minimal gastrointestinal irritation and maximal absorption, the essential oils should be evenly dispersed in a vehicle.

Poisoning

If poisoning from essential oil ingestion is suspected, either telephone a general practitioner or take the person to a hospital accident and emergency department. If the person is showing severe signs of poisoning (such as loss of consciousness) telephone the emergency service. Gastric lavage is often undertaken in such cases, but do not try to induce vomiting yourself, unless advised to by a health professional. See page 240 for further details.

Pregnancy

It is likely that all essential oils cross the placenta, and therefore reach the fetus, after administration to a pregnant woman. There are many oils that should therefore not be taken in oral doses, and it is recommended that oils are only taken orally with great caution throughout pregnancy. Equal caution is recommended for rectal and vaginal administration, since the amount of essential oil reaching the fetus could easily be as high as from oral dosing. A maximum dilution of 2% for massage is suggested, and there are a number of oils which should be avoided altogether. As well as potential risks to the fetus, there are a very small number of oils which might induce abortion in oral doses. See page 108.

Premature babies

Since premature babies have very thin and sensitive skin it might be prudent to avoid the external use of essential oils altogether, unless there are very important benefits to be gained by the baby. Even vaporising oils into the air might be best avoided, again unless there are important benefits to be gained.

Skin

Certain essential oils are best avoided on hypersensitive, inflamed or diseased skin, in order to guard against inflammation. See page 229. Also see 'Undiluted oils'.

Skin diseases

Special caution is needed when applying essential oils to diseased or damaged skin. This is because it is more prone to allergic reactions, as well as being more readily permeable to essential oils. Certain oils should be avoided, or used with caution; see page 229.

Sunbathing

Certain essential oils should not be applied to the skin in certain concentrations up to 12 hours prior to sunbathing. This is to avoid phototoxic reactions. See page 229.

Thyroid disease

Certain essential oils should not be taken orally by people with thyroid disease. See page 231.

Undiluted oils

As a general rule undiluted essential oils are not used in massage, nor are they applied to eyes, diseased skin, mucous membrane, or broken skin (cuts, wounds, surgery, etc.). This is to avoid causing inflammation. However, there are instances in which the benefits may outweigh the risks. Such instances could include: bites and stings, burns, herpes simplex, herpes zoster, leg ulcers, malignant ulceration (fungating carcinomas), mouth ulcers, neutropenic ulceration, pressure sores, tinea, verrucae, warts, etc. In many of these instances medical supervision would be advisable.

14

Safety guidelines

This chapter contains a summary of the advice and conclusions of the book. It largely comprises a series of recommendations addressed to different groups: aromatherapists, essential oil suppliers and consumers.

AROMATHERAPISTS

- Aromatherapists should check with new clients to find out if they have ever experienced skin reactions to essential oils or perfumes.
- Aromatherapists should know how to take a proper case history in order to ascertain whether a new client has a condition, or is taking a drug, which might contraindicate particular essential oils.
- Aromatherapists should always record medication being taken by a client, as well as any negative reactions to the treatment.
- Aromatherapists should know how to patch test, and should use patch testing in appropriate cases.
- Aromatherapists should ensure that all undiluted (single or mixed) essential oils given to clients are adequately labelled and are in bottles with integral drop-dispensers.
- Aromatherapists should not advise clients to ingest essential oils, unless the therapist is also a primary care practitioner, such as a medical doctor or medical herbalist, who has a thorough knowledge of essential oil toxicology.
- Also see advice below under 'The consumer'.

We should probably add that, in order to practice safely an aromatherapist should have sufficient knowledge of clinical medicine to know when to refer a client, or avoid a particular procedure.

Notification

In order that the true toxic potential of aromatherapy can be assessed, it is vital that a notification system be put in place, akin to the 'yellow card' system used by GPs. This would involve therapists notifying a central body of any adverse reactions to essential oils which they observe in clients, giving full details.

Practitioners' safety

The exposure of aromatherapists to essential oil vapours is generally more varied and more prolonged than the client's. Inhalation of oil vapours represents a particular risk for busy aromatherapists who work in cramped conditions with little or no ventilation. This can give rise to headaches, fatigue or nausea. Good ventilation is perhaps the most important factor in preventing such occupational side-effects for aromatherapists. For safety in practice, our recommendations are:

- Good ventilation in treatment room
- Air treatment room thoroughly in between clients
- Minimum of 5 minutes for fresh-air break in between clients
- Wash hands in between appointments
- Keep and dispense oils in a separate room.

Similar guidelines will apply to others who work with essential oils on a daily basis, such as bottlers and packers. Contact, via inhalation and skin, should be minimised.

THE CONSUMER
General safety guidelines

- Keep essential oils where small children cannot reach them.
- Use less rather than more and, if necessary, gradually increase the amount applied.
- Do not take essential oils internally unless prescribed by a medical doctor or medical herbalist.

- Do not purchase undiluted essential oils unless the bottles have integral drop-dispensers and/or child-resistant tops.
- Avoid essential oils which are poorly labelled (e.g. only the name of the oil on the label) and which have no safety guidelines (e.g. 'keep out of reach of children').
- Do not apply *undiluted* essential oils to mucous membranes (eyes, mouth, vagina, rectum).
- If phototoxic oils are used at over recommended levels (see p. 85) the areas of skin to which they have been applied should not be exposed to UV light (daylight or sunbeds) for a minimum of 12 hours.

Avoiding degraded oils

See page 10 for a full explanation of why avoiding degraded oils is important.

- Do not use essential oils if they are in clear bottles.
- Small amounts of essential oil remaining in large bottles should be transferred to smaller bottles, in order to reduce the risk of oxidation.
- Do not use essential oils which are more than 1 year old (after purchase/opening) or 2 years if they have been kept in a refrigerator.

OVER THE COUNTER

There are four important elements which impinge on the safety of oils as they are sold to the public:

- Packaging
- Labelling
- Restricted oils
- Awareness.

Packaging

It cannot be over-emphasised that many essential oils could be life-threatening if a child were to drink them. Integral drop-dispensers are a great help in limiting the quantity of oil which a child is likely to ingest accidentally.

The use of child-resistant bottle caps is also of enormous benefit in helping prevent accidents in young children. Even though children sometimes manage to open these bottles, this measure could potentially save lives.

Essential oils in open-topped bottles probably represent the biggest single risk in aromatherapy today. We would like to see essential oils sold only in bottles with integral drop-dispensers and/or child-proof closures.

Labelling

We would like to see a move towards clearer and more informative packaging on essential oil bottles. From a safety standpoint, useful information could include:

- A use-by date, or other indication of useful lifespan
- The botanical name of the originating plant, if relevant
- The part of the plant the oil came from, if relevant
- The percentage concentration of essential oil, if the product is diluted.

The use-by date will help prevent skin reactions from oxidised essential oils, as well as ensuring that the oil is used when still fresh and potent. The botanical name could avoid confusion between different essential oils with the same common name. The plant part will often not be necessary, but could be very important in cases such as cinnamon leaf and bark, where the bark oil is considerably more hazardous than the leaf oil with regard to skin reactions. The percentage dilution will help the user of the oil to know how much to use, and would also help doctors in case of the oil being accidentally ingested.

Recommended cautions for all undiluted oils

The following are suggestions for inclusion on the packaging of undiluted essential oils:

- 'For external use only'
- 'Keep out of reach of children'
- 'If ingested by a child, seek urgent medical help'
- 'Essential oils can irritate if used undiluted'.

However, there are practical problems in having a large number of words on what is usually a very small label.

Awareness

There are undoubtedly some oils which present negligible risk, but it is worrying that so many oils are sold in open-mouthed bottles in shops where there is no suitably trained supervision. We would

Table 14. 1 Potentially hazardous essential oils and their toxic components

Oil	Toxic component
Almond (bitter, unrectified)	Hydrocyanic acid
Armoise	Thujone
Artemisia arborescens	Thujone
Basil (high estragole)	Estragole
Birch (sweet)	Methyl salicylate
Boldo	Ascaridole
Buchu (B. crenulata)	Pulegone
Cade (unrectified)	Benzo[a]pyrene
Calamus	Asarone
Camphor (brown)	Safrole
Camphor (yellow)	Safrole
Cassia	Cinnamaldehyde
Cinnamon bark	Cinnamaldehyde
Costus	Costuslactone
Elecampane	Alantolactone
Fig leaf[A]	?
Horseradish	Allyl isothiocyanate
Lanyana	Thujone
Melaleuca bracteata	Methyleugenol
Mustard	Allyl isothiocyanate
Pennyroyal	Pulegone
Ravensara anisata	Estragole
Sage (Dalmatian)	Thujone
Sassafras	Safrole
Snakeroot	Methyleugenol
Tansy	Thujone
Tarragon	Estragole and methyleugenol
Tea[A]	?
Thuja	Thujone
Verbena	?
Wintergreen	Methyl salicylate
Wormseed	Ascaridole
Wormwood	Thujone

like to see appropriate training for all retail staff who sell essential oils, even where packaging and labelling of the products is of a good standard.

Restricted oils

Because of their potentially hazardous nature we recommend that the oils listed in Table 14.1 should not be sold, *in undiluted form*, over the counter, or be available to the general public in any other way. It is not our intention to restrict the availability of any essential oils to practitioners, perfumers, food flavourists or anyone else who will know how to use them safely and responsibly.

WHAT TO DO IN A CASE OF ESSENTIAL OIL POISONING

If a child (or an adult) ingests an unknown or excessive quantity of essential oil do one of the following:

- Telephone a general practitioner.
- Take the person to a hospital accident and emergency department, taking the bottle with you.

If you discover, some time after the event, that ingestion took place, and the person is showing severe signs of poisoning (such as loss of consciousness) you should telephone the ambulance emergency service.

Whoever you speak to will want to know what was ingested, and how much. The doctor may contact one of the national poisons units. One of their main functions is to give advice to doctors in such situations. Do not try to contact a poisons unit directly.

As in any case of poisoning, prompt action could very well result in the person's life being saved. Many of the cases of poisoning referred to in this book only survived because they received rapid intervention. Gastric lavage is often undertaken, but do not try to induce vomiting yourself, unless advised to do so by a health professional.

Glossary

Absolute: The aromatic material extracted from a plant using solvent extraction. Essential oils are only produced by distillation or (in the case of citrus oils) cold pressing. Absolutes tend to be more viscous than essential oils, and are more likely to be coloured.

Adduct: Most often used in the context of DNA—'DNA adduct'. A molecule attached to DNA, which alters its function.

Alkylating agent: A chemical which can permanently alter the chemical structure of another, such as DNA, by attaching a carbon chain. Alkylating agents are potential carcinogens.

Anosmic: A person or animal having no sense of smell.

Anthelmintic: A remedy which destroys intestinal worms.

Anticholinergic: Antagonising (counteracting) acetylcholine, a neurotransmitter found in the CNS and periphery.

Ataxia: Loss of control over voluntary movements.

Bioavailability: In this text, bioavailability is taken to mean the proportion of a substance which reaches the systemic circulation after administration. It is generally lower after dermal or oral dosing than after injection (depending on the substance, however) and is, by definition, 100% after intravenous administration.

Biotransformation: The changing of one chemical substance into another, which may be biologically active or inactive, by the body's metabolism.

Bradycardia: Slowing of heartbeat.

Carboxylic acids: These are carbon-containing acids very common in foods, but relatively unusual in essential oils. Acetic acid, found in vinegar, and citric acid, found in citrus fruits, are two common examples from foodstuffs. Valeric acid, cinnamic, anisic, citronellic and phenylacetic acids are all found in essential oils. Like all carboxylic acids they contain the COOH functional group.

Chemotypes: Varieties of a species of aromatic plant which are of identical appearance but which produce essential oils with significant differences in chemical composition and hence toxicity. The various chemotypes of a species will usually possess quite different major components. See basil, oregano and thyme profiles for examples of chemotypes.

Clonic: Convulsions characterised by repetitive muscular contraction.

Conjugation: The addition of a water-soluble chemical group to a molecule taken into the body so as to aid excretion into the urine. 'Phase II' metabolism.

Cultivar: A plant variety produced from a naturally occurring species that has been developed and maintained by cultivation.

Cyanosis: Bluish colouration of the skin and mucous membranes due to inadequate oxygenation of the blood.

Cytochrome P_{450}:   A collective name for a group of metabolic enzymes found in many tissues, but most prevalently in the liver. They are responsible for the oxidation (Phase I metabolism) of foreign molecules.

Cytotoxic: Most cytotoxic drugs interfere with the metabolism and replication of DNA. They can do this in several ways, but all have the potential for great toxicity, as well as the ability to provoke malignant change, somewhat paradoxically. Some cytotoxic drugs work through other mechanisms. Cytotoxic activity which has been discovered in essential oil molecules is generally of a much lower order than that of anti-cancer drugs, and cytotoxic essential oils would not be expected to be toxic as used in aromatherapy.

Ecchymosis: Effusion of blood into tissue, giving the appearance of a bruise.

Electrophile: A molecule which has a high affinity for electron-rich chemicals, such as DNA and protein. Metabolic electrophiles (e.g. epoxides) are very toxic, especially to the liver.

Epoxidation: The formation of epoxides.

Epoxide: Phase I metabolism produces epoxides from certain molecules. They are highly reactive and therefore toxic, most importantly to the liver.

Exchange transfusion: The gradual removal of most or all of the recipient's blood with the simultaneous transfusion of an equal volume of normal blood.

Excoriation: The stripping away of the surface layer of the skin.

Fetotoxicity: A general term for toxicity to the developing fetus, strictly speaking after the embryo stage.

Fraction: A group of constituents of an essential oil, with similar volatility. Splitting an essential oil into fractions is known as fractionation.

Functional groups: In essential oils, usually oxygen-containing combinations of atoms within a terpenoid structure which contribute to the oils' chemical and aromatic characteristics. For example: alcohol, ester, ketone, aldehyde.

Furanocoumarin: Also known as furocoumarin or psoralen. A cyclic structure containing oxygen, commonly found in cold-pressed citrus oils. Frequently associated with phototoxicity on exposure to UV light.

Gastric lavage: Washing out of the stomach, a procedure often used in cases of poisoning.

Genotoxic: Substances which interfere with the structure and/or function of cellular DNA are known as genotoxic (i.e. toxic to genes). It is possible to look for such abnormal cellular changes using cultures of human or animal cells. Genotoxicity indicates possible, but not definite, carcinogenesis.

Glutathione (reduced): A sulphur-containing substance in the liver responsible for detoxifying highly reactive metabolites. Once glutathione is depleted, the toxicity of the reactive metabolites is greatly enhanced.

Haemolytic anaemia: Destruction of the red blood cells.

Hepatocarcinogen: A chemical which can assist the development of liver cancer. Alk(en)ylbenzenes, such as asarone, safrole and estragole are potential hepatocarcinogens.

Hepatocyte: Liver cell.

Hepatotoxicity: Toxicity to the liver in a general sense.

Histopathological: Damaging to tissue.

Hydroxylation: The addition of an -OH functional group to a carbon chain, often at a C=C double bond.

Hyperkeratosis: Thickening of the upper dead cell layer in the epidermis.

Hyperplasia: Any condition in which there is an increase in the number of cells in a part. Benign prostatic hyperplasia is common in older men, and results in enlargement of the prostate.

Hyperpyrexic: Suffering from a severely raised temperature (over 41°C).

Induction: A chemically-induced increase in enzyme activity.

Isomers: There are many different types of isomer. In the chemistry of essential oils, the types of isomer usually encountered are:

Optical isomers: the isomers are non-identical mirror images of one another, rather like a pair of gloves. The isomers can rotate the plane of plane-polarised light; one optical isomer will rotate it clockwise (the *d*-isomer), the other anti-clockwise (the *l*-isomer). Despite their chemical similarities, optical isomers can smell very different (e.g. *d*- and *l*-carvone).

Geometric isomers: the atoms in a molecule are joined up in the same order, but take up different positions in space (e.g. *trans-trans*- and *cis-trans*-farnesol. See pp. 13 and 14 for diagrams of isomers).

LD$_{50}$: This stands for 'lethal dose 50%', and is the traditionally accepted method of acute toxicity testing. Different doses of test substance are administered to groups of matched animals (one dose level per group) usually rats or mice. The dosage that kills 50% of a group is taken as the LD$_{50}$ value for the test substance. LD$_{50}$ values are expressed in grams of test substance per kilogram of body weight. The toxicity of the test substance may then be extrapolated from test animal to human.

Lymphoid hyperplasia: Overgrowth of the lymphatic system.

Maximation test: Test for skin sensitisation. The test substance is applied repeatedly to a patch of skin until a maximal response is obtained.

Metabolite: Substance produced as a result of a biochemical reaction in the body, sometimes by way of detoxifying a chemical.

Microsomal enzymes: Enzymes, including the cytochromes P_{450}, which are located in sub-cellular structures called microsomes.

Mutagenic: Mutagenic chemicals can alter the genetic messages in DNA by damaging it. They can thus effectively switch genes on or off, or change what they do. They are often associated with cancer causation. Mutagenic tests can be carried out in bacteria or fungi.

Myeloid: Pertaining to the bone marrow.

Myelosuppression: Suppressing the production of blood cells in the bone marrow.

Necrosis: Tissue death due to insufficient flow of oxygenated blood, or to direct action of a corrosive chemical. May lead to gangrene.

Neurotoxic: Having a destructive or toxic effect on nervous tissue.

Nidation: Implantation: the embedding of the fertilised ovum in the uterine mucous membrane.

Occlusion: The covering of the skin with an impermeable material which prevents evaporation of volatile substances from the skin, and enhances absorption.

Olfactory epithelium: The small area of mucous membrane at the top of the nose, into which the olfactory nerve endings project.

Oncogenesis: A general term for cancer formation.

Orthostatic: Orthostatic hypotension is hypotension which only occurs when the person stands up.

Oxidation: The addition of oxygen, or the removal of electrons or hydrogen from an organic molecule.

Percutaneous: By diffusion through the skin. (In contrast to subcutaneous, which refers to injection through the skin.).

Photocarcinogenesis: The initiation of cancer by UV light, or by a chemical in the presence of UV light.

Phototumorigenesis: The initiation of tumours by UV light, or by a chemical in the presence of UV light.

Polar: Term used to describe a molecule which has separate regions with distinct partial positive and negative electrical charges. Polar molecules are generally water-soluble and fat-insoluble.

Polycyclic: A molecular structure with several rings of atoms joined together.

Potentiate: To give power to, or increase (the effect of). Methyl salicylate, for instance, potentiates the anticoagulant action of warfarin.

Psoralen: See furanocoumarin.

Psychotropic: Affecting the brain and influencing mood and behaviour.

Reactive: Likely to undergo chemical reaction, possibly resulting in a pharmacological or toxicological effect.

Sesquiterpenoids: These are essential oil constituents based on $1\frac{1}{2}$ terpene units (the sesquiterpene unit) and which contain functional groups such as aldehyde, ester or ketone. They are relatively unusual in essential oils. Examples include alantolactone and patchouli alcohol.

Subcutaneous: Beneath the skin.

Substrate: Compound which is acted upon by an enzyme.

Teratogenic: Causing developmental abnormalities in the fetus.

Teratology: The study of the prediction of developmental abnormalities in fetuses caused by chemical agents.

Terpenoids: Essential oil constituents based on the terpene hydrocarbon skeleton, but which contain functional groups such as aldehyde, ester or ketone. Most essential oil constituents which are not terpene hydrocarbons are terpenoids.

Tumorigenic: Tumour-inducing.

ABBREVIATIONS

CNS	central nervous system
CT	chemotype
EC	European Community
et al	and others (usually authors)
FCF	furanocoumarin-free
FDA	Food and Drug Administration
FFPA	free from prussic acid
g	grams
g/kg	grams (of substance) per kilogram of body weight
IFRA	International Fragrance Association
i.p.	intraperitoneal (into the abdominal cavity)
i.v.	intravenous
mg	milligrams (one thousandth of a gram)
μg	micrograms (one millionth of a gram)
ml	millilitres
ml/kg	millilitres (of substance) per kilogram of body weight
NCA	not commercially available
ng	nanograms (one thousand millionth of a gram)
ppm	parts per million
ppb	parts per billion
RIFM	Research Institute for Fragrance Materials
s.c.	subcutaneous
<>	'found in' (see NOTES p. xiii)
>	'Over', e.g. '> 5 g/kg' means '5 g/kg or over'.
<	'Less than', e.g. '< 1 g/kg' means 'less than 1 g/kg'

References

[1] Falk-Filipsson A et al 1993 d-Limonene exposure to humans by inhalation: uptake, distribution, elimination, and effects on the pulmonary function. Journal of Toxicology and Environmental Health 38: 77-88

[2] Jirovetz L et al 1992 Analysis of fragrance compounds in blood samples of mice by gas chromatography, mass spectrometry, GC/FTIR and GC/AES after inhalation of sandalwood oil. Biochemical Chromatography 6(3): 133-134

[3] Guenther E 1949–1952 The essential oils. Vols. 1–6. D. Van Nostrand, New York

[4] Evans W C 1989 Trease and Evans' Pharmacognosy, 13th edn. Baillière Tyndall, London, ch 27

[5] Reynolds J E F (ed) 1993 Martindale: The extra pharmacopoeia. The Pharmaceutical Press, London

[6] Harborne J B, Baxter H (eds) 1993 Phytochemical dictionary, a handbook of bioactive compounds from plants. Taylor & Francis, London

[7] Parry E J 1922 The chemistry of essential oils and artificial perfumes, 4th edn. Scott, Greenwood, London, vols 1 and 2

[8] Rodricks J V 1992 Calculated risks: understanding the toxicity and human health risks of chemicals in our environment. Cambridge University Press, Cambridge

[9] List P H, Hörhammer L 1976 Hager's Handbuch der Pharmazeutischen Praxis. Springer-Verlag, Berlin, vols 2–6

[10] Grandjean P 1990 Skin penetration. Hazardous chemicals at work. Taylor & Francis, London p 3–4

[11] Schilcher H 1985 Effects and side-effects of essential oils. In: Baerheim-Svendsen A, Scheffer J J C (eds) Essential oils and aromatic plants. Martinus Nijhoff, Dordrecht p 217–231

[12] Tisserand R 1985 The essential oil safety data manual. Tisserand Aromatherapy Institute, Brighton

[13] Gilman A G, Goodman L S, Gilman A 1980 The pharmacological basis of therapeutics, 6th edn. Baillière Tyndall, London, p 31

[14] Gerarde H W 1960 Toxicology and biochemistry of aromatic hydrocarbons. Elsevier, London, p 59

[15] Wester R C, Noonan P K 1980 Relevance of international models for percutaneous absorption. International Journal of Pharmaceutics 7: 99–110

[16] Bowman W C, Rand M J 1980 Textbook of pharmacology, 2nd edn. Blackwell Scientific Publications, Oxford

[17] Phillips J C et al 1976 Studies on the absorption, distribution and excretion of citral in the rat and mouse. Food and Cosmetics Toxicology 14: 537–540

[18] Schroder V, Vollmer H 1932 The excretion of thymol, carvacrol, eugenol, and guaiacol and the distribution of these substances in the organism. Naunyn-Schmeidebergs Arch Exp Path Pharmak 168: 331–353

[19] Nuttall N 1992 Sweet oils may put women in peril. The Times, 1st March

[20] Boyland E, Chasseaud L F 1970 The effect of some carbonyl compounds on rat liver glutathione levels. Biochemical Pharmacology 19: 1526–1528

[21] Goeger D E, Anderson K E 1991 Coumarin-induced changes in d-aminolaevulinic acid synthase and cytochrome P_{450} in chick embryo liver. Food and Chemical Toxicology 29(3): 145–151

[22] Moorthy B 1991 Toxicity and metabolism of R-(+)-pulegone in rats: its effects on hepatic cytochrome P_{450} in vivo and in vitro. Journal of the Indian Institute of Science 71(1): 76–78

[23] Messiha F S 1990 Effect of almond and anise oils on mouse liver alcohol dehydrogenase, aldehyde dehydrogenase and heart lactate dehydrogenase isoenzymes. Toxicology Letters 54(2–3): 183–188

[24] Roffey S J et al 1990 Hepatic peroxisomal and microsomal induction by citral and linalool in rats. Food and Chemical Toxicology 28(6): 403–408

[25] Moorthy B et al 1989 Hepatotoxicity of pulegone in rats: its effects on microsomal enzymes. Toxicology 55: 327–337

[26] Parke D V, Rahman H 1970 The induction of hepatic microsomal enzymes by safrole. Proceedings of the Biochemical Society 119: 53P–54P

[27] Woo W S et al 1983 Effect of naturally occurring coumarins on the activity of drug metabolizing enzymes. Biochemical Pharmacology 32: 1800–1803

[28] Hoskins A 1984 The occurrence, metabolism and toxicity of cinnamic acid and related compounds. Journal of Applied Toxicology 4: 283–292

[29] Le Bourhis B 1970 Identification de quelques métabolites du trans-anéthole chez l'homme, le lapin et le rat. Annales Pharmaceutiques Francaises 28: 355–361

[30] Sangster S A et al 1987 The metabolic disposition of (methoxy-C)-labelled trans-anethole, estragole and p-propylanisole in human volunteers. Xenobiotica 17: 1223–1232

[31] Jori A et al 1969 Effect of essential oils on drug metabolism. Biochemical Pharmacology 18(9): 2081–2085

[32] Borchert P et al 1973 The metabolism of the naturally occurring hepatocarcinogen safrole to l-hydroxysafrole and the electrophilic reactivity of l-acetoxysafrole. Cancer Research 33: 575–589

[33] Strolin-Benedetti M, Le Bourhis B 1972 Répartition dans l'organisme et élimination du trans-anéthole. C R Hebd Seanc Acad Sci Paris 274(D): 2378–2381

[34] Franchomme P, Pénoël D 1990 L'aromatherapie exactement. Jollois, Limoges

[35] Ishida T et al 1989 Terpenoid biotransformation in mammals V. Metabolism of (+)-citronellal, (+/-)-7-hydroxycitronellal, citral, (-)-perillaldehyde, (-)-myrtenal, cuminaldehyde, thujone, and (+/-)-carvone in rabbits. Xenobiotica 19: 843–855

[36] Ishida T et al 1977 Biotransformation of terpenoids in mammals; biotransformation of 3-carene and related compounds in rabbits. Tetrahedron Letters 28: 2437–2440

[37] Ritschel W A et al 1977 Pharmacokinetics of coumarin and its 7-hydroxy-metabolites upon intravenous and peroral administration of coumarin in man. European Journal of Clinical Pharmacology 12: 457–461

[38] Spector W S (ed) 1956 Handbook of toxicology vol. 1. Acute toxicities. W B Saunders, Philadelphia

[39] Hohenwallner W, Klima J 1971 In vivo activation of glucuronyl transferase in rat liver by eucalyptol. Biochemical Pharmocology 20(12): 3463–3472

[40] Parke D V et al 1974 Effect of linalool on hepatic drug-metabolizing enzymes in rats. Biochemical Society Transactions 2: 615–618

[41] Chadha A, Madyastha K M 1984 Metabolism of geraniol and linalool in the rat and effects on liver and lung microsomal enzymes. Xenobiotica 14: 365–374

[42] Madyastha K M, Chadha A 1986 Metabolism of 1,8-cineole in rat: its effects on liver and lung microsomal cytochrome P_{450} systems. Bulletin of Environmental Contamination and Toxicology 37: 759–766

[43] Parke D V, Rahman H 1969 The effects of some terpenoids and other dietary anutrients on hepatic drug-metabolizing enzymes. Proceedings of the Biochemical Society 113: 12P

[44] Jori A et al 1970 Effect of eucalyptol (1,8-cineole) on the metabolism of other drugs in rats and in man. European Journal of Pharmacology 9: 362–366

[45] Ioannides C et al 1981 Safrole: its metabolism, carcinogenicity and interactions with cytochrome P_{450}. Food and Cosmetics Toxicology 19: 657–666

[46] Jori A et al 1972 On the inducing activity of eucalyptol. Journal of Pharmacy and Pharmacology 24: 464–469

[47] Wade A E et al 1968 Alteration of drug metabolism in rats and mice by an environment of cedarwood. Pharmacology 1: 317–328

[48] Regan J W, Bjeldanes L F 1976 Metabolism of (+)-limonene in rats. Journal of Agricultural and Food Chemistry 24: 377–380

[49] Kodama R et al 1974 Studies on the metabolism of d-limonene II. The metabolic fate of d-limonene in rabbit. Xenobiotica 4: 85–95

[50] Parke D V et al 1974 The absorption, distribution and excretion of linalool in the rat. Biochemical Society Transactions 547th Meeting, London 2: 612–615

[51] Chidgey M A et al 1987 Studies on benzyl acetate III. The percutaneous absorption and disposition of methylene-C benzyl acetate in the rat. Food and Chemical Toxicology 25: 521–525

[52] Moorthy B et al 1989 Metabolism of a monoterpene ketone, R-(+)-pulegone—a hepatotoxin in rat. Xenobiotica 19: 217–224

[53] Private communication from Dr Sharon Hotchkiss, Dept of Toxicology, St Mary's Hospital, London

[54] Madyastha K M, Srivatsan V 1987 Metabolism of b-myrcene in vivo and in vitro: its effects on rat liver microsomal enzymes. Xenobiotica 17: 539–549

[55] Kovar K A et al 1987 Blood levels of 1,8-cineole and locomotor activity of mice after inhalation and oral administration of rosemary oil. Planta Medica 53(4): 315–318

[56] Chidgey M A, Caldwell J 1986 Studies on benzyl acetate I. Effects of dose size and vehicle on the plasma pharmacokinetics and metabolism of (methyline-14C) benzyl acetate in the rat. Food and Chemical Toxicology 24: 1257–1265

[57] Asakawa Y et al 1986 Terpenoid biotransformation in mammals IV. Biotransformation of (+)-longifolene, (-)-caryophyllene, (-)-caryophyllene oxide, (-)-cyclocolorenone, (+)-nootkatone, (-)-elemol, (-)- abietic acid and (+)-dehydroabietic acid in rabbits. Xenobiotica 16: 753–767

[58] Ritschel W A et al 1976 Pharmacokinetics of coumarin upon i.v. administration in man. Arzneimittel Forschung 26: 1382–1387

[59] Wislocki P G et al 1976 The metabolic activation of the carcinogen 1´-hyroxysafrole in vivo and in vitro and the electrophilic reactivities of possible ultimate carcinogens. Cancer Research 36: 1686–1695

[60] Fry J 1985 Common diseases, their nature, incidence and care. M T P, London

[61] Solheim E, Scheline R R 1980 Metabolism of alkenebenzene derivatives in the rat III. Elemicin and isoelemicin. Xenobiotica 10: 371–380

[62] Zangouras A et al 1981 Dose-dependent conversion of estragole in the rat and mouse to the carcinogenic metabolite 1´-hydroxyestragole. Biochemical Pharmacology 30: 1383–1386

[63] Ishida T et al 1981 Terpenoids biotransformation in mammals III. Biotransformation of a-pinene, b-pinene, pinane, delta-3-carene, carane, myrcene, and p-cymene in rabbits. Journal of Pharmaceutical Sciences 70: 406–414

[64] Walder A et al 1983 p-Cymene metabolism in rats and guinea pig. Xenobiotica 13: 503–512

[65] Sangster S A et al 1984 Metabolism of anethole I. Pathways of metabolism in the rat and in the mouse. Food and Chemical Toxicology 22: 695–706

[66] Rommelt H et al 1987 Pharmakokinetik ätherische Öle nach Inhalation mit einer terpenhaltigen Salbe. Zeitschrift für Phytotherapie 9: 14–16

[67] Asano M, Yamakasa T 1950 The fate of branched chain fatty acids in animal body. I. A contribution to the problem of 'Hildebrandt Acid'. Journal of Biochemistry 37: 321–327

[68] Le Bourhis B 1968 Recherches préliminaires sur le métabolisme du trans-anéthole. Annales de Biologie Clinique 26: 711–715

[69] Grisk A, Fisher W 1969 Zur pulmonalen Ausscheidung von Cineol, Menthol und Thymol bei Ratten nach rektaler Applikation. Zschr Ärztl Fortbild 63(4): 233–236

[70] Solheim E, Scheline R R 1973 Metabolism of alkenebenzene derivatives in the rat. I. p-methoxyallylbenzene (estragole) and p-methoxypropenylbenzene (anethole). Xenobiotica 3: 493–510

[71] Leibman K C, Ortiz E 1973 Mammalian metabolism of terpenoids I. Reduction and hydroxylation of camphor and related compounds. Drug Metabolism and Disposition 1: 543–551

[72] Horning M G et al 1974 GC-MS studies of the metabolism of safrole, an hepatocarcinogen in the rat and guinea pig. Toxicology and Applied Pharmacology 29: 89

[73] Kozam G, Mantell G M 1978 The effect of eugenol on oral mucous membrane. Journal of Dental Research 57: 954–957

[74] Pirilä V, Siltanen E 1957 On the eczematous agent in oil of turpentine. Proceedings of the International Congress on Occupational Health—Helsinki (January 6) 3: 400–402

[75] Guin J D et al 1984 The effect of quenching agents on contact urticaria caused by cinnamic aldehyde. Journal of the American Academy of Dermatology 10(1): 45–51

[76] Rudzki E et al 1976 Sensitivity to 35 essential oils. Contact Dermatitis 2: 196-200

[77] Naganuma M et al 1985 A study of the phototoxicity of lemon oil. Archives of Dermatological Research 278: 31–36

[78] Fujii T et al 1972 Studies on compounded perfumes for toilet goods. On the non-irritative compounded perfumes for soaps. Yukagaku 21: 904–908

[79] Marzulli F N, Maibach H I 1970 Perfume phototoxicity. Journal of the Society of Cosmetic Chemists 21: 695–715

[80] Meyer J 1970 Accidents dus à un cosmétique de bronzage à base d'essence de bergamote. Bulletin du Société Francaise de Dermatologie et de Syphiligraphie 77(6): 881–884

[81] Greif N 1967 Cutaneous safety of fragrance materials as measured by the maximization test. American Perfumer and Cosmetics 82: 54–57

[82] Caporale G et al 1967 Skin photosensitizing activity of some methylpsoralens. Experientia 23: 985–986

[83] Foussereau J et al 1967 Contact dermatitis from laurel I. Clinical aspects. Transactions of St John's Hospital Dermatological Society 53: 141–146

[84] Pirilä V et al 1964 On the chemical nature of the eczematogenic agent in oil of turpentine IV. The primary irritant effect of terpenes. Dermatologica 128: 16–21

[85] Papa C M, Shelly W B 1964 Menthol hypersensitivity. Journal of the American Medical Association 189: 546–548

[86] Calvery H O et al 1946 The metabolism and permeability of normal skin. Physiological Reviews 26: 495–540

[87] Wester R C, Maibach H I 1975 Percutaneous absorption in the Rhesus monkey compared to man. Toxicology and Applied Pharmacology 32: 394–398

[88] Hotchkiss S A M et al 1992 Percutaneous absorption of benzyl acetate through rat skin *in vitro* 2. Effect of vehicle and occlusion. Food and Chemical Toxicology 30(2): 145–153

[89] Hotchkiss S A M et al 1990 Percutaneous absorption of benzyl acetate through rat skin *in vitro* 1. Validation of an *in vitro* model against *in vivo* data. Food and Chemical Toxicology 28(6): 443–447

[90] Wester R C, Maibach H I 1983 Cutaneous pharmacokinetics: 10 stages to percutaneous absorption. Drug Metabolism Reviews 14: 169–205

[91] Bronaugh R L et al 1990 *In vivo* percutaneous absorption of fragrance ingredients in rhesus monkeys and humans. Food and Chemical Toxicology 28(5): 369–373

[92] Franz T J 1975 Percutaneous absorption. On the relevance of *in vitro* data. Journal of Investigative Dermatology 64: 190–195

[93] Scheuplein R J, Blank I H 1971 Permeability of the skin. Physiological Reviews 51(4): 702–747

[94] Scheuplein R J, Ross L 1970 Effects of surfactants and solvents on the permeability of the epidermis. Journal of the Society of Cosmetic Chemists 21: 853–873

[95] Wepierre J et al 1968 Percutaneous absorption and removal by the body fluids of 14C ethyl alcohol 3H perhydrosqualene and 14C-p-cymene. European Journal of Pharmacology 3: 47–51

[96] Valette G, Sobrin E 1963 Absorption percutanée de diverses huiles animales ou végétales. Pharmaceutica Acta Helvetiae 38: 710–716

[97] Valette G, Cavier R 1954 Absorption percutanée et constitution chimique. Cas des hydrocarbures, des alcools et des esters. Arch Int Pharmacodyn Ther 97: 232–240

[98] Macht D 1938 The absorption of drugs and poisons through the skin and mucous membranes. Journal of the American Medical Association 110: 409–414

[99] Brown E W, Scott W O 1934 Absorption of methyl salicycate by human skin. Journal of Pharmacology and Experimental Therapeutics 50: 32–50

[100] Heng M C 1987 Local necrosis and interstitial nephritis due to topical methyl salicylate and menthol. Cutis 39(5): 442–444

[101] Rudzki E, Grzywa Z 1976 Sensitizing and irritating properties of star anise oil. Contact Dermatitis 2(6): 305–308

[102] Cheminat A et al 1981 Allergic contact dermatitis to costus: removal of haptens with polymers. Acta Dermato-venereologica (Stockholm) 61: 525–529

[103] Parys B T 1983 Chemical burns resulting from contact with peppermint oil mar: a case report. Burns 9(5): 374–375

[104] Hanau D et al 1983 The influence of limonene on induced delayed hypersensititiviy to citral in guinea pigs I. Histological study. Acta Dermato-venereologica (Stockholm) 63(1): 1–7

[105] Opdyke D L J 1977 Safety testing of fragrances: problems and implications. Clinical Toxicology 10(1): 61–77

[106] Hasimoto M et al 1972 Effect of different routes of administration of cedrene on hepatic drug metabolism. Biochemical Pharmacology 21: 1514–1517

[107] Uehleke H, Brinkschulte-Freitas M 1979 Oral toxicity of an essential oil from myrtle and adaptive liver stimulation. Toxicology 12(3): 335–342

[108] Cohen A J 1979 Critical review of toxicology of coumarin with special reference to interspecies differences in metabolism and hepatotoxic response and their significance to man. Food and Cosmetics Toxicology 17: 277–289

[109] Olowe S A, Ransome-Kuti O 1980 The risk of jaundice in glucose-6-phosphate dehydrogenase deficient babies exposed to menthol. Acta Paediatrica Scandinavica 69(3): 341–345

[110] Gordon W P et al 1982 Hepatotoxicity and pulmonary toxicity of pennyroyal oil and its constituent terpenes in the mouse. Toxicology and Applied Pharmacology 65(3): 413–424

[111] Madyastha P et al 1985 *In vivo* and *in vitro* destruction of rat liver cytochrome P-450 by a monoterpene ketone, pulegone. Biochemical and Biophysical Research Communications 128: 921–927

[112] Gordon W P et al 1987 The metabolism of the abortifacient terpene, (R)-(+)-pulegone to a proximate toxin, menthofuran. Drug Metabolism and Disposition 15(5): 589–594

[113] Mizutani T et al 1987 Effects of drug metabolism modifiers on pulegone-induced hepatotoxicity in mice. Research Communications in Chemical Pathology and Pharmacology 58(1): 75–83

[114] Jackson G M et al 1987 Comparison of the short-term hepatic effects of orally administered citral in Long-Evans Hooded and Wistar Albino rats. Food and Chemical Toxicology 25: 505–513

[115] Thomassen D et al 1988 Contribution of menthofuran to the hepatotoxicity of pulegone: assessment based on

matched area under the curve and on matched time course. The Journal of Pharmacology and Experimental Therapeutics 244: 825–829

[116] Lake B G et al 1989 Studies on the mechanism of coumarin-induced toxicity in rat hepatocytes: comparison with dehydrocoumarin and other coumarin metabolites. Toxicology and Applied Pharmacology 97: 311–323

[117] McClanahan R H et al 1989 Metabolic activation of (R)-(+)-pulegone to a reactive enonal that covalently binds to mouse liver proteins. Chemical Research Toxicology 2: 349–355

[118] Madyastha K M, Moorthy B 1989 Pulegone mediated hepatotoxicity: evidence for covalent binding of R(+)-(14C)pulegone to microsomal proteins *in vitro*. Chemico-biological Interactions 72: 325–333

[119] Thomassen D et al 1990 Menthofuran-dependent and independent aspects of pulegone hepatotoxicity: roles of glutathione. The Journal of Pharmacology and Experimental Therapeutics 253(2): 567–572

[120] Fennell T R et al 1984 Characterization of the major hepatic DNA adduct of the strong hepatocarcinogen 1'-hydroxy-2',3'-dehydroestragole (Hodhe) in Male B6C3F1 mice. Proceedings of the American Association of Cancer Research 25: 88

[121] Bradley P R (ed) 1992 British herbal compendium. British Herbal Medicine Association, Bournemouth

[122] Phillips D H et al 1984 ^{32}P-post-labelling analysis of DNA adducts formed in the livers of animals treated with safrole, estragole and other naturally-occurring alkenylbenzenes II. Newborn male B6C3F1 mice. Carcinogenesis 5(12): 1623–1628

[123] Phillips D H et al 1981 Structures of the DNA adducts formed in mouse liver after administration of the proximate hepatocarcinogen 1'-hydroxyestragole. Cancer Research 41: 176–186

[124] Webb D R et al 1990 Assessment of the subchronic oral toxicity of *d*-limonene in dogs. Food and Chemical Toxicology 28(10): 669–676

[125] Lehman-McKeeman L D et al 1989 *d*-Limonene-induced male-rat-specific nephrotoxicity: evaluation of the association between *d*-limonene and alpha$_{2u}$-globulin. Toxicology and Applied Pharmacology 99: 250–259

[126] Webb D R et al 1989 Acute and subchronic nephrotoxicity of *d*-limonene in Fischer 344 rats. Food and Chemical Toxicology 27(10): 639–650

[127] Kanerva R L, Alden C L 1987 Review of kidney sections from a subchronic *d*-limonene oral dosing study conducted by the National Cancer Institute. Food and Chemical Toxicology 25: 355–358

[128] Kanerva R L et al 1987 Comparison of short-term renal effects due to oral administration of decalin or *d*-limonene in young adult male Fischer-344 rats. Food and Chemical Toxicology 8(6): 345–353

[129] Woo D C, Hoar R M 1972 Apparent hydronephrosis as a normal aspect of renal development in late gestation of rats: the effect of methyl salicylate. Teratology 6: 191–196

[130] Holmes G 1971 Urinary calculi in Fiji Indians—the curry kidney. Medical Journal of Australia 2: 755–756

[131] Gray T J B et al 1972 Biochemical and pathological differences in hepatic response to chronic feeding of safrole and butylated hydroxytoluene. Proceedings of the Biochemical Society 130: 91P–92

[132] Dhalla N S et al 1961 Effect of Acorus oil *in vitro* on the respiration of rat brain. Journal of Pharmaceutical Sciences 50(7): 580–582

[133] Sampson W K, Hernandez G 1939 Experimental convulsions in the rat. The Journal of Pharmacology and Experimental Therapeutics 65: 275–280

[134] Wenzel D G, Ross C R 1957 Central stimulating properties of some terpenones. Journal of the American Pharmaceutical Association 46: 77–82

[135] Merkulova O S 1957 Reflex mechanism of camphor and pyramidone experimental epilepsy. Dokl Akad Nauk USSR 112: 968–971

[136] Weiss G 1960 Hallucinogenic and narcotic-like effect of powdered myristica (nutmeg). Psychiatric Quarterly 34: 346–356

[137] Early D F 1961 Pennyroyal: a rare case of epilepsy. Lancet 281: 580–581

[138] Shulgin A T 1966 Possible implication of myristicin as a psychotropic substance. Nature 210: 380–384

[139] Cvetko B et al 1973 Elektroencefalografske spremembe v epilepticnem statusu sprozenem z oljem morskega pelina. Neuropsihijatrija 21(3–4): 297–300

[140] Arditti J et al 1978 Trois observations d'intoxication par des essences végétales convulsivantes. Annales Medicales de Nancy 17: 371–374

[141] O'Mullane N M et al 1982 Adverse CNS effects of menthol-containing olbas oil. Lancet 1: 1121

[142] Thorup I et al 1983 Short term toxicity study in rats dosed with peppermint oil. Toxicology Letters 19: 211–215

[143] Olsen P, Thorup I 1984 Neurotoxicity in rats dosed with peppermint oil and pulegone. Archives of Toxicology Supplement 7: 408–409

[144] Steinmetz M D et al 1985 Action d'huiles essentielles de sauge, thuya, hysope et de certains constituants, sur la respiration de coupes de cortex cérébral *in vitro*. Plantes Medicinales et Phytotherapie 19: 35–47

[145] Steinmetz M D et al 1987 Actions de l'huile essentielle de romarin et de certains de ses constituants (eucalyptol et camphre) sur le cortex cérébral de rat *in vitro*. Journal de Toxicologie Clinique et Expérimentale 7(4): 259–271

[146] Beattie R T 1968 Nutmeg as a psychoactive agent. British Journal of Addiction 63: 105–109

[147] Weil A T 1966 The use of nutmeg as a psychotropic agent. Bulletin on Narcotics 18(4): 15–23

[148] Kalbhen D A 1971 Nutmeg as a narcotic. A contribution to the chemistry and pharmacology of nutmeg (*Myristica fragrans*). Angewandte Chemie International Edition 10: 370–374

[149] Boissier J R et al 1967 Action psychotrope expérimentale des anétholes isomères *cis* et *trans*. Therapie 22: 309–323

[150] Forrest J E, Heacock R A 1972 Nutmeg and mace, the psychotropic spices from *Myristica fragrans*. Lloydia 35: 440–449

[151] Butler T C 1971 The distribution of drugs. In: La Du B N et al (eds) Fundamentals of drug metabolism and drug disposition. Williams & Wilkins, Baltimore

[152] Bowman W C, Rand M J 1982 Textbook of pharmacology, 2nd edn. Blackwell Scientific Publications, Oxford, ch 40

[153] Gleason M N et al 1984 Clinical toxicology of commercial products, 5th edn. Williams & Wilkins, Baltimore

[154] Miller et al 1979 The metabolic activation of safrole and related naturally occurring alkylbenzenes in relation to carcinogenesis by these agents. In: Naturally occurring carcinogens—mutagens and modulators of carcinogenesis. Proceedings of the 9th International Symposium of the Princess Takamatsu Cancer Research Fund, Tokyo University, Park Press, Baltimore

[155] Arctander S 1960 Perfume and flavor materials of natural origin. Elizabeth, New Jersey

[156] Leung A Y 1980 Encyclopedia of common natural ingredients used in food, drugs and cosmetics. John Wiley, New York

[157] Anon 1968 The Merck Index, 8th edn. Merck, New Jersey

[158] Budavari S (ed) 1989 The Merck Index, 11th edn. Merck, New Jersey

[159] Grieve M 1978 A modern herbal. Penguin Books, Harmondsworth

[160] Ford R A 1991 The toxicology and safety of fragrances. In: Muller P M, Lamparsky D (eds) Perfumes—art, science & technology. Elsevier, London

[161] Vesselinovitch S D et al 1979 Transplacental and lactational carcinogenesis by safrole. Cancer Research 39: 4378–4380

[162] Tisserand R 1977 The art of aromatherapy. C W Daniels, Saffron Walden

[163] Farley D R, Howland V 1980 The natural variation of the pulegone content in various oils of peppermint. Journal of the Science of Food and Agriculture 31: 1143–1151

[164] Anon 1975 Safe and unsafe herbs in herbal teas. Department of Health Education and Welfare, Public Health Service Food and Drug Administration, Washington DC

[165] Neale A 1893 Case of death following blue gum (Eucalyptus globulus) oil. The Australasian Medical Gazette 12: 115–116

[166] Braithwaite P F 1906 A case of poisoning by pennyroyal: recovery. British Medical Journal 2: 865

[167] Jones C O 1913 A case of poisoning by pennyroyal. British Medical Journal 2: 746

[168] Anon 1978 Fatality and illness associated with consumption of pennyroyal oil—Colorado. Morbidity and Mortality Weekly Report 27: 511–513

[169] Vallance W B 1955 Pennyroyal poisoning—a fatal case. Lancet 2: 850–851

[170] Sullivan J B et al 1979 Pennyroyal oil poisoning and hepatotoxicity. Journal of the American Medical Association 242: 2873–2874

[171] Gunby P 1979 Plant known for centuries still causes problems today. Journal of the American Medical Association 241: 2246–2247

[172] Gold J, Gates W 1980 Herbal abortifacients. Journal of the American Medical Association 243: 1365–1366

[173] Macht D I 1913 The action of so-called emmenagogue oils on the isolated uterine strip. Journal of Pharmacology and Experimental Therapeutics 4: 547–553

[174] Gunn J W C 1921 The action of the 'emmenagogue' oils on the human uterus. The Journal of Pharmacology and Experimental Therapeutics 16: 485–489

[175] Datnow M M 1928 An experimental investigation concerning toxic abortion produced by chemical agents. Journal of Obstetrics and Gynaecology 35: 693–724

[176] Papavassiliou M J et al 1937 Rue as an abortifacient and poison. Ann Med Légale Criminol Police Sci 17: 993–999

[177] Patior A et al 1938 Note sur l'action de l'essence de rue sur l'organisme animal. C R Soc Biol 127: 1324–1325

[178] Renaux J, La Barre J 1941 Les effets ocytociques de rue et de sabine. Acta Biol Belg 1: 334–335

[179] Takacs E, Warkany J 1968 Experimental production of congenital cardiovascular malformations in rats by salicylate poisoning. Teratology 1: 109–111

[180] Pyun J S 1970 Effect of methyl salicylate on developing rat embryos. Ch'oesin Uihak 13(7): 63–72

[181] Collins T F X et al 1971 Effect of methyl salicylate on rat reproduction. Toxicology and Applied Pharmacology 18: 755–765

[182] Abramovici A 1972 The teratogenic effect of cosmetic constituents on the chick embryo. Advances in Experimental and Medical Biology 27: 161–174

[183] Pecevski J et al 1981 Effect of oil of nutmeg on the fertility and induction of meiotic chromosome rearrangements in mice and their first generation. Toxicology Letters 7: 239–244

[184] Abramovici A, Rachmuth-Roizman P 1983 Molecular structure – teratogenicity relationships of some fragrance additives. Toxicology 29: 143–156

[185] McCloskey S E et al 1986 Toxicity of benzyl alcohol in adult neonatal mice. Journal of Pharmaceutical Sciences 75(7): 702–705

[186] Neubert D et al 1987 Principles and problems in assessing prenatal toxicity. Archives of Toxicology 60: 238–245

[187] Pages N et al 1989 Les échantillons commerciaux de 'sabine' (rameaux feuilles et huile essentielle) sont-ils tératogènes? Étude chez la souris. Plantes Medicinales et Phytotherapie 23(3): 186–192

[188] Kong Y et al 1989 Antifertility principle of Ruta graveolens. Planta Medica 55: 176–178

[189] Pages N et al 1989 Teratogenic effect of Plectranthus fruticosus essential oil in mice. Journal of Ethnopharmacology 27(3): 94–96

[190] Pages N et al 1989 Teratological evaluation of Juniperus sabina essential oil in mice. Planta Medica 55: 144–146

[191] Mantovani A et al 1989 Pre-natal (segment II) toxicity study of cinnamic aldehyde in the Sprague-Dawley rat. Food and Chemical Toxicology 27(12): 781–786

[192] Fournier G et al 1989 Étude d'échantillons commerciaux de sabine: rameaux, feuilles et huile essentielle. Plantes Medicinales et Phytotherapie 23: 169–179

[193] Pages N et al 1990 Les huiles essentielles et leurs propriétés tératogènes potentielles: exemple de l'huile essentielle d'eucalyptus étude préliminaire chez la souris. Plantes Medicinales et Phytotherapie 24(1): 21–26

[194] Pages N et al 1992 Potential teratogenicity in mice of the essential oil of Salvia lavandulifolia Vahl. Study of a fraction rich in sabinyl acetate. Phytotherapy Research 6(2): 80–83

[195] Nath D et al 1992 Commonly used Indian abortifacient plants with special reference to their

teratologic effects in rats. Journal of Ethnopharmacology 36(2): 147–154

[196] Gaworski C L et al 1992 Developmental toxicity evaluation of inhaled citral in Sprague-Dawley rats. Food and Chemical Toxicology 30(4): 269–277

[197] Pages N et al 1988 Teratological evaluation of *Plectranthus fruticosus* leaf essential oil. Planta Medica 54: 296–298

[198] Forschmidt P 1979 Teratogenic activity of flavor additives. Teratology 19(2): 26A

[199] Toaff M E et al 1979 Selective oocyte degeneration and impaired fertility in rats treated with the aliphatic monoterpene, citral. Journal of Reproduction and Fertility 55: 347–352

[200] Maickel R P, Snodgrass W R 1973 Physicochemical factors in maternal–fetal distribution of drugs. Toxicology and Applied Pharmacology 26: 218–230

[201] del Castillo J et al 1975 Marijuana, absinthe and the central nervous system. Nature 253: 365–366

[202] Hendriks H et al 1981 Pharmacological screening of valerenal and some other components of the essential oil of *Valeriana officinalis*. Planta Medica 42: 62–68

[203] Lennox W G, Gibbs F A 1936 Effect on the electro-encephalogram of drugs and conditions which influence seizures. Archives of Neurology and Psychiatry 36: 1236–1250

[204] Macht D I et al 1921 Sedative properties of some aromatic drugs and fumes. Journal of Pharmacology and Experimental Therapeutics 18(5): 361–372

[205] Wabner D 1993 Purity and pesticides. International Journal of Aromatherapy 5(2): 27–29

[206] Kudrzycka-Bieloszabska F W et al 1966 Pharmacodynamic properties of oleum chamomillae and oleum millefolii. Dissertationes Pharmaceuticae et Pharmacologicae (Pol) 18(5): 449–454

[207] Shipochliev T 1968 Pharmacological study of a group of essential oils II. Vet Med Nauki 5(10): 87–92

[208] Seto T A, Keup W 1969 Effects of alkylmethoxybenzene and alkylmethylenedioxybenzene essential oils on pentobarbital and ethanol sleeping time. Arch Int Pharmacodyn Ther 180: 232–241

[209] Atanassova-Shopova S, Roussinov K 1970 Experimental studies on certain effects of the essential oil of *Salvia sclarea* L. on the central nervous system. Izv Inst Fiziol Bulg Akad Nauk 13: 89–95

[210] Atanassova-Shopova S, Roussinov K S 1970 On certain central neurotropic effects of lavender essential oil. Izv Inst Fiziol Bulg Akad Nauk 13: 69–77

[211] Binet L et al 1972 Recherches sur les propriétés pharmacodynamiques (action sédative et action spasmolytique) de quelques alcools terpéniques aliphatiques. Annales Pharmaceutiques Francaises 30: 611–616

[212] Wagner H, Sprinkmeyer L 1973 Pharmacological effect of balm spirit. Deutsche Apotheker-Zeitung 113(30): 1159–1166

[213] Atanassova-Shopova S et al 1973 On certain central neurotropic effects of lavender essential oil. II communication: studies on the effects of linalool and of terpineol. Izv Inst Fiziol Sof 15: 149–156

[214] Lawrence B M 1979 Essential oils 1976–1978. Allured Publishing, Wheaton

[215] Lawrence B M 1981 Essential oils 1979–1980. Allured Publishing, Wheaton

[216] Lawrence B M 1989 Essential oils 1981–1987. Allured Publishing, Wheaton

[217] Lawrence B M 1993 Essential oils 1988–1991. Allured Publishing, Wheaton

[218] Rossi T et al 1988 Sedative, anti-inflammatory and anti-diuretic effects induced in rats by essential oils of varieties of *Anthemis nobilis*: a comparative study. Pharmacological Research Communications 20(5): 71–74

[219] Houghton P J 1988 Biological activity of valerian and related plants. Journal of Ethnopharmacology 22(Feb–Mar): 121–142

[220] Delaveau P et al 1989 Sur les propriétés neuro-dépressive de l'huile essentielle de lavande. C R Soc Biol 183(4): 342–348

[221] Guillemain J et al 1989 Neurosedative effects of essential oil of *Lavandula angustifolia* Mill. Annales Pharmaceutiques Francaises 47(6): 337–343

[222] Rakieten N, Rakieten M L 1957 The effect of *l*-menthol on the systemic blood pressure. Journal of the American Pharmaceutical Association 46(2): 82–84

[223] Agshikar N V, Abraham G J 1972 Pharmacology and acute toxicity of essential oil extracted from *Zanthoxylum budrunga*. Indian Journal of Medical Research 60(5): 757–762

[224] Caujolle F, Franck C 1945 Pharmacodynamic actions of clary sage and condiment sage. C R Soc Biol 139: 1109–1110

[225] Clerc A et al 1934 Experiments on dogs with certain substances which lower the surface tension of the blood. C R Soc Biol 116: 864–867

[226] Painter J C et al 1971 Nutmeg poisoning—a case report. Clinical Toxicology 4(1): 1–4

[227] Weil A T 1965 Nutmeg as a narcotic. Economic Botany 19: 194–217

[228] Briggs C J, McLaughlin L D 1974 Low-temperature thin-layer chromatograph for detection of polybutene contamination in volatile oils. Journal of Chromatography 101(2): 403–407

[229] Belanger A 1989 Residues of azinphos-methyl, cypermethrin, benomyl and chlorothalonil in monarda and peppermint oil. Acta Horticulturae 249: 67–73

[230] Dikshith T S et al 1989 Pesticide residues in edible oils and oil seeds. Bulletin of Environmental Contamination and Toxicology 42(1): 50–56

[231] McCord J A, Jervey L P 1962 Nutmeg (myristicin) poisoning: case report. South Carolina Medical Journal 58: 436–439

[232] Hughes R F 1932 A case of methyl salicylate poisoning. Canadian Medical Association Journal 27: 417–418

[233] Mack R B 1982 Toxic encounters of the dangerous kind. The nutmeg connection. North Carolina Medical Journal 43(5): 439

[234] Millet Y 1981 Toxicity of some essential plant oils. Clinical and experimental study. Clinical Toxicology 18(12): 1485–1498

[235] Gronka P A et al 1969 Camphor exposures in a packaging plant. American Industrial Hygiene Association Journal 30: 276–279

[236] Thomas J G 1962 Peppermint fibrillation. Lancet (27 Jan): 222

[237] Davison C et al 1961 On the metabolism and toxicity of methyl salicylate. Journal of Pharmacology and Experimental Therapeutics 132: 207–211

[238] Cushny A R 1908 Therapeutical and pharmacological section: nutmeg poisoning. Proceedings of the Royal Society of Medicine 1(3): 39–44

[239] Dale H 1909 Therapeutical and pharmacological section: note on nutmeg poisoning. Proceedings of the Royal Society of Medicine 2(3): 69–74

[240] Weiss J et al 1973 Camphorated oil intoxication during pregnancy. Pediatrics 52: 713–714

[241] Riggs J et al 1965 Camphorated oil intoxication in pregnancy. Obstetrics and Gynaecology 25: 255–258

[242] Macht D 1921 The action of so-called emmenagogue oils on the isolated uterus. Journal of the American Medical Association 61: 105–107

[243] Bellman M H 1973 Camphor poisoning in children. British Medical Journal 2: 177

[244] Smith A, Margolis G 1954 Camphor poisoning. American Journal of Pathology 30: 857–869

[245] Koppel C et al 1988 Hemoperfusion in acute camphor poisoning. Intensive Care Medicine 14(4): 431–433

[246] Brown W I et al 1982 Fatal benzyl alcohol poisoning in a neonatal intensive care unit. Lancet 1: 1250

[247] Clark T L 1924 Fatal case of camphor poisoning. British Medical Journal 1: 467

[248] Wilkinson H F 1991 Childhood ingestion of volatile oils (letter). Medical Journal of Australia 154(6): 430–431

[249] Temple W A et al 1991 Management of oil of citronella poisoning. Clinical Toxicology 29(2): 257–262

[250] Melis K et al 1990 Accidental nasal eucalyptol and menthol instillation. Acta Clin Belg Suppl 13: 101–102

[251] Pilapil V R 1989 Toxic manifestations of cinnamon oil ingestion in a child. Clinical Pediatrics 28(6): 276

[252] Grande G A, Dannewitz S R 1987 Symptomatic sassafras oil ingestion. Veterinary and Human Toxicology 6: 447

[253] Buechel D W et al 1983 Pennyroyal oil ingestion: report of a case. Journal of the American Osteopathic Association 82(10): 793–794

[254] Lovejoy F H 1982 Fatal benzyl alcohol poisoning in neonatal intensive care units. American Journal of Disease in Childhood 136: 974–975

[255] Patel S, Wiggins J 1980 Eucalyptus oil poisoning. Archives of Disease in Childhood 5: 405–406

[256] Aronow R 1976 Camphor poisoning. Journal of the American Medical Association 235: 1260

[257] Phelan W J III 1976 Camphor poisoning: over-the-counter dangers. Pediatrics 57: 428–431

[258] Pack W K et al 1972 Über eine tödliche Blausäurevergiftung nach dem Genuß bitterer Mandeln (*Prunus amygdalus*). Z Rechtsmedizin 70: 53–54

[259] Breuninger H et al 1970 Zur nasalen Anwendung ätherischer Öle im Säuglins- und Kleinkindesalter. Z Laryngol Rhinolotol Otol 49(12): 800–804

[260] Kloss J L, Boeckman C R 1967 Methyl salicylate poisoning. Ohio State Medical Journal 63: 1064–1065

[261] Gurr F W, Scroggie J G 1965 Eucalyptus oil poisoning treated by dialysis and mannitol infusion. Australian Annals of Medicine 14: 238–249

[262] Payne R B 1963 Nutmeg intoxication. New England Journal of Medicine 269: 36–38

[263] Jacobziner H, Raybin H W 1962 Methyl salicylate poisoning. New York State Journal of Medicine (Feb 1): 403–405

[264] Mant A K 1961 A case of poisoning by oil of citronella. Medicine, Science and the Law—Association Proceeding VI 1–2: 170–171

[265] Graham D C 1961 Methyl salicylate: a lethal hazard in the home. Canadian Medical Association Journal 84: 960

[266] Green R C Jr 1959 Nutmeg poisoning. Journal of the American Medical Association 171: 1342–1344

[267] Tauscher J W, Polich J J 1959 Treatment of pine oil poisoning by exchange transfusions. Journal of Pediatrics 55: 511–515

[268] Adams J T et al 1957 A case of methyl salicylate intoxication treated by exchange transfusion. Journal of the American Medical Association 165: 1563–1565

[269] Craig J O 1953 Poisoning by the volatile oils in childhood. Archives of Disease in Childhood 28: 475–483

[270] Craig J O, Frase M S 1953 Accidental poisoning in childhood. Archives of Disease in Childhood 28: 259–267

[271] Mele A 1952 Acute fatal poisoning with chenopodium oil. Folia Medica 35: 955–963

[272] Rubin M B et al 1949 Ingestion of poisons—survey of 250 children. Clinical Proceedings Childrens Hospital 5: 57–73

[273] Eimas A 1938 Methyl salicylate poisoning in an infant. Report of a patient with partial necropsy. Journal of Pediatrics 13: 550–554

[274] Krober F 1936 A case of severe poisoning with chenopodium oil following its therapeutic use. Deutsche Medizinische Wochenschrift 62: 1759

[275] Klingensmith W R 1934 Poisoning by camphor. Journal of the American Medical Association 102: 2182–2183

[276] Ibru-Maar A 1932 Fatal poisoning with oil of chenopodium. Dt Z Ges Gerichtl Medizin 20: 158–160

[277] Sewell J S 1925 Poisoning by eucalyptus oil. British Medical Journal 1: 922

[278] McPherson J 1925 The toxicology of eucalyptus oil. Medical Journal of Australia 2: 108–110

[279] Kirkness W R 1910 Poisoning by oil of eucalyptus. British Medical Journal 1: 261

[280] Foggie W E 1911 Eucalyptus oil poisoning. British Medical Journal 1: 359–360

[281] Zani F et al 1991 Studies on the genotoxic properties of essential oils with *Bacillus subtilis* rec-assay and salmonella/microsome reversion assay. Planta Medica 57(3): 237–241

[282] Roe F J C, Peirce W E H 1960 Tumor promotion by citrus oils: tumors of the skin and urethral orifice in mice. Journal of the National Cancer Institute 24: 1389–1403

[283] Roe F J C 1959 Oil of sweet orange: a possible role in carcinogenesis. British Journal of Cancer 13: 92–93

[284] Peirce W E H 1961 Tumor-promotion by lime oil in the mouse forestomach. Nature 189: 497–498

[285] Taylor J M et al 1967 Toxicity of oil of calamus (Jammu variety). Toxicology and Applied Pharmacology 10: 405

[286] Borchert P et al 1973 1´-hydroxysafrole, a proximate carcinogenic metabolite of safrole in rat and mouse. Cancer Research 33: 590–600

[287] Wislocki P G et al 1977 Carcinogenic and mutagenic activities of safrole, 1′-hydroxysafrole and some known or possible metabolites. Cancer Research 37: 1883–1891

[288] Ishidate M Jr et al 1984 Primary mutagenicity screening of food additives currently used in Japan. Food and Chemical Toxicology 22: 623–636

[289] Randerath K et al 1984 ^{32}P-post-labelling analysis of DNA adducts formed in the livers of animals treated with safrole, estragole and other naturally-occurring alkenylbenzenes. I. Adult female CD-1 mice. Carcinogenesis 5(12): 1613–1622

[290] Stevenson C S 1937 Oil of wintergreen (methyl salicylate) poisoning. Report of three cases, one with autopsy, and a review of the literature. American Journal of Medical Science 193: 772–788

[291] Howes A J et al 1990 Structure-specificity of the genotoxicity of some naturally occurring alkenylbenzenes determined by the unscheduled DNA synthesis assay in rat hepatocytes. Food and Chemical Toxicology 28(8): 537–542

[292] Swanson A B et al 1979 The mutagenicities of safrole, estragole, eugenol, trans-anethole, and some of their known or possible metabolites for Salmonella typhimurium mutants. Mutation Research 60: 143–153

[293] Howrie D L et al 1985 Candy flavoring as a source of salicylate poisoning. Paediatrics 75(5): 869–871

[294] To L P et al 1982 Mutagenicity of trans-anethole, estragole, eugenol and safrole in the Ames Salmonella typhimurium assay. Bulletin of Environmental Contamination and Toxicology 28: 647–654

[295] Sezikawa J, Shibamoto T 1982 Genotoxicity of safrole-related chemicals in microbial test systems. Mutation Research 101: 127–140

[296] Damhoeri A et al 1985 In vitro mutagenicity tests on capsicum pepper, shallot and nutmeg oleoresins. Agricultural and Biological Chemistry 49: 1519–1520

[297] Zolotovich G et al 1967 Cytotoxic effect of carbonyl compounds isolated from essential oils. Comptes Rendus de L'Académie Bulgare des Sciences 20: 1213–1216

[298] Silyanowska K et al 1967 Cytotoxic effect of esters isolated from essential oils. Comptes Rendus de L'Académie Bulgare des Sciences 20: 1337–1339

[299] Nachev Ch et al 1967 Cytotoxic effect of alcohols isolated from essential oils. Comptes Rendus de L'Académie Bulgare des Sciences 20: 1081–1084

[300] Stoll A, Seebeck E 1951 Chemical investigations on alliin, the specific principle of garlic. Advances in Enzymology and Related Areas of Molecular Biology 11: 377–400

[301] Homburger F, Boger E 1968 The carcinogenicity of essential oils, flavors and spices: a review. Cancer Research 28: 2372–2374

[302] Schmid E 1969 Über die cytogenetische Wirkung dreier Monoterpene, eines Anthrachinons und eines Tropolons auf die Mitose von Vicia faba. Genetica 40: 65–83

[303] Longnecker D S et al 1990 Evaluation of promotion of pancreatic carinogenesis in rats by benzyl acetate. Food and Chemical Toxicology 28(10): 665–668

[304] Truhaut R et al 1989 Chronic toxicity/carcinogenicity study of trans-anethole in rats. Food and Chemical Toxicology 27: 11–20

[305] Ames B N et al 1987 Ranking possible carcinogenic hazards. Science 236: 271–280

[306] Anthony A et al 1987 Metabolism of estragole in rat and mouse and influence of dose size on excretion of the proximate carcinogen 1′-hydroxyestragole. Food and Chemical Toxicology 25: 799–806

[307] Elegbede J A et al 1986 Mouse skin tumor promoting activity of orange peel oil and d-limonene: a re-evaluation. Carcinogenesis 7(12): 2047–2049

[308] Caldwell J 1991 Basil & methyl chavicol—statement on new data. International Journal of Aromatherapy 3(1): 6

[309] Miller E C et al 1983 Structure–activity studies of the carcinogenicities in the mouse and rat of some naturally occurring and synthetic alkenylbenzene derivatives related to safrole and estragole. Cancer Research 43: 1124–1134

[310] Miller J A et al 1982 The metabolic activation and carcinogenicity of alkenylbenzenes that occur naturally in many spices. Carcinogens and Mutagens in the Environment 1: 83–96

[311] Green N R, Savage J R 1978 Screening of safrole, eugenol, their ninhydrin positive metabolites and selected secondary amines for potential mutagenicity. Mutation Research 57: 115–121

[312] Drinkwater N R et al 1976 Hepatocarcinogenicity of estragole (1-allyl-4-methoxybenzene) and 1′-hydroxyestragole in the mouse and mutagenicity of 1′-acetoxyestragole in bacteria. Journal of the National Cancer Institute 57: 1323–1331

[313] Hagan E C et al 1965 Toxic properties of compounds related to safrole. Toxicology and Applied Pharmacology 7: 18–24

[314] Rompelberg C J M et al 1993 Effects of the naturally occurring alkenylbenzenes eugenol and trans-anethole on drug-metabolising enzymes in the rat liver. Food and Chemical Toxicology 31(9): 637–645

[315] Roe F J C, Field W E H 1965 Chronic toxicity of essential oils and certain other products of natural origin. Food and Cosmetic Toxicology 3: 311–324

[316] Field W E H, Roe F J C 1965 Tumor promotion in the forestomach epithelium of mice by oral administration of citrus oils. Journal of the National Cancer Institute 35: 771–787

[317] Dickens F, Jones H E H 1965 Further studies on the carcinogenic action of certain lactones and related substances in the rat and mouse. British Journal of Cancer 19: 392–403

[318] Delgado I F et al 1993 Peri- and postnatal developmental toxicity of beta-myrcene in the rat. Food and Chemical Toxicology 31(9): 623–628

[319] Long E L, Jenner P M 1963 Esophageal tumors produced in rats by the feeding of dihydrosafrole. Federation Proceedings of the Federation of American Societies for Experimental Biology 2: 275

[320] Long E L et al 1963 Liver tumors produced in rats by feeding safrole. Archives of Pathology 75: 595–604

[321] Power F B, Salway A H 1908 Chemical examination and physiological action of nutmeg. American Journal of Pharmacy 80: 563–580

[322] Anon 1973 Evaluation of the health aspects of garlic and oil of garlic as a food ingredient. Federation of American Societies for Experimental Biology PB-223 838: 1–13

[323] Anon 1972 GRAS (generally recognised as safe) food ingredients—garlic. Food and Drug Research Labs Inc PB 221 219: 1–43

[324] Leach E H, Lloyd J P F 1956 Citral poisoning. Proceedings of the Nutrition Society 15: xv–xvi

[325] Millet Y et al 1979 Étude expérimentale des propriétés toxiques convulsivantes des essences de sauge et d'hysope du commerce. Revue d'Electroencephalographie et de Neurophysiologie Clinique 1: 12–18

[326] Jori A, Briatico G 1973 Effect of eucalyptol on microsomal enzyme activity of foetal and newborn rats. Biochemical Pharmacology 22: 543–544

[327] Madsen C et al 1986 Short-term toxicity study in rats dosed with menthone. Toxicology Letters 32(1–2): 147–152

[328] Powers K A, Beasley V R 1985 Toxicological aspects of linalool: a review. Veterinary and Human Toxicology 27(6): 484–486

[329] Thorup I et al 1983 Short term toxicity study in rats dosed with pulegone and menthol. Toxicology Letters 19(3): 207–210

[330] Honda G et al 1986 Isolation of sedative principles from Perilla frutescens. Chemical and Pharmaceutical Bulletin Tokyo 34(4): 1672–1677

[331] Habersang S et al 1979 Pharmacological studies of chamomile constituents IV. Studies on the toxicity of (-)-bisabolol. Planta Medica 37: 115–123

[332] Wilson B J 1979 Naturally occurring toxicants of foods. Nutrition Reviews 37: 305–312

[333] Wilson B J et al 1977 Perilla ketone: a potent lung toxin from the mint plant, Perilla frutescens Britton. Science 197: 573–574

[334] Powers M F et al 1961 A study of the toxic effects of cinnamon oil. Pharmacologist 3: 62

[335] Anon 1974 Oil of rue—proposed affirmation of GRAS status with specific limitations as direct human food ingredient. Federal Register 39(185): 34215

[336] Hagan E C et al 1967 Food flavorings and compounds of related structure II. Subacute and chronic toxicity. Food and Cosmetics Toxicology 5: 141–157

[337] Eickholt T H, Box R H 1965 Toxicities of peppermint and Pycnanthemun albescens oils, Fam. Labiateae. Journal of Pharmaceutical Sciences 54: 1071–1072

[338] Jenner P M et al 1964 Food flavorings and compounds of related structure I. Acute oral toxicity. Food and Cosmetics Toxicology 2: 327–343

[339] Taylor J M 1964 A comparison of the toxicity of some allyl, propenyl, and propyl compounds in the rat. Toxicology and Applied Pharmacology 6: 378–387

[340] Aydelotte M B 1963 The effects of vitamin A and citral on epithelial differentiation in vitro 2. The chick oesophageal and corneal epithelia and epidermis. Journal of Embryology and Experimental Morphology 11: 621–635

[341] Webb W K, Hansen W H 1963 Chronic and subacute toxicology and pathology of methyl salicylate in dogs, rats and rabbits. Toxicology and Applied Pharmacology 5: 576–587

[342] Wolf I J 1935 Fatal poisoning with oil of chenopodium in a negro child with sickle cell anemia. Archives of Pediatrics 52: 126–130

[343] Abbot D D et al 1961 Chronic oral toxicity of oil of sassafras and safrol. Pharmacologist 3(73): 62

[344] Homburger F et al 1961 Toxic and possible carcinogenic effects of 4-allyl-1,2-methylenedioxy-benzene (safrole) in rats on deficient diets. Medicina Experimentalis: 4: 1–11

[345] Von Oettingen W F 1960 The aliphatic acids and their esters: toxicity and potential dangers. The saturated monobasic aliphatic acids and their esters. AMA Archives of Industrial Health 21: 28–65

[346] Caujolle F, Meynier D 1958 Toxicité de l'estragol et des anétholes (cis et trans). C R Hebd Seanc Acad Sci Paris 246: 1465–1468

[347] Hazleton L W et al 1956 Toxicity of coumarin. Journal of Pharmacology and Experimental Therapeutics 118: 348–358

[348] Leach E H, Lloyd J P F 1956 Experimental ocular hypertension in animals. Transactions of the Ophthalmological Societies of the United Kingdom 76: 453–460

[349] Mahindru S N 1992 Indian plant perfumes. Metropolitan, New Delhi

[350] Sharma J D, Dandiya P C 1962 Studies in Acorus calamus Part VI. Pharmacological actions of asarone and beta-asarone on cardiovascular system and smooth muscles. Indian Journal of Medical Research 50: 61–65

[351] Dandiya P C, Sharma J D 1962 Studies on Acorus calamus. Part V. Pharmacological actions of asarone and beta-asarone on central nervous system. Indian Journal of Medical Research 50: 46–60

[352] Chandhoke N, Ghatak B J R 1969 Studies on Tagetes minuta: some pharmacological actions of the essential oil. Indian Journal of Medical Research 57: 864–876

[353] Caujolle F, Franck C 1945 Sur l'action pharmacodynamique de l'essence d'hysope. C R Soc Biol 139: 1111–1112

[354] Pinto-Scognamiglio W 1967 Connaissances actuelles sur l'activité pharmacodynamique de la thuyone, aromatisant naturel d'un emploi étendu. Boll Chim Farm 106: 292–300

[355] Teuscher E et al 1989 Components of essential oils as membrane active compounds. Planta Medica 55: 660–661

[356] Cavallito C J, Bailey J H 1944 Allicin, the antibacterial principle of Allium sativum I. Isolation, physical properties and antibacterial action. Journal of the American Chemical Society 66: 1950–1951

[357] Rice K C, Wilson R S 1976 (-)-3-Isothujone, a small nonnitrogenous molecule with antinociceptive activity in mice. Journal of Medicinal Chemistry 19(8): 1054–1057

[358] Cowan J W et al 1967 Antithyroid activity of onion volatiles. Australian Journal of Biological Sciences 20: 683–685

[359] Salji J P et al 1971 The antithyroid activity of Allium volatiles in the rat—in vitro studies. European Journal of Pharmacology 16: 251–253

[360] Zondek B et al 1938 Phenol methyl esters as estrogenic agents. Biochemical Journal 32: 641–645

[361] Albert-Puleo M 1980 Fennel and anise as estrogenic agents. Journal of Ethnopharmacology 2: 337–344

[362] Papageorgiou C et al 1983 Allergic contact dermatitis to garlic (Allium sativum L.). Identification of the allergens: the role of mono-, di-, and trisulfides present in garlic. Archives of Dermatological Research 275: 229–234

[363] Fenwick G R, Hanley A B 1985 The genus *Allium*— part 3. CRC Critical Revues in Food Science and Nutrition 23(1): 1–73

[364] George K et al 1974 Mode of action of garlic oil. Effect on oxidative phosphorylation in hepatic mitochondria of mice. Biochemical Pharmacology 23(5): 931–936

[365] Co-operative Group for Essential Oil of Garlic 1986 The effect of essential oil of garlic on hyperlipemia and platelet aggregation—an analysis of 308 cases. Journal of Traditional Chinese Medicine 6(2): 117–120

[366] Barrie S et al 1987 Effects of garlic oil on platelet aggregation. Journal of Orthomolecular Medicine 2(1): 15–21

[367] Truitt E B et al 1963 Evidence of monoamine oxidase inhibition by myristicin and nutmeg. Proceedings of the Society for Experimental Biology and Medicine 112: 647–650

[368] Truitt E B Jr et al 1961 The pharmacology of myristicin. Journal of Neuropsychiatry 2: 205–210

[369] Truitt E B 1967 The pharmacology of myristicin and nutmeg. Public Health Service Publications Washington 1645: 215–222

[370] Rasheed A et al 1984 Pharmacological influence of nutmeg and nutmeg constituents on rabbit platelet function. Planta Medica 50: 222–226

[371] Truitt E B Jr, Ebersberger E M 1962 Evidence of monoamine oxidase inhibition by myristicin and nutmeg *in vivo*. Federation Proceedings of the Federation of American Societies for Experimental Biology 21: 418

[372] Nutrition International 1990 First World Congress on the Health Significance of Garlic and Garlic Constituents 1–48

[373] Boullin D J 1981 Garlic as a platelet inhibitor. Lancet 1(4 April): 776–777

[374] Devasagayam T P A et al 1982 Diallyl disulphide-induced changes in microsomal enzymes of suckling rats. Indian Journal of Experimental Biology 20: 430–432

[375] Hikino H 1986 Antihepatotoxic actions of *Allium sativum* bulbs. Planta Medica 0(3): 163–171

[376] Janku I et al 1960 Das diuretische Prinzip des Wacholders. Arch Exptl Pathol Pharmacol 238: 112–113

[377] Christensen B V, Lynch H J 1937 A comparative study of the pharmacological actions of natural and synthetic camphor. Journal of the American Pharmaceutical Association 26: 786–96

[378] Saratikov A S et al 1957 Camphor–adrenaline synergism. (Russian), Farmak Toks 20(5): 84–90

[379] Fournier G et al 1993 Contribution to the study of *Salvia lavandulifolia* essential oil: potential toxicity attributable to sabinyl acetate. Planta Medica 59: 96–97

[380] Aqel M B 1991 Relaxant effect of the volatile oil of *Romarinus officinalis* on tracheal smooth muscle. Journal of Ethnopharmacology 33(1, 2): 57–62

[381] Northover B J, Verghese J 1962 The pharmacology of certain terpene alcohols and oxides. Journal of Scientific and Industrial Research 21C: 342–345

[382] Mathias C G T et al 1980 Contact urticaria from cinnamic aldehyde. Archives of Dermatology 116: 74–76

[383] Bhargava A K et al 1967 Pharmacological investigation of the essential oil of *Daucus carota*. The Indian Journal of Pharmacy 28(4): 127–129

[384] Wijesekera R O 1978 The chemistry and technology of cinnamon. CRC Critical Revues in Food Science and Nutrition 10: 1–30

[385] Clegg R J et al 1980 Inhibition of hepatic cholesterol synthesis and S-3-hydroxy-3-methylglutaryl CoA reductase by mono and bycyclic monoterpenes administered *in vivo*. Biochemical Pharmacology 29: 2125–2127

[386] Patoir A et al 1938 Action toxique de l'essence de sabine et de l'armoise sur l'organisme. C R Soc Biol 127: 1325–1326

[387] Lysenko L V Pharmacological study of geraniol, carvone, linalool and thymol. (Russian) (Journal title unknown): 176–178

[388] De Smet P A G M, Keller K, Hänsel R, Chandler R F (eds) 1992 Adverse effects of herbal drugs, vol. 1. Springer-Verlag, Heidelberg

[389] De Smet P A G M, Keller K, Hänsel R, Chandler R F (eds) 1993 Adverse effects of herbal drugs, vol. 2. Springer-Verlag, Heidelberg

[390] Allen W T 1897 Note on a case of supposed poisoning by pennyroyal. Lancet 1: 1022–1023

[391] Kaiser R et al 1975 Analysis of buchu leaf oil. Journal of Agriculture and Food Chemistry 23(5): 943–950

[392] Sparks T 1985 Cinnamon oil burn. The Western Journal of Medicine 142(6): 835

[393] Perry P A et al 1990 Cinnamon oil abuse by adolescents. Veterinary and Human Toxicology 32(2): 162–164

[394] Chandler R F 1986 An inconspicuous but insidious drug. Canadian Pharmaceutical Journal 119: 563–566

[395] Duke J A 1985 Handbook of medicinal herbs. CRC Press, Boca Raton

[396] Buechel D W et al 1983 Pennyroyal oil ingestion: report of a case. Journal of the American Osteopathic Association 793: 129

[397] Anon 1983 British herbal pharmacopoeia. British Herbal Medicine Association, Bournemouth

[398] Anon 1934 The British pharmaceutical codex. The Pharmaceutical Press, London

[399] Lane B W et al 1991 Clove oil ingestion in an infant. Human and Experimental Toxicology 10: 291–294

[400] Mizutani T et al 1991 Hepatotoxicity of eugenol and related compounds in mice depleted of glutathione: structural requirements for toxic potency. Research Communications in Chemical Pathology and Pharmacology 73(1): 87–95

[401] Thompson D C et al 1991 Metabolism and cytotoxicity of eugenol in isolated rat hepatocytes. Chemico-biological Interactions 77: 137–147

[402] Hartnoll G et al 1993 Near fatal ingestion of oil of cloves. Archives of Disease in Childhood 69: 392–393

[403] Isaacs G 1983 Permanent local anaesthesia and anhidrosis after clove oil spillage. Lancet 1(April 16): 882

[404] Hills J M, Aaronson P I 1991 The mechanism of action of peppermint oil on gastrointestinal smooth muscle. Gastroenterology 101: 55–65

[405] Schafer K et al 1986 Effect of menthol on cold receptor activity. Journal of General Physiology 88: 757–776

[406] Watson H R et al 1978 New compounds with the menthol cooling effect. Journal of the Society of Cosmetic Chemists 29: 185–200

[407] Sidell N et al 1990 Menthol blocks dihydropyridine-insensitive Ca^{2+} channels and induces neurite

outgrowth in human neuroblastoma cells. Journal of Cellular Physiology 142(2): 410–419

[408] Allenby C F et al 1984 Diminution of immediate reactions to cinnamic aldehyde by eugenol. Contact Dermatitis 11: 322–323

[409] Schwartz R H 1990 Cinnamon oil: kids use it to get high. Clinical Pediatrics 29(3): 196

[410] Homburger F et al 1971 Inhibition of murine subcutaneous and intravenous benzo(rst)pentaphene carcinogenesis by sweet orange oils and *d*-limonene. Oncology 25: 1–10

[411] Elegbede J A et al 1984 Inhibition of DMBA-induced mammary cancer by the monoterpene *d*-limonene. Carcinogenesis 5(5): 661–664

[412] Elegbede J A et al 1986 Regression of rat primary mammary tumours following dietary *d*-limonene. Journal of the National Cancer Institute 76(2): 323–325

[413] Gold L S et al 1992 Rodent carcinogens: setting priorities. Science 258: 261–265

[414] Von Skramlik E V 1959 Über die Giftigkeit und Verträglichkeit von ätherischen Ölen. Pharmazie 14: 435–445

[415] Webb W K, Hansen W H 1963 Chronic and subacute toxicology and pathology of methyl salicylates in dogs, rats and rabbits. Toxicology and Applied Pharmacology 5: 576–587

[416] Blackmon W P, Curry H B 1957 Camphor poisoning— report of a case occurring during pregnancy. Journal of the Florida Medical Association 43(10): 999–1000

[417] Wilson J G 1973 Present status of drugs as teratogens in man. Teratology 7: 3–15

[418] Forbes P D et al 1977 Phototoxicity testing of fragrance raw materials. Food and Cosmetics Toxicology 15(1): 55–60

[419] Opdyke D L J 1973 Monographs on fragrance raw materials. Food and Cosmetics Toxicology 11

[420] Opdyke D L J 1974 Monographs on fragrance raw materials. Food and Cosmetics Toxicology 12

[421] Opdyke D L J 1975 Monographs on fragrance raw materials. Food and Cosmetics Toxicology 13

[422] Opdyke D L J 1976 Monographs on fragrance raw materials. Food and Cosmetics Toxicology 14

[423] Opdyke D L J 1977 Monographs on fragrance raw materials. Food and Cosmetics Toxicology 15

[424] Opdyke D L J 1978 Monographs on fragrance raw materials. Food and Cosmetics Toxicology 16

[425] Opdyke D L J 1979 Monographs on fragrance raw materials. Food and Cosmetics Toxicology 17

[426] Opdyke D L J 1982 Monographs on fragrance raw materials. Food and Cosmetics Toxicology 20

[427] Opdyke D L J 1983 Monographs on fragrance raw materials. Food and Cosmetics Toxicology 21

[428] Opdyke D L J 1988 Monographs on fragrance raw materials. Food and Chemical Toxicology 26

[429] Opdyke D L J 1992 Monographs on fragrance raw materials. Food and Chemical Toxicology 30

[430] International Fragrance Association 1992 Code of practice. Latest amendments, June 1992. International Fragrance Association, Geneva

[431] Cadeac M, Meunier A 1891 Contribution à l'étude physiologique de l'intoxication par le vulnéraire. Nouvelles preuves des propriétés épileptisantes de l'essence d'hyssop. Bull Acad Méd 43: 261–264

[432] Millet Y et al 1980 Étude de la toxicité d'huiles essentielles végétales du commerce: essence d'hysope et de sauge. Médecine Légale, Toxicologie 23(1): 9–21

[433] Verhulst H L et al 1961 Communications from the national clearinghouse for poison control centres: camphor. American Journal of Disease in Childhood 101: 536

[434] Janssens J et al 1990 Nutmeg oil: identification and quantification of its most active constituents as inhibitors of platelet aggregation. Journal of Ethnopharmacology 29: 179–188

[435] Papavassiliou M J 1935 Sur deux cas d'intoxication par la sabine la perméabilité placentaire a l'essence de sabine. Société de Médecine Légale 15: 778–781

[436] Patior A et al 1938 Action toxique de l'essence de sabine et de l'armoise sur l'organisme. C R Soc Biol 127: 1325–1326

[437] Jacobziner H, Raybin H W 1962 Camphor poisoning. Archives of Pediatrics 79: 28–30

[438] Van Lookeren J 1939 Vergiftiging door oleum chenopodii. Nederlandisch Tijdschrit Voor Geneeskunde 83: 5472–5476

[439] Turner R 1993 Absinthe—the green fairy. International Journal of Aromatherapy 5(2): 24–26

[440] Nedbal J 1967 Der Einfluss der Geruchsreize auf die Anfallsbereitschaft der Epilepsie. Zeitschrift Ärzneimittel Fortbild 61(1): 21–23

[441] Young A R et al 1988 Inhibition of UV radiation-induced DNA damage by a 5-methoxypsoralen tan in human skin. Pigment Cell Research 1: 350–354

[442] Young A R et al 1991 Photoprotection and 5-MOP photochemoprotection from UVR-induced DNA damage in humans: the role of skin type. Journal of Investigative Dermatology 97(5): 942–948

[443] Buchbauer G 1991 Aromatherapy: evidence for sedative effects of the essential oil of lavender after inhalation. Zeitschrift für Naturforschung 46C: 1067–1072

[444] Jäger W et al 1992 Percutaneous absorption of lavender oil from a massage oil. Journal of the Society of Cosmetic Chemists 43: 49–54

[445] Hasheminejad G, Caldwell J 1994 Genotoxicity of the alkenylbenzenes *alpha*- and *beta*-asarone, myristicin and elemicin as determined by the UDS assay in cultured rat hepatocytes. Food and Chemical Toxicology 32(3): 223–232

[446] Czygan F-C 1987 Warnung vor unkritischem Gebrauch von Wacholderbeeren. Zeitschrift für Phytotherapie 8: 10

[447] Orafidiya L O 1993 The effect of autoxidation of lemongrass oil on its antibacterial activity. Phytotherapy Research 7: 269–271

[448] Hamond P W 1906 Nutmeg poisoning. British Medical Journal 2: 778

[449] Reekie J S 1909 Nutmeg poisoning. Journal of the American Medical Association 52: 62

[450] Marsden K 1991 The aromatherapy phenomenon. International Journal of Aromatherapy 3(4): 6–8

[451] Davis P 1988 Aromatherapy an A–Z. C W Daniel, Saffron Walden

[452] Marshall A D, Caldwell J 1992 Influence of modulators of epoxide metabolism on the cytotoxicity of *trans*-anethole in freshly isolated rat hepatocytes. Food and Chemical Toxicology 30(6): 467–473

[453] Wiseman R W et al 1987 Structure–activity studies of the hepatocarcinogenicities of alkenylbenzene derivatives related to estragole and safrole on administration to preweanling male C57BL/6J x C3H/HeJF1 mice. Cancer Research 47: 2275–2283

[454] Ramos-Ocampo V E 1988 Mutagenicity and DNA-damaging activity of calamus oil, asarone isomers, and dimethoxypropenylbenzene analogues. The Philippine Entomologist 7(3): 275–291

[455] Anon 1988 Maximum limits for certain undesirable substances present in foodstuffs as consumed as a result of the use of flavourings. Council directive, 15. 7. 88. Official Journal of the European Communities 1(L): 184/61–184/65

[456] Anon 1992 The flavourings in food regulations 1992. Statutory Instruments No. 1971

[457] Malingré T M et al 1973 The presence of cannabinoid components in the essential oil of *Cannabis sativa* L. Pharmaceutisch Weekblad 108: 549–552

[458] Malingré T M et al 1975 The essential oil of *Cannabis sativa*. Planta Medica 28: 56–61

[459] Jirovetz L et al 1990 Determination of lavender oil fragrance compounds in blood samples. Fresenius J Anal Chem 338: 922–923

[460] Foussereau M J 1963 L'allergie de contact à l'huile de laurier. Bull Soc Derm Syph 70: 698–701

[461] Foussereau J et al 1967 Contact dermatitis from laurel II. Chemical aspects. Transactions of St John's Hospital Dermatological Society 53: 147–153

[462] Bouhlal K et al 1988 Le cade en dermatologie. Parfums, Cosmetiques, Aromes 83: 73–82

[463] Schoket B et al 1990 Formation of DNA adducts in the skin of psoriasis patients, in human skin in organ culture, and in mouse skin and lung following topical application of coal-tar and juniper tar. Journal of Investigative Dermatology 94(2): 241–246

[464] Agrawal O P et al 1980 Antifertility effects of fruits of *Juniperus communis*. Planta Medica 39: 98–101

[465] Prakash A O et al 1985 Anti-implantation activity of some indigenous plants in rats. Acta Europaea Fertilitatis 16(6): 441–448

[466] Zheng G et al 1992 Myristicin: a potential cancer chemoprotective agent from parsley leaf oil. Journal of Agricultural and Food Chemistry 40: 107–110

[467] Zheng G et al 1992 Inhibition of benzo[a]pyrene-induced tumorigenesis by myristicin, a volatile aroma constituent of parsley leaf oil. Carcinogenesis 13(10): 1921–1923

[468] Hotchkiss S 1994 How thin is your skin? New Scientist 141(1910): 24–27

[469] Bär Von F, Griepentrog F 1967 Die Situation in der gesundheitlichen Beurteilung der Aromatisierungsmittel für Lebensmittel. Medizin und Ernährung 8: 244–251

[470] Dayton L 1993 T-shirts find their place in the sun. New Scientist 139(1888): 19

[471] Foussereau J et al 1975 Contact allergy to *Frullania* and *Laurus nobilis*: cross-sensitization and chemical structure of the allergens. Contact Dermatitis 1: 223–230

[472] Sherry C J, Burnett R E 1978 Enhancement of ethanol-induced sleep by whole oil of nutmeg. Experientia 34(4): 492–493

[473] Mathela C S et al 1992 Reinvestigation of *Skimmia laureola* essential oil. Indian Perfumer 36(3): 217–222

[474] Lavy G 1987 Nutmeg intoxication in pregnancy. Journal of Reproductive Medicine 32: 63–64

[475] Swales N J, Caldwell J 1992 Cytotoxicity and depletion of glutathione (GSH) by cinnamaldehyde in rat hepatocytes. Human and Experimental Toxicology 10: 488–489

[476] Koren G (ed) 1990 Maternal–fetal toxicology—a clinicians' guide. Marcel Dekker, New York

[477] Anon 1993 Condom concern. International Journal of Aromatherapy 5(1): 4

[478] Swanson A B et al 1981 The side-chain epoxidation and hydroxylation of the hepatocarcinogens safrole and estragole and some related compounds by rat and mouse liver microsomes. Biochem Biophys Acta 673: 504

[479] Chan V S W, Caldwell J 1992 Comparative induction of unscheduled DNA synthesis in cultured rat hepatocytes by allylbenzenes and their 1´-hyroxy metabolites. Food and Chemical Toxicology 30(10): 831–836

[480] Caldwell J et al 1990 Comparative studies on the metabolism of food: case examples in the safety evaluation of the allylbenzene natural flavors. Nutritional Biochemistry 1: 402–409

[481] Ashwood-Smith M J et al 1980 5-methoxypsoralen, an ingredient in several suntan preparations, has lethal, mutagenic and clastogenic properties. Nature 285(5764): 407–409

[482] Averbeck D et al 1990 Genotoxicity of bergapten and bergamot oil in *Saccharomyces cerevisiae*. Journal of Photochemistry and Photobiology 7: 209–229

[483] Young A R et al 1990 Phototumorigenesis studies of 5-methoxypsoralen in bergamot oil: evaluation and modification of risk of human use in an albino mouse skin model. Journal of Photochemistry and Photobiology 7: 231–250

[484] Dubertret L et al 1990 The photochemistry and photobiology of bergamot oil as a perfume ingredient: an overview. Journal of Photochemistry and Photobiology 7: 362–365

[485] Zaynoun S T et al 1977 A study of bergamot and its importance as a phototoxic agent. Contact Dermatitis 3: 225–239

[486] Gloxhuber C 1970 Prüfung von Kosmetik-Grundstoffen auf fototoxische Wirkung. Journal of the Society of Cosmetic Chemists 21: 825–833

[487] Morlière P et al 1991 Photoreactivity of 5-geranoxypsoralen and lack of photoreaction with DNA. Photochemistry and Photobiology 53(1): 13–19

[488] Baker J B E 1960 The effects of drugs on the foetus. Pharmacological Reviews 12: 37–90

[489] Tomatis L 1979 The predictive value of rodent carcinogenicity tests in the evaluation of human risks. Annual Review of Pharmacology and Toxicology 19: 511–530

[490] Private communication from Dr Tim Betts, Consultant Neuropsychiatrist, Queen Elizabeth Psychiatric Hospital, Birmingham

[491] Musajo L et al 1966 Skin-photosensitising furocoumarins: photochemical interaction between DNA and -O^{14}CH$_3$ bergapten (5-methoxy-psoralen). Photochemistry and Photobiology 5: 739–745

[492] Zaynoun S T et al 1977 A study of oil of bergamot and its importance as a phototoxic agent. I. Characterisation

and quantification of the photoactive component. British Journal of Dermatology 96: 475–482

[493] Zajdela F, Bisagni E 1981 5-methoxypsoralen, the melanogenic additive in sun-tan preparations, is tumorigenic in mice exposed to 365nm u.v. radiation. Carcinogenesis 2(2): 121–127

[494] Young A R et al 1983 A comparison of the phototumorigenic potential of 8-MOP and 5-MOP in hairless albino mice exposed to solar stimulated radiation. British Journal of Dermatology 108: 507–518

[495] Sambuco C P et al 1987 Protective value of skin tanning induced by ultraviolet radiation plus a sunscreen containing bergamot oil. Journal of the Society of Cosmetic Chemists 38: 11–19

[496] Grube D D 1977 Photosensitizing effects of 8-methoxypsoralen on the skin of hairless mice. II. Strain and spectral differences for tumorigenesis. Photochemistry and Photobiology 25: 269–276

[497] Fournier G et al 1990 Contribution à l'étude des huiles essentielles de différentes espèces de juniperus. J Pharm Belg 45(5): 293–298

[498] Abel G 1987 Chromosomenschädigende Wirkung von beta-asaron in menschlichen Lymphocyten. Planta Medica 53: 251–253

[499] Brophy J J 1993 The essential oils of the genus Doryphora. Journal of Essential Oil Research 5(6): 581–586

[500] Barni B, Barni I 1967 La intossicazione da apiolo. I. Rassegna casistica. Zacchia 3(2): 197–221

[501] Amerio A et al 1968 La nefrotapia da apiolo. Minerva Nefrologica 15(1): 49–70

[502] Marozzi E, Farneti A 1968 Recente esperienza in tema di avvelenamento da apiolo e sostanze correlate. Zacchia 4(4): 563–580

[503] Colalillo R 1974 Avvelenamento da apiolo in 7 donne e 2 bambini. Rivista Tossicologia Sperimental 18(2): 125–130

[504] Marshall A D, Caldwell J 1993 The cytotoxicity and genotoxicity of anethole 1,2-epoxide, a primary metabolite of the food flavour trans-anethole. Human and Experimental Toxicology 12: 427–428

[505] Littleton F 1990 Warfarin and topical salicylates. Journal of the American Medical Association 263: 2888

[506] Bartek M J et al 1972 Skin permeability in vivo: comparison in rat, rabbit, pig and man. Journal of Investigative Dermatology 58: 114–123

[507] Abramovici A et al 1985 Benign hyperplasia of ventral prostate in rats induced by a monoterpene (preliminary report). The Prostate 7: 389–394

[508] Servadio C et al 1986 Early stages of the pathogenesis of rat ventral prostate hyperplasia induced by citral. European Urology 12: 195–200

[509] Sandbank M et al 1988 Sebaceous gland hyperplasia following topical application of citral: an ultrastructural study. The American Journal of Dermatopathology 10(5): 415–418

[510] Geldof A A et al 1992 Estrogenic action of commonly used fragrant agent citral induces prostatic hyperplasia. Urological Research 20: 139–144

[511] Tenenbein M 1990 Maternal–fetal toxicology: a clinician's guide. Dekker, New York

[512] Zhu L et al 1994 The Cinnamomum species in China: resources for the present and future. Perfumer and Flavorist 19(4): 17–22

[513] Musajo L et al 1953 L'attività fotodinamica delle cumarine naturali. La Chimica e L'Industria 35: 13–15

[514] Musajo L et al 1954 L'activité photodynamique des coumarines naturelles. Bull Soc Chim Biol 36(9): 1213–1224

[515] Pathak M A, Fitzpatrick T B 1959 Bioassay of natural and synthetic furocoumarins (psoralens). Journal of Investigative Dermatology 32: 509–518

[516] Ashwood-Smith M J et al 1983 Photobiological activity of 5,7-dimethoxycoumarin. Experientia 39: 262–264

[517] Anon 1992 Lemon burn warning. International Journal of Aromatherapy 4(4): 4

[518] Bourrel C et al 1993 Chemical analysis, bacteriostatic and fungistatic properties of the essential oil of elecampane (Inula helenium L.). Journal of Essential Oil Research 5: 411–417

[519] Forbes R J 1970 A short history of the art of distillation. E J Brill, Leiden

[520] Cox D et al 1989 The rarity of liver toxicity in patients treated with coumarin (1,2-benzopyrone). Human Toxicology 8: 501–506

[521] Le Bourhis B, Soenen A-M 1973 Recherches sur l'action psychotrope de quelques substances aromatiques utilisées en alimentation. Food and Cosmetics Toxicology 11: 1–9

[522] Williams A C, Barry B W 1989 Essential oils as novel human skin penetration enhancers. International Journal of Pharmaceutics 57: R7–R9

[523] Hardisty R M, Weatherall D J 1982 Blood and its disorders, 2nd edn. Blackwell Scientific Publications, Oxford

[524] Hazleton L W et al 1956 Toxicity of coumarin. Journal of Pharmacology and Experimental Therapeutics 118: 348–358

[525] Booth A N et al 1959 Urinary metabolites of coumarin and o-coumaric acid. Journal of Biological Chemistry 234(4): 946–948

[526] Shilling W H et al 1969 Metabolism of coumarin in man. Nature 221: 664–665

[527] Gangoli S D et al 1974 Studies on the metabolism and hepatotoxicity of coumarin in the baboon. Biochemical Society Transactions 2: 310–312

[528] Evans J G et al 1979 Two-year toxicity study on coumarin in the baboon. Food and Cosmetics Toxicology 17: 187–193

[529] Feuer G 1974 The metabolism and biological actions of coumarins. Progress in Medicinal Chemistry 10: 85–158

[530] D'Aprile F 1928 Studio clinico-sperimentale sull'intossicazione da apiolo. Annali di Ostetricia e Ginecologia 50: 1204–1227

[531] Patoir A et al 1936 Le rôle abortif de l'apiol. Paris Médical 3: 442–446

[532] Hermann K et al 1956 Death from apiol used as an abortifacient. Lancet 1: 937–939

[533] Lowenstein L, Ballew D H 1958 Fatal acute haemolytic anaemia, thrombocytopenic purpura, nephrosis and hepatitis resulting from ingestion of a compound containing apiol. Canadian Medical Association Journal 78: 195–198

[534] Critchley M (ed) 1984 Butterworths Medical Dictionary. Butterworths, London

[535] Mann J 1992 Murder magic and medicine. Oxford University Press, Oxford

[536] Åkesson H O, Wålinder J 1965 Nutmeg intoxication. The Lancet 1: 1271–1272

[537] Panayotopoulos D J, Chisholm D D 1970 Hallucinogenic effect of nutmeg. British Medical Journal 1: 754

Recommended Reading

The following publications are particularly recommended as sources of detailed information in the areas indicated.

ESSENTIAL OIL SAFETY

• The fragrance raw materials monographs (see references 418–429) in which the Research Institute for Fragrance Materials reports the results of safety testing on both essential oils and chemicals used in fragrances. The monographs generally include data on acute oral and dermal toxicity, skin irritation and sensitisation, and phototoxicity. A new volume is published every few years. The publishers, Pergamon Press, are now known as Elsevier Science.

North America: Elsevier Science Publishing Co. Inc., 655 Avenue of the Americas, New York, NY 10010 USA Tel: (212) 989-5800 Fax: (212) 633-3990

Rest of the World: Elsevier Science Ltd., The Boulevard, Langford Lane, Kidlington, Oxford, OX5 1GB UK Tel: 01865-843000 Fax: 01865-843010

• An ongoing series of monographs on the safety of herbs and essential oils. (Some of the monographs cover non-aromatic plants.) At the time of writing two volumes have been published.

De Smet P A G M, Keller K, Hänsel R, Chandler R F (eds) 1992, 1993 Adverse effects of herbal drugs. Springer-Verlag, Heidelberg, vol 1 (1992), Vol 2 (1993)

ESSENTIAL OIL COMPOSITION

• An ongoing series of monographs on the composition of essential oils by Brian M Lawrence. At the time of writing four volumes have been published.

Essential Oils 1976–1978 (1979)
Essential Oils 1979–1980 (1981)
Essential Oils 1981–1987 (1989)
Essential Oils 1988–1991 (1993)
Allured Publishing Corporation, 362 S. Schmale Road, Carol Stream, IL 60188 USA Tel: (708) 653-2155 Fax: (708) 653–2192

ESSENTIAL OIL RESEARCH

• Edited by the authors of this volume, the International Journal of Aromatherapy is published quarterly and contains summarised reports on current essential oil research, a refereed section, and items of general interest to aromatherapists.

The International Journal of Aromatherapy, Aromatherapy Publications, PO Box 746, Hove, East Sussex, BN3 3XA, UK Tel: 01273-772479 Fax: 01273-329811

• Edited by Brian Lawrence, the quarterly Journal of Essential Oil Research publishes only refereed articles, the great majority of which are on essential oil composition.

The Journal of Essential Oil Research, Allured Publishing Corporation, 362 S. Schmale Road, Carol Stream, IL 60188 USA Tel: (708) 653-2155 Fax: (708) 653 2192

Botanical index

For cross–referral between the Botanical Index and the General Index: page numbers given in the Botanical Index refer only to locations in the text where botanical names are given. For fuller information specific oils may be sought in the General Index under their common names.

General index